New Frontiers in Zoology

New Frontiers in Zoology

Editor: Simon Benson

R CALLISTO
REFERENCE

www.callistoreference.com

Callisto Reference,
118-35 Queens Blvd., Suite 400,
Forest Hills, NY 11375, USA

Visit us on the World Wide Web at:
www.callistoreference.com

ISBN: 978-1-64116-251-7 (Hardback)

Cataloging-in-Publication Data

New frontiers in zoology / edited by Simon Benson.
    p. cm.
Includes bibliographical references and index.
ISBN 978-1-64116-251-7
1. Zoology. 2. Animals. 3. Biology. 4. Natural history. I. Benson, Simon.
QL45.2 .N49 2020
590--dc23

Table of Contents

Preface

Every book is initially just a concept; it takes months of research and hard work to give it the final shape in which the readers receive it. In its early stages, this book also went through rigorous reviewing. The notable contributions made by experts from across the globe were first molded into patterned chapters and then arranged in a sensibly sequential manner to bring out the best results.

Zoology is the branch of biology concerned with the study of animals, including their embryology, structure, habits, distribution, classification and evolution. Some of the current areas of zoological investigation include anatomy, histology, physiology, embryology, teratology and ethology. The field of zoology, particularly the study of morphology and physiology, underwent a significant development due to the Darwinian theory of evolution by natural selection. This facilitated the classification of animals on a genealogical basis and attempts to determine their genetic relationships. Modern classification system for animals is based on the three-domain system of archaea, bacteria and eukaryota. Animals are further classified under the categories of domain, kingdom, phylym, class, order, family, genus and species. This book traces the progress of zoology and highlights some of its key concepts and applications. It is a valuable compilation of topics, ranging from the basic to the most complex advancements in this field. The extensive content of this book provides the readers with a thorough understanding of the subject.

It has been my immense pleasure to be a part of this project and to contribute my years of learning in such a meaningful form. I would like to take this opportunity to thank all the people who have been associated with the completion of this book at any step.

Editor

Seasonal habitat selection of the red deer (*Cervus elaphus alxaicus*) in the Helan Mountains, China

Mingming Zhang[1], Zhensheng Liu[1,2] & Liwei Teng[1,2,3]

[1] College of Wildlife Resources, Northeast Forestry University, No.26 Hexing Road, Xiangfang District, Harbin 150040, P.R. China.
[2] Key Laboratory of Conservation Biology, State Forestry Administration, No.26 Hexing Road, Xiangfang District, Harbin 150040, P.R. China.
[3] Corresponding author. E-mail: tenglw@gmail.com

ABSTRACT. We studied the seasonal habitat selection of the red deer, *Cervus elaphus alxaicus* Bobrinskii & Flerov, 1935, in the Helan Mountains, China, from December 2007 to December 2008. Habitat selection varied widely by season. Seasonal movements between high and low elevations were attributed to changes in forage availability, alpine topography, the arid climate of the Helan Mountains, and potential competition with blue sheep, *Pseudois nayaur* (Hodgson, 1833). The use of vegetation types varied seasonally according to food availability and ambient temperature. Red deer used montane coniferous forest and alpine shrub and meadow zones distributed above 2,000 m and 3,000 m in summer, alpine shrub and meadows above 3,000 m in autumn, being restricted to lower elevation habitats in spring and winter. The winter habitat of *C. elaphus alxaicus* was dominated by *Ulmus glaucescens* Franch. and *Juglans regia* Linnaeus, deciduous trees, and differed from the habitats selected by other subspecies of red deer. *Cervus elaphus alxaicus* preferred habitats with abundant vegetation coverage to open habitats in winter, but the reverse pattern was observed in summer and autumn. Red deer preferred gentle slopes (<10°) but the use of slope gradient categories varied seasonally. Red deer avoidance of human disturbance in the Helan Mountains varied significantly by season. Information on red deer habitat selection can help understand the factors affecting seasonal movements and also support decision making in the management and conservation of red deer and their habitats.

KEY WORDS. Habitat use; migration; seasonal movement; ungulates.

The red deer, *Cervus elaphus alxaicus* Bobrinskii & Flerov, 1935 is an endemic subspecies in China, distributed in the central region of the Helan Mountains, which range from the border between Alxa League of the Inner Mongolia Autonomous Region to the Ningxia Hui Autonomous Region in west-central China. The Helan Mountains support the only known population of this subspecies (WANG *et al.* 1999). Its population is small and is isolated from other subspecies of red deer (ZHANG *et al.* 1999). Historically, the survival of *C. e. alxaicus* has been threatened by the deterioration of habitats on the Alxa Plateau, and its reduced population has since been isolated in the relatively poor ecological environment of the Helan Mountains for a long period. Population estimates for *C. e. alxaicus* in the Helan Mountains ranged from 850 to 1,060 in 1983, increasing to 1,705 ± 523 in 2005, after the creation of the Ningxia Helan Mountain Nature Reserve in 1982 (ZHANG *et al.* 2006). While *C. e. alxaicus* is a threatened subspecies, *C. elaphus* is categorized as a species of Least Concern in the World Conservation Union's Red List of Threatened Animals (LOVARI *et al.* 2008).

Habitat selection by animals is considered as an optimization process that involves factors such as food supply, conspecific population density, body size, competitors, predators, and landforms (MORRISON *et al.* 1998). Information on which resources are preferred or avoided by organisms improves our understanding of how they meet their requirements for survival and reproduction (MANLY *et al.* 2002). The distribution and availability of trophic resources are important factors that affect habitat selection. In most temperate habitats, food is scarce during the winter months and abundant in spring and early summer (MOEN 1976, SCHMITZ 1991). This forces animals in temperate regions to adapt to seasonal changes in food supply. Ungulates that inhabit temperate and boreal regions often exhibit cyclical seasonal movements between summer and winter ranges in response to environment factors (e.g., snow conditions, food availability), social constraints and predation risk (FRYXELL & SINCLAIR 1988, PÉPIN *et al.* 2008). Regular, round-trip movements between seasonal home ranges (WHITE & GARROTT 1990) have evolved to enable animals to avoid undesirable conditions at a particular time of the year (Vaughan *et al.* 2000). Generally, seasonal movement patterns of ungulates include short-distance movements, dispersal, and migration (Grovenburg *et al.* 2009). Seasonal migration of cervids involves dispersal to areas of lower elevation, particularly in winter, when the environment is less hospitable at higher elevations (Albon & Langvatn 1992, Igota *et al.* 2004, Pépin *et al.* 2008). Mixed strategies of migration have, however, been found among cervids that inhabit temperate and boreal mountainous regions (Igota *et al.* 2004, Brinkman *et al.* 2005, Grovenburg *et al.* 2009).

Despite numerous studies of other subspecies of red deer, little is known about the seasonal habitat use or movement patterns of *C. e. alxaicus* during different seasons. The objective of this study was to: 1) compare habitats used by *C. e. alxaicus* during different seasons to document differences in habitat selection; and 2) examine environmental variables that affect the seasonal movement of *C. e. alxaicus*.

MATERIAL AND METHODS

This study was conducted over four seasons from December 2007 to December 2008 in the Helan Mountain region, which is located between the eastern Yinchuan plain in Ningxia Hui Autonomous Region and the western Alxa Plateau in Inner Mongolia Autonomous Region (105°44'-106°42'E, 38°21'-39°22'N) (Fig. 1). The Helan Mountain region, located in northwestern China, is on the transitional zone between steppe and desert regions of central Asia (Takhtajan 1986). It generally lies at 2,000-3,000 masl, with a maximum elevation of 3,556 masl. It covers an area of 2,740 km² [including Ningxia Helan Mountain National Nature Reserve (2,063 km²) and Inner Mongolia Helan Mountain National Nature Reserve (677 km², Fig. 1)], with a north-south length of about 250 km and an east-west width of about 20-40 km (Z.S. Liu unpublished data).

The region has a typical continental climate, characterized by cool and dry conditions, with annual mean temperature of -0.9°C and mean annual rainfall of 420 mm. The local climate is influenced by the topography of the Helan Mountain, the low temperature center of the northern Ningxia. The maximum monthly mean temperature is 11.9°C, in July, and the minimum is -14.2°C, in January, 8.8 ~9.8°C lower than in Yinchuan, the capital city of Ningxia Hui Autonomous Region. Precipitation varies seasonally, with 62% falling as rain in summer. There is little precipitation in winter, about 10.1% of the annual total. Snow cover is limited in the Helan Mountains (Geng & Yang 1990). The vegetation distribution is strongly influenced by moisture conditions. The elevation differential between the Helan Mountains and the plains to the south and east is about 2100 m, which creates an elavational climatic gradient that results in the formation of four elevational vegetation zones. The mountain steppe zone (MS) occurs at 1,400-1,600 masl and covers an area of 1,241 km² dominated by *Stipa breviflora* Griseb., *Ajania fruticulosa* (Ledeb.) Poljak,

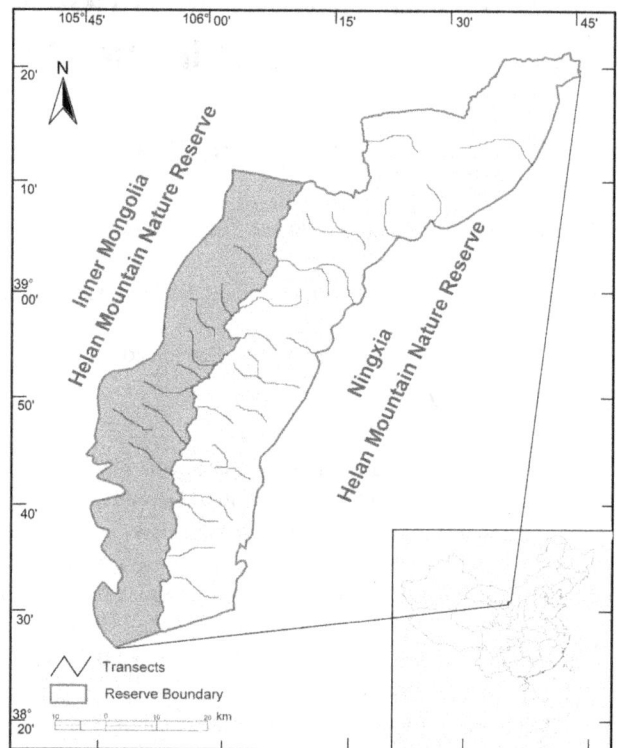

Figure 1. Location and distribution of the study area and transects in the Helan Mountain region, China. Grey shading represents Helan Mountain Nature Reserve in Inner Mongolia Autonomous Region, while the unshaded area represents the reserve in Ningxia Hui Autonomous Region.

Ptilagrostis pelliotii (Danguy) Grubov.), *Oxytropis aciphylla* Ledeb.), *Convolvulus gortschakovii* Schrenk ex Fisch. & C.A. Mey., and *Salsola laricifolia* Turcz. ex Litv. The open mountain forest and steppe zone (MOFS) (1,600-2,000 masl, 1,155 km²) is dominated by *Ulmus glaucescens* Franch., *Prunus mongolica* Maxim., *Stipa grandis* P.A. Smirn., and *S. bungeana* Trin. The mountain coniferous forest zone (MCF) (1,900-3,000 masl, 319 km²) is dominated by *Picea crassifolia* Kom., *Pinus tabulaeformis* Carrière, *Juniperus rigida* Siebold & Zucc., and *Potentilla parvifolia* Fisch. ex Lehm. The alpine shrub and meadow zone (ABM) (3,000-3,556 masl, 23 km²) is dominated by *Salix cupularis* var. *lasilogyne* Rehd., *Caragana jubata* (Pall.) Poir., *Kobresia* spp., *Polygonum viviparum* Linnaeus and *Arenaria* spp. (Jiang *et al.* 2000, Di 1986). Mammals found in the area include insectivores: Daurian hedgehog, *Mesechinus dauuricus* (Sundeyall, 1842); carnivores: red fox, *Vulpes vulpes* (Linnaeus, 1758); artiodactyls: blue sheep, alpine musk deer: *Moschus chrysogaster* (Hodgson, 1839); lagomorphs: Daurian pika, *Ochotona dauurica* (Pallas, 1776); and chiropterans and rodents. Blue sheep and *C. e. alxaicus* are the two dominant ungulate species in the Helan Mountain region (Liu 2009).

From December 2007 to December 2008, we carried out four surveys, one per season, to determine the distribution of *C. e. alxaicus* throughout the Helan Mountain area. Each seasonal survey sampled 32 line transects established along the valleys and each survey took about one month to sample all topographic types. Differences in topographic relief prevented walking along the transects at the same velocity and length, therefore, transects ranged in length from 4.5 to 8.5 km, for a total of 350 km traversing the whole study area and covering all four elevational vegetation zones. The distance between any two transects was at least 2 km, to ensure the independence of each transect.

Because of the rarity of *C. e. alxaicus* in the area and their high sensitivity to anthropogenic disturbance (JEPPESEN 1987), we documented habitat use mainly by recording fresh signs, such as evidence of bedding or the presence of feces. We also observed deer in order to record their activities. It was easy to differentiate the signs left by *C. e. alxaicus* from those of other ungulates (e.g., *P. nayaur*, *M. chrysogaster*) in the Helan Mountains, because *C. e. alxaicus* is larger than *P. nayaur* and *M. chrysogaster*, and the size and shape of their feces are different (CHANG & XIAO 1988). To ensure the accu-

racy of data, we recorded fresh feces only (3-5 days old as estimated by the color and water content). When deer were observed during line transect surveys, we first used telescopes to observe their feeding or bedding behavior, without disturbing the animals. After deer had departed the area, we carried out detailed sampling.

We recorded terrestrial coordinates according to a Global Positioning System (GPS) after a feeding or bedding habitat was identified. We then established a plot with five sample quadrants (Fig. 2) to collect data of 18 topographic and biological variables (Table I) using the methods described by LIU *et al.* (2002). The distance between any two plots was at least 500 m to ensure the independence of each plot (Fig. 2).

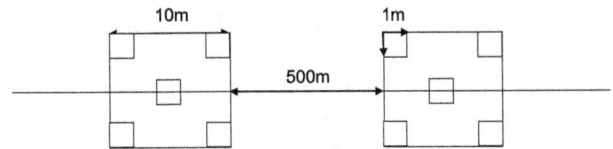

Figure 2. Diagramatic presentation of the survey transects, plots and sample quadrats used in this study.

Table I. Variables collected in feeding and bedding habitat plots used by *Cervus elaphus alxaicus* in the Helan Mountain region, China.

Variables	Categorization and Criterion	Abbreviation
Altitude (m)	The altitude of the plot accorded to GPS	AL
Vegetation types	Mountain steppe zone (MS); Mountain open forest and steppe zone (MOFS); Mountain coniferous forest zone (MCF); Alpine bush and meadow zone (ABM)	VT
Topography	Categorized by the slope and fault of a hillside, divided into 5 levels: Smooth undulating slope; Moderately broken slope; Distinctly broken slope; Scree/landslide; Cliff	TO
Dominant tree	The tree covers 70% of the density in the 10×10 m plot. It usually was *Ulmus glaucescens*, *Ziziphus jujube*, *Salix* spp., *Juniperus rigida*, *Pinus tabulaeformis*, *Picea crassifolia* Mixture or Open land with no tree	DT
Tree density (trees/100 m²)	The total number of trees in the 10×10 m plot	TD
Tree height (m)	The mean height of trees in the 10×10 m plot	TH
Distance to the nearest tree (m)	Distance from the center of the 10×10 m plot to the nearest tree	DtT
Shrub density (trees/100 m²)	The number of shrubs in the 10×10 m plot	SD
Shrub height (m)	The mean height of shrubs in the 10×10 m plot	SH
Distance to the nearest shrub (m)	Distance from the center of the 10×10 m plot to the nearest shrub	DtS
Herb coverage (%)	The mean herb coverage of the 5 sample quadrats in the 10×10 m plot	HC
Slope gradient (°)	Slope gradient of the hillside where the spot located measured with military compass	SG
Slope location	A visual assessment of the site location relative to the macroslope which is usually from valley bottom to ridge top, classed as: lower slope (includes valley bottom and flat), middle slope and upper slope (includes ridge top)	SL
Slope aspect	Aspect was surveyed to eight compass points, translated as 0°, 45°, 90°, 135°, 180°, 225°, 270° and 315° from North, as 0° is equivalent to 360°. And the slope aspect was grouped into 3 main directions: sunny slope (135°~225°), partial shade slope (45°~135° and 225°~315°) or shady slope (315°~45°)	SA
Distance to water resource (m)	The distance from the spot to the nearest water resource	DtW
Distance to human disturbance (m)	The distance from the spot to the nearest place of human activity such as highway, road and shelter forest station, etc	DtH
Distance to bare rock (m)	The distance from the spot to the nearest bare rock	DtR
Hiding cover (%)	The coverage of the hiding conditions. Percent hiding cover was determined by visually estimating the percent of a deer or a substitute (a 1 m stick) obscured at 30 m in the four cardinal directions (KUNKEL & PLETSCHER 2001)	HiC

Generally, research on resource selection requires a comparison of the habitats used by *C. e. alxaicus* (observed plots) with those that are available (expected plots, assuming no differential selection by deer). To provide comparison plots for the analysis of habitat selection, 617 randomly located plots were surveyed. The random plots were established along the survey transects, in areas with no obvious evidence of *C. e. alxaicus* use. The distance between random plots and occupied plots or any two random plots was at least 500 m to reduce the possibility of overlap between used and unused plots. Comparison plots were surveyed in each vegetation zone according to the proportion of used plots in each zone. Data were recorded for comparison plots using the same methods used in the occupied plots.

Data were analyzed to quantify habitat selection by *C. e. alxaicus* by season. To assess seasonal differences between the 18 factors recorded at used plots and comparison plots, we used a chi-square goodness-of-fit test within classified categories for each variable (Marcum & Loftsgaarden 1980). P-values less than 0.05 were considered statistically significant. Bonferroni confidence intervals were calculated by the following formula to identify variables that indicate preference or avoidance, following the method developed by Neu *et al.* (1974) and Byers *et al.* (1984). $(p_i - r_i) \pm Z_{1-a/2k} \times \sqrt{p_i(1-p_i)/n_i + r_i(1-r_i)/m_i}$; where, n_i is the number of comparison plots in category i, and p_i is the proportion of the comparison plots that fall in category i; m_i is the total number of plots used by *C. e. alxaicus*, r_i is the proportion of plots used by *C. e. alxaicus* in category i; $Z_{1-a/2k}$ is the upper standard normal table value corresponding to a probability tail area of $1-a/2k$; a is the level of significance; and k is the number of categories tested. The confidence intervals indicated that *C. e. alxaicus* showed avoidance (marked "-") of category i when $(p_i - r_i) - Z_{1a/2k} \times \sqrt{p_i(1-p_i)/n_i + r_i(1-r_i)/m_i} > 0$; whereas *C. e. alxaicus* showed preference (marked "+") for category i when $(p_i - r_i) + Z_{1a/2k} \times \sqrt{p_i(1-p_i)/n_i + r_i(1-r_i)/m_i} < 0$; and *C. e. alxaicus* showed no obvious selection (marked " = ") for category i when $(p_i - r_i) - Z_{1a/2k} \times \sqrt{p_i(1-p_i)/n_i + r_i(1-r_i)/m_i} < 0$ and $(p_i - r_i) + Z_{1a/2k} \times \sqrt{p_i(1-p_i)/n_i + r_i(1-r_i)/m_i} > 0$.

Data for non-numeric ecological factors (VT, TO, DT, SL and SA; see Table I) were examined with chi-square tests. Data for the remaining numeric ecological variables were initially analyzed with one-sample Kolmogorov-Smirnov tests to determine if they were normally distributed. The normally distributed data were analyzed with independent-samples t-tests, while the non-normally distributed data were analyzed with Kruskal-Wallis H tests.

RESULTS

A total of 602 plots used by *C. e. alxaicus* (observed plots) were recorded and compared among the four vegetation zones (Fig. 3). Across the whole study period, 209 used plots in the coniferous forest mountain zone were measured, and this type was the most common vegetation type, followed by open mountain forest and steppe zone of 169 plots, mountain steppe zone of 108 plots and alpine shrub and meadow zone of 106 plots (Fig. 3).

Across the entire study area in spring, we sampled 181 plots used by deer (Fig. 3) and 181 comparison plots (Appendix 1). Deer selected habitats characterized by gentle (<10°), undulating, sunny slopes in spring. Deer preferred habitats with 4-6 m high trees, near dense shrubs (> 10 trees/100 m²) taller than 1.3 m, with high herb coverage of more than 50%, good hiding conditions (hiding cover lower than 50%) (Fig. 4), and distant from bare rock. However, no significant preference was shown during spring with respect to dominant tree species (x^2 = 9.65, df = 10, p > 0.05), distance to the nearest tree (x^2 = 4.02, df = 2, p > 0.05), altitude (x^2 = 0.59, df = 3, p > 0.05), distance to water resource (x^2 = 0.59, df = 2, p > 0.05), or distance to human disturbance (x^2 = 0.00, df = 2, p > 0.05) (Appendix 1).

In summer, 146 plots were used by deer (Fig. 3) and 167 comparison plots (Appendix 1) were surveyed. Deer preferred habitats with gentle, undulating slopes (<20°) on the south side of the MCF zone above 2,000 m and the ABM zone above 3,000 m, respectively, during summer. Habitats used in summer were on lower slopes near dense stands of trees (> 4 tree/100 m²) with mixed tree species of 4-6 m height, near dense shrub stands (> 10 tree/100 m²) taller than 1.3 m, with high herb coverage (> 80%), good hiding condition (hiding coverage lower than 50%) (Fig. 4) and far from bare rock (> 50 m). Distance to human disturbance did not affect habitat use by deer in summer (Appendix 1).

A total of 144 plots used by deer (Fig. 3) and 138 comparison plots (Appendix 1) were surveyed in autumn. Autumn habitat use was similar to that during summer: *C. e. alxaicus* preferred habitats with gentle (<10°), undulating and partially shaded slopes in the ABM zone, at elevations above 3,000 m in autumn. They also preferred habitats with more gentle slopes, high tree density (> 4 trees/100 m²), mixed tree species of 4-6 m height, dense shrubs (> 10 trees/100 m²), high herb coverage (> 80%) good hiding condition (hiding coverage lower than 50%) (Fig. 4), and distant from bare rock (> 100 m). There was no significant difference between the used plots and comparison plots in autumn with respect to distance to the nearest tree (x^2 = 0.36, df = 2, p > 0.05), shrub height (x^2 = 4.00, df = 2, p > 0.05), distance to water resource (x^2 = 2.95, df = 2, p > 0.05), or distance to human disturbance (x^2 = 5.41, df = 2, p > 0.05) (Appendix 1).

131 plots used by *C. e. alxaicus* (Fig. 3) and 131 comparison plots (Appendix 1) were surveyed in winter. Deer preferred winter habitats with gentle (<10°), undulating, sunny slopes, lower slopes with high herb coverage (> 50%), and average hiding condition (hiding coverage lower than 75%) (Fig. 4). However, deer maintained a distance of more than 1.5 m from shrubs

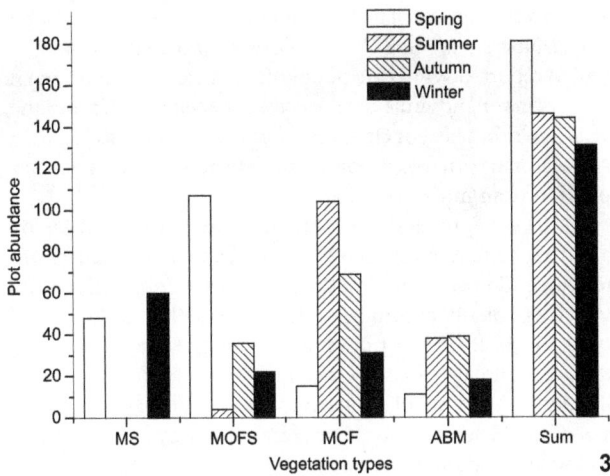

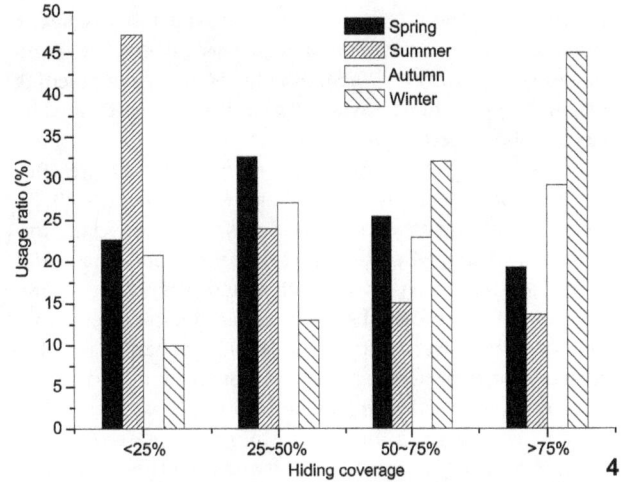

Figures 3-4. (3) Abundance and proportion of plots used by *Cervus elaphus alxaicus* in different vegetation type zones among four seasons, in the Helan Mountain region, China. (MS) Mountain steppe zone, (MOFS) Mountain open forest and steppe zone, (MCF) Mountain coniferous forest zone, (ABM) Alpine bush and meadow zone. (4) Seasonal changes in the usage ratio of hiding coverage on *C. e. alxaicus* in the Helan Mountain region, China.

and 50-100 m from bare rock. Deer showed no significant preference with respect to vegetation type ($x^2 = 0.90$, df = 3, p > 0.05), shrub height ($x^2 = 2.40$, df = 2, p > 0.05), altitude ($x^2 = 1.59$, df = 3, p > 0.05), distance to water source ($x^2 = 0.14$, df = 2, p > 0.05), or distance to human disturbance ($x^2 = 2.13$, df = 2, p > 0.05) (Appendix 1).

There were statistically significant differences in the use of five non-numeric ecological factors by season. *Cervus elaphus alxaicus* selected habitats by season based on vegetation type ($x^2 = 58.611$, df = 3, p < 0.001), topography ($x^2 = 969.841$, df = 3, p < 0.001), dominant tree species ($x^2 = 820.947$, df = 10, p < 0.001), slope aspect ($x^2 = 74.355$, df = 2, p < 0.001) and slope location ($x^2 = 143.096$, df = 2, p < 0.001). There were also statistically significant differences in the use of 13 numeric ecological factors by season (Table II).

DISCUSSION

Cervus elaphus alxaicus in the Helan Mountains displayed a pattern of seasonal elevational migration similar to that of other red deer subspecies in mountainous areas (ALBON & LANGVATN 1992, JARNEMO 2008, PÉPIN *et al.* 2008). However, we observed differences between the habitat selection of *C. e. alxaicus* and that of other subspecies. HUTTO (1985) described the mechanisms determining habitat selection as: geographic restrictions, genetic evolution, influence of experience, and settlement decisions following exploration.

Vegetation type determines the composition and distribution of deer forage, and is determined by soil type, climate, sunlight, topography, landform and many microhabitat factors. The heterogeneous distribution of biotic and abiotic fac-

tors in environments leads to spatial heterogeneity in vegetation types. Physiological and ecological requirements of deer are met to varying degrees by different vegetation types. Thus the geographic and seasonal variation in vegetation types affect red deer habitat selection. In the Helan Mountains, *C. e. alxaicus* range annually from the mountain steppe zone below 1,600 m to alpine shrub and meadow zone above 3,000 m (LIU 2009). Deer preferred habitats in the montane coniferous forest zone (> 2,000 masl) and alpine shrub and meadow zone (> 3,000 masl), in summer, and those in the alpine shrub and meadow zone (> 3,000 m) in autumn. Deer showed no preference for any vegetation type in spring or winter (Appendix 1), using habitats in proportion to their availability during those seasons. IGOTA *et al.* (2004) reported that deer rarely change their summer home ranges for breeding and nursing of offspring. We predicted that some *C. e. alxaicus* might migrate down from the alpine shrub and meadow zone during winter and spring, while some might stay at high altitude. Further study with GPS collars on *C. e. alxaicus* has been conducted to test the prediction.

CUI *et al.* (2007) and CHANG *et al.* (2010) reported that *U. glaucescens*, *Populus davidiana* Dode, *P. monglica*, *Potentilla* spp., Graminoids (*Stipa* spp., *Poa* spp.), *Caragana* spp. were important in the winter diet of *C. e. alxaicus*. In summer, *C. e. alxaicus* ate 18 plant species of 11 families, which were mostly *Salix microstachya* var. *bordensis* (Nakai) C.F. Fang, *P. davidiana*, *U. glaucescens* and *Agropyron cristatum* (L.) Gaertn. which were mainly distributed in the mountainous open forest and steppe zone, ranging from 1,600 m to 2,000 m altitude, and *C. e. alxaicus* of migratory group migrated down from alpine shrub and meadow zone during winter and spring (Table II) as the amount of food resources declined. Food availability is clearly critical for the nutrition of ungulates, especially for fawns and females (PETTORELLI *et al.* 2005). Adaptation to varying feeding conditions throughout seasons or years was also confirmed by GROOT & HAZEBROEK (1995).

Habitat selection by deer is determined by the presence of both food and hiding cover (BORKOWSKI & UKALSKA 2008). The role of hiding cover in habitat use may be especially important in winter, when cervids reduce their food intake and survive, to a large extent, using their fat reserves (PUTMAN 1988). PEEK *et al.* (1982) studied the role of cover in habitat selection. Two cover types were recognized: thermal cover, in which a forest overstory protects against weather and sun, and hiding (security) cover used to escape and avoid predators and humans. *C. e. alxaicus* in the Helan Mountains preferred habitats near cover to open habitats in winter, but the reverse was found in summer and autumn (Fig. 4). Habitats dominated by *U. glaucescens* and *Juglans regia* Linnaeus were frequently used in winter (Appendix 1). *C. e. alxaicus* are likely to be under low predation risk except from poaching in the Helan Mountains. Therefore, selection of habitat by deer mainly reflects their food requirements and need for protection against severe weather. The habitat selection process presumably results from demands to maximize their energy efficiency while minimizing their movements when searching for food, water and cover (QIAO *et al.* 2006). It differed from other subspecies of red deer, such as *Cervus elaphus xanthopygus* Milne-Edwards, 1867 in northeast China and *Cervus elaphus nelsoni* (Bailey, 1935) in the Rocky Mountains, which preferred coniferous forest and sapling trees and shrubs, feeding on shrub and epicormics in winter (UNSWORTH *et al.* 1998). *C. e. alxaicus* preferred montane coniferous forest and alpine shrub and meadow zones in summer and autumn, but preferred mountain steppe and mountain open forest and steppe zones in winter and spring, similar to *Cervus elaphus scoticus* Lönnberg, 1906 (WELCH *et al.* 1990).

The special alpine topography and arid climate in the Helan Mountain region were the two important factors explaining the habitat selection of *C. e. alxaicus*. Given that precipitation is low in the Helan Mountain region, especially in winter, and the snow coverage is also very low (GENG & YANG 1990), the snow depth and coverage are not important factors initiating the habitat selection and movement of *C. e. alxaicus*. The local population of blue sheep is widespread and numbers more than 10,000 (LIU 2009, LIU *et al.* 2007b). Blue sheep inhabit the mountain open forest and steppe zone, and probably compete with deer for forage, especially in winter and early spring, when food is scarce (LIU *et al.* 2007a). We recorded migration of *C. e. alxaicus* during winter and early spring to the mountain open forest and steppe zone from higher altitude areas (Appendix 1). A reasonable assumption is that blue sheep density, to a certain degree, has a negative influence on *C. e. alxaicus* density, which might also explain deer movements to higher el-

evations in summer. Further study on the niche overlap between blue sheep and *C. e. alxaicus* is underway. AGER *et al.* (2003) reported sympatric populations of Rocky Mountain elk (*C. e. nelsoni*) and mule deer, *Odocoileus hemionus* (Rafinesque, 1817), in northeastern Oregon, where elk exhibited strong daily and seasonal patterns of movements and habitat use under competition from mule deer.

Topography and slope gradient were important factors affecting habitat selection by *C. e. alxaicus*. Generally, deer preferred gentle (<10°), undulating slopes (Appendix 1), but this varied seasonally according to food availability and temperature (Fig. 5). Deer selection of lower slopes was also caused by the predominantly steep topography of the Helan range, which includes only a small proportion of the total area as gentle slopes (DI 1986). *C. e. alxaicus* preferred sunny slopes with direct solar radiation, especially during the cold winter and spring (Fig. 5).

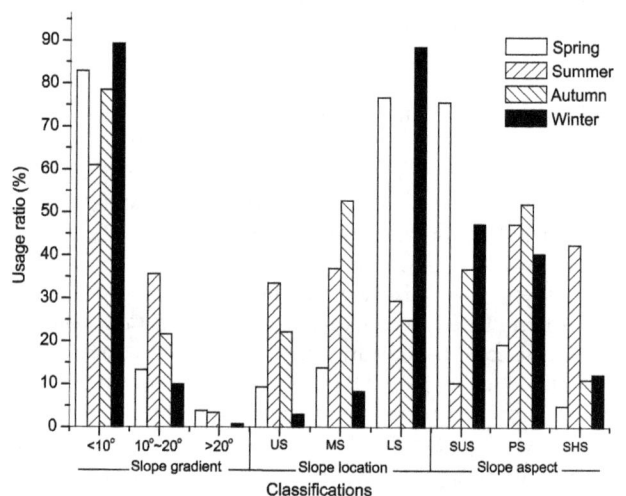

Figure 5. Seasonal changes in the effect of slope characteristics on *Cervus elaphus alxaicus*. (US) Upper Slope, (MS) Middle Slope, (LS) Lower Slope, (SUS) Sunny Slope, (PS) Partial shade slope, (SHS) Shady slope.

Pressure from predators and human disturbance are often considered to be important factors that influence ungulate behavior and habitat selection (BONENFANT *et al.* 2004, BORKOWSKI 2004). However, historically, the main predators of *C. e. alxaicus*, the snow leopard, *Uncia uncia* (Schreber,1775); the gray wolf, *Canis lupus* Linnaeus, 1758; and lynx, *Lynx lynx* (Linnaeus, 1758), became extinct during the 1980s in the Helan mountain region (WANG & SCHALLER 1996). Numerous studies have demonstrated that deer behavior and habitat use are mostly influenced by humans (BORKOWSKI & UKALSKA 2008), including hunting, other human activities (LICOPPE 2006), travel routes and road traffic (YOST & WRIGHT 2001). JIANG et al. (2007) reported that C. e. xanthopygus in northeastern China showed some behavioral plasticity in response to human influences, and the effect of habitat loss or fragmentation caused by human activities are expected to be great (JAEGER et al. 2005). However, in the Helan Mountain region, we found that the deer neither preferred nor avoided human disturbance (p > 0.05) (Appendix 1), and they profited from the programs of Forest-Grass Conservation Project and Shelterbelt Forestry Project and Returning Husbandry to Forestry (Grass) launched in 1996 (ZHAO et al. 2000) and 1999 (ZHAO et al. 2004), respectively. Wild animal hunting and livestock grazing have been forbidden in most of the Helan Mountain region and habitats are generally well-preserved. Therefore, C. e. alxaicus has not suffered much human disturbance, apart from regular ranger, activities whose impact is probably negligible. However, in our data, there was a significant amount of seasonal variation in deer avoidance of human disturbance in the Helan Mountains (Table II). The migration of C. e. alxaicus individuals to lower altitudes for foraging in winter and spring has brought deer closer to areas that are impacted by humans. Deer preferred habitats far from human disturbance at higher elevations during summer and autumn. During our survey, human-related C. e. alxaicus fatalities were low compared to those of other studies (LICOPPE 2006, PÉPIN et al. 2008); therefore, limited persecution and food availability may have facilitated and encouraged C. e. alxaicus individuals to use human-impacted areas in winter and spring.

Many studies on ungulate seasonal movement and habitat selection are based on data provided by GPS-collars or radio-collars, which are very precise. However, because of the difficulties involved in capturing C. e. alxaicus and the rugged topographic conditions in the Helan Mountains, we conducted our surveys by comparing use versus availability of habitats to quantify the seasonal habitat selection of C. e. alxaicus. This study was based on several years of observation and one year of data collection, but the biodiversity and natural conditions in the Helan Mountains have remained stable for years due to its continental climate. Understanding habitat selection by the red deer and its strategies of seasonal movement can be used for decision making on appropriate management and conservation measures (GUISAN & THUILLER 2005) for C. e. alxaicus and its habitat.

ACKNOWLEDGMENTS

Our project was financially supported by the National Nature Science Foundation of China (#30970371), Program for New Century Excellent Talents in University (#NCET-08-0753), the Fundamental Research Funds for the Central Universities (#DL09CA03, #DL13EA01-01), and the Optional Project of Ningxia Hui Autonomous Region (2011-017). We are grateful for the support of Ningxia and Inner Mongolia the Helan Mountain National Nature Reserve. We would like to thank Wang Zhaoding for his great efforts and expertise in conducting field research. We also thank two anonymous reviewers and Tom Dahmer for valuable review of this manuscript.

LITERATURE CITED

AGER, A.A.; B.K. JOHNSON; J.W. KERN & J.G. KIE. 2003. Daily and seasonal movements and habitat use by female rocky mountain elk and mule deer. Journal of Mammalogy 84 (3): 1076-1088.

ALBON, S.D. & R. LANGVATN. 1992. Plant phenology and the benefits of migration in a temperate ungulate. Oikos 65: 502-513.

Table II. Characteristics of 13 ecological factors in the habitats used by Cervus elaphus alxaicus during the year in the Helan Mountain region, China. Significant P-values: * p ≤ 0.001.

Ecological factors	Spring (mean ± SE)	Summer (mean ± SE)	Autumn (mean ± SE)	Winter (mean ± SE)	Kruskal-Wallis H Test χ^2
TD (individuals/100 m²)	2.06 ± 5.065	8.89 ± 10.258	3.13 ± 5.634	2.90 ± 4.642	58.051*
TH (m)	1.91 ± 2.050	3.09 ± 2.561	1.66 ± 2.423	1.90 ± 1.879	3.319*
DtT (m)	13.28 ± 11.243	9.83 ± 11.437	17.01 ± 10.845	12.81 ± 11.129	60.433*
SD (individuals/100 m²)	12.08 ± 10.954	19.20 ± 13.947	12.29 ± 11.127	6.70 ± 10.808	76.191*
SH (m)	0.95 ± 1.012	0.90 ± 0.567	0.64 ± 0.611	0.64 ± 3.499	71.366*
DtS (m)	3.24 ± 3.782	1.51 ± 2.143	3.29 ± 3.795	6.30 ± 4.252	78.617*
HC (%)	67.84 ± 23.194	82.40 ± 23.873	88.56 ± 17.755	66.98 ± 24.664	126.156*
SG (°)	5.55 ± 6.151	10.21 ± 10.977	6.40 ± 3.350	4.61 ± 6.081	75.136*
AL (m)	1,903.62 ± 427.717	2,512.13 ± 375.169	2,512.46 ± 493.990	1,890.61 ± 570.243	203.163*
DtW (m)	1,060.92 ± 1351.187	1,176.24 ± 863.845	1,510.63 ± 1221.745	1,327.50 ± 1431.247	26.920*
DtH (m)	2,933.31 ± 1907.579	5,380.26 ± 3124.869	7,255.49 ± 3343.345	3,134.27 ± 1913.725	158.822*
DtR (m)	120.04 ± 185.693	186.05 ± 168.370	247.47 ± 222.321	75.61 ± 217.001	178.969*
HiC (%)	47.36 ± 27.710	35.80 ± 28.239	52.68 ± 27.888	67.25 ± 24.434	86.234*

BONENFANT, C.; E.L. LEIF; A. MYSTERUD; R. LANGVATN; N.C. STENSETH; J.M. GAILLARD & F. KLEIN. 2004. Multiple causes of sexual segregation in European red deer: enlightenments from varying breeding phenology at high and low latitude. **Proceedings of the Royal Society B Biological Sciences 271**: 883-892.

BORKOWSKI, J. & J. UKALSKA. 2008. Winter habitat use by red and roe deer in pine-dominated forest. **Forest Ecology and Management 255**: 468-475.

BORKOWSKI, J. 2004. Distribution and habitat use by red and roe deer following a large forest fire in South-western Poland. **Forest Ecology and Management 201**: 287-293.

BRINKMAN, T.J.; C.S. DEPERNO; J.A. JENKS; B.S. HAROLDSON & R.G. OSBORN. 2005. Movement of female white-tailed deer: effects of climate and intensive row-crop agriculture. **The Journal of Wildlife Management 69** (3): 1099-1111.

BYERS, C.R.; K. STEINHORST & P.R. KRAUSMAN. 1984. Clarification of a technique for analysis of utilization-availability data. **The Journal of Wildlife Management 48**: 1050-1053.

CHANG, H. & Q.Z. XIAO. 1988. Selection of winter habitat of Red deer in Dailing region. **Acta Theriologica Sinica 8** (2): 81-88.

CHANG, Y.; M.M. ZHANG; Z.S. LIU; T.H. HU & Z.G. LI. 2010. Summer Dies of Sympatric Blue Sheep (*Pseudois nayaur*) and Red Deer (*Cervus elaphus alxaicus*) in the Helan Mountains, China. **Acta Ecologica Sinica 30** (6): 1486-1493.

CUI, D.Y.; Z.S. LIU; X.M. WANG; H. ZHAI; T.H. HU & Z.G. LI. 2007. Winter food-habits of red deer in the Helan Mountains, China. **Zoological Research 28** (4): 383-388.

DI, W.Z. 1986. **Plantae vasculares the Helan Mountain**. Xi'an, Northwestern University Press.

FRYXELL, J.M. & A.R.E. SINCLAIR. 1988. Causes and consequences of migration by large herbivores. **Trends in Ecology & Evolution 3** (9): 237-241.

GENG, K. & Z.R. YANG. 1990. Climatic characteristics and climatic landforms in Helan Mountain. **Yantai Teacher's College Journal (Natural Science Edition) 6** (2): 49-56.

GROOT, B. & E. HAZEBROEK. 1995. Ingestion and diet composition of red deer (*Cervus elaphus L.*) in the Netherlands from 1954 till 1992. **Mammalia 59**: 187-195.

GROVENBURG, T.W.; J.A. JENKS; R.W. KLAVER; C.C. SWANSON; C.N. JACQUES & D. TODRY. 2009. Seasonal movements and home ranges of white-tailed deer in north-central South Dakota. **Canadian Journal of Zoology 87** (10): 876-885.

GUISAN, A. & W. THUILLER. 2005. Predicting species distribution: offering more than simple habitat models. You have full text access to this content. **Ecology Letters 8** (9): 993-1009.

HUTTO, R.L. 1985. **Habitat selection in birds**. New York, Academic Press, 558p.

IGOTA, H.; M. SAKURAGI; H. UNO; K. KAJI; M. KANEKO; R. AKAMATSU & K. MAEKAWA. 2004. Seasonal migration patterns of female sika deer in eastern Hokkaido, Japan. **Ecological Research 19** (2): 169-178.

JAEGER, J.A.G; J. BOWMAN; J. BRENNAN; L. FAHRIG; D. BERT; J. BOUCHARD; N. CHARBONNEAU; K. FRANK; B. GRUBER & K.T. VON TOSCHANOWITZ.

2005. Predicting when animal populations are at risk from roads: an interactive model of road avoidance behavior. **Ecological Modelling 185** (2-4): 329-348.

JARNEMO, A. 2008. Seasonal migration of male red deer (*Cervus elaphus*) in southern Sweden and consequences for management. **European Journal of Wildlife Research 54** (2): 327-333.

JEPPESEN, J.L. 1987. Impact of human disturbance on home range, movements and activity of red deer (*Cervus elaphus*) in a Danish environment. **Danish Review of Game Biology 13** (2): 35-38.

JIANG, G.S.; J.Z. MA & M.H. ZHANG. 2007. Effects of human disturbance on movement, foraging and bed selection in red deer *Cervus elaphus xanthopygus* from the Wandashan Mountains, northeastern China. **Acta Theriologica 52** (4): 435-446.

JIANG, Y.; M. KANG; S. LIU; L.S. TIAN & M.D. LEI. 2000. A study on the vegetation in the east side of Helan Mountain. **Plant Ecology 149** (2): 119-130.

KUNKEL, K. & D.H. PLETSCHER. 2001. Winter hunting patterns of wolves in and Near Glacier National Park, Montana. **The Journal of Wildlife Management 65** (3): 520-530.

LICOPPE, A.M. 2006. The diurnal habitat used by red deer (*Cervus elaphus L.*) in the Haute Ardenne. **European Journal of Wildlife Research 52** (3): 164-170.

LIU, Z.S. 2009. **Notes of vertebrates in the Helan Mountain**. Yinchuan, Ningxia People's Publishing House.

LIU, Z.S.; X.M. WANG; Z.G. LI; D.Y. CUI & X.Q. LI. 2007a. Feeding habitats of blue sheep (*Pseudois nayaur*) during winter and spring in the Helan Mountains, China. **Frontiers of Biology in China 2** (1): 100-107.

LIU, Z.S.; X.M. WANG; Z.G. LI; H. ZHAI & T.H. HU. 2007b. Distribution and abundance of blue sheep in the Helan Mountains, China. **Chinese Journal of Zoology 42** (3): 1-8.

LIU, Z.S.; J.P. WU & L.W. TENG. 2002. Time budget and behavior pattern of semi free *Cervus nippon* in spring. **Chinese Journal of Ecology 21** (6): 29-32.

LOVARI, S.; J. HERRERO; J. CONROY; T. MARAN; G. GIANNATOS; M. STUBBE; S. AULAGNIER; T. JDEIDI; M. MASSETI; I. NADER; K. DE SMET & F. CUZIN. 2008. *Cervus elaphus*. In: IUCN 2010 (Ed). **IUCN Red List of Threatened Species. Version 2010.4**. Available online at: http://www.iucnredlist.org/apps/redlist/details/41785/0 [Accessed: 12/X/2011].

MANLY, B.F.J.; L.L. MCDONALD; D.L. THOMAS; T.L. MCDONALD & W.P. ERICKSON. 2002. **Resource selection by animals: statistical design and analysis for field studies**. London, Chapman & Hall.

MARCUM, C.L. & D.O. LOFTSGAARDEN. 1980. A nonmapping technique for studying habitat preferences. **The Journal of Wildlife Management 44** (4): 963-968.

MOEN, A.N. 1976. Energy conservation by white-tailed deer in the winter. **Ecology 57**: 192-198.

MORRISON, M.L.; B.G. MARCOT & R.W. MANNAN. 1998. **Wildlife-habitat relationships: concepts and applications**. Madison, The University of Wisconsin Press.

NEU, C.W.; C.R. BYERS & J.M. PEEK. 1974. A technique for analysis of utilization-availability data. **The Journal of Wildlife Management 38** (3): 541-545.

PEEK, J.M.; M.D. SCOTT; L.J. NELSON & D.J. PIERCE. 1982. Role of cover in habitat management for big game in northwestern United States. **Transactions of North American Wildlife and Natural Resources Conference 47**: 363-373.

PÉPIN, D.; C. ADRADOS; G. JANEAU; J. JOACHIM & C. MANN. 2008. Individual variation in migratory and exploratory movements and habitat use by adult red deer (*Cervus elaphus*) in a mountainous temperate forest. **Ecological Research 23** (2): 1005-1013.

PETTORELLI, N.; A. MYSTERUD; N.G. YOCCOZ; R. LANGVATN & N.C. STENSETH. 2005. Importance of climatological downscaling and plant phenology for red deer in heterogeneous landscapes. **Proceedings of the Royal Society Biological Sciences 272**: 2357-2364.

PUTMAN, R. 1988. **The Natural History of Deer.** London, Christopher Helm.

QIAO, J.F.; W.K. YANG & X.Y. GAO. 2006. Natural diet and food habitat use of the Tarim red deer, *Cervus elaphus yarkandensis*. **Chinese Science Bulletin 51** (Supp. I): 147-152.

SCHMITZ, O.J. 1991. Thermal constrains and optimization of winter feeding and habitat choice in white-tailed deer. **Ecography 14**: 104-111.

TAKHTAJAN, A. 1986. **Floristic regions of the world.** Berkeley, University of California Press, 62p.

UNSWORTH, J.W.; L. KUCK; E.O. GARTON & B.R. BUTTERFIELD. 1998. Elk habitat selection on Clearwater national forest, Idaho. **The Journal of Wildlife Management 62** (4): 1255-1263.

VAUGHAN, T.A.; J.M. RYAN & N.J. CZAPLEWSKI. 2000. **Mammalogy.** Orlando, Saunders College Publishing, 4th ed.

WANG, S. 1998. **China red data book of endangered animals: mammals.** Beijing, Science Press.

WANG, X.M.; M. LI; S.Y. TANG; Z.X. LIU; Y.G. LI & H.L. SHENG. 1999. The Study of resource and conservation of artiodactyls in the Helan Mountain. **Chinese Journal of Zoology 34** (5): 26-29.

WANG, X.M. & G.B. SCHALLER. 1996. Status of large mammals in Inner Mongolia, China. **Journal of East China Normal University 6** (Special Issue): 94-104.

WELCH, D.; B.W. STAINES; D.C. CATT & D. SCOTT. 1990. Habitat usage by red (*Cervus elaphus*) and roe deer (*Capreolus capreolus*) in a Scottish Sitka Apruce plantation. **Journal of Zoology 221** (3): 453-476.

WHITE, C.G. & R.A. GARROTT. 1990. **Analysis of Wildlife Radio-Tracking Data.** San Diego, Academic Press.

YOST, A.C. & R.G. WRIGHT. 2001. Moose, Caribou,and Grizzly Bear distribution in relation to road traffic in Denali National Park, Alaska. **Arctic 54** (1): 41-48.

ZHANG, X.L.; Z.G. LI; Z. LI; Y.X. MA; T.S. ZHANG & H. ZHAI. 2006. Studies on population quantity and dynamics of red deer in spring for Helanshan Mountain of Ningxia. **Journal of Ningxia University (Natural Science Edition) 27** (3): 263-265.

ZHANG, X.L.; Z.G. LI; H.J. LÜ & H.L. GUO. 1999. Studies on ecological habits and population dynamics of Ningxia red deer. **Ningxia Journal of Agriculture and Forestry Science and Technology** (Supp. I): 22-27.

ZHAO, C.L.; Z.G. LI; H.J. LÜ; T. LI; T.H. HU; H. ZHAI; H.L. WANG; Y.X. LI; Z.C. WANG; Z.L. CHANG; R.F. JIAO; H.Y. SHI & H.P. XU. 2000. Vegetation cover degree monitoring in the Helanshan Mountain project area of Sino-Germany cooperation Ningxia shelter-forest project. **Ningxia Journal of Agriculture and Forestry Science and Technology** (Supp. I): 6-14.

ZHAO, Y.S.; P. SUN; X.Q. ZHOU & D.J. CUI. 2004. The role of closing hills for reforestation on eco-environment in the Helanshan Mountains. **Inner Mongolia Forestry Investigation and Design 27** (4): 7-9.

Appendix 1. The proportion of 18 ecological factors in used (Obs) plots by *Cervus elaphus alxaicus* and comparison (Exp) plots during different seasons in the Helan Mountain region, China. Selection was determined according to the method of BYERS *et al.* (1984). Factors: (AL) Altitude, (VT) Vegetation types, (TO) Topography, (DT) Dominant tree, (TD) Tree density, (TH) Tree height, (DtT) Distance to the nearest tree, (SD) Shrub density, (SH) Shrub height, (DtS) Distance to the nearest shrub, (HC) Herb coverage, (SG) Slope gradient, (SL) Slope location, (SA) Slope aspect, (DtW) Distance to water resource, (DtH) Distance to human disturbance, (DtR) Distance to bare rock, (HiC) Hiding cover.

Factors	Item	Spring			Summer			Autumn			Winter		
		Obs	Exp	χ^2(Sig)	Obs	Exp	χ^2(Sig)	Obs	Exp	χ^2(Sig)	Obs	Exp	χ^2(Sig)
VT	MS	48(=)	45	p < 0.05	0(−)	45	p < 0.01	0(−)	26	p < 0.01	60(=)	56	p > 0.05
	MOFS	107(=)	108		4(−)	28		36(=)	47		22(=)	26	
	MCF	15(=)	17		104(+)	71		69(=)	48		31(=)	31	
	ABM	11(=)	11		38(+)	23		39(+)	17		18(=)	18	
TO	Smooth undulating slope	156(+)	12	p < 0.01	82(+)	8	p < 0.01	120(+)	7	p < 0.01	112(+)	44	p < 0.01
	Moderately broken slope	19(−)	42		64(+)	16		24(+)	5		18(−)	43	
	Distinctly broken slope	6(−)	70		0(−)	97		0(−)	83		0(−)	36	
	Scree/landslide	0(−)	26		0(−)	10		0(−)	21		0(=)	5	
	Cliff	0(−)	31		0(−)	36		0(−)	22		1(=)	3	
DT	*Ulmus glaucescens*	54(=)	54	p > 0.05	0(−)	23	p < 0.01	9(=)	17	p < 0.01	47(=)	54	p < 0.01
	Ziziphus jujuba var.*spinosa*(Bunge)Hu	2(=)	0		0(=)	0		0(=)	1		5(=)	0	
	Salix babylonica©	3(=)	0		0(=)	1		0(=)	1		2(=)	0	
	Juniperus rigida	11(=)	16		20(=)	19		11(=)	12		8(=)	8	
	Pinus tabulaeformis	1(=)	3		1(=)	4		3(=)	1		2(=)	9	
	Picea crassifolia	4(=)	8		32(=)	27		18(=)	13		0(−)	10	
	Prunus armeniaca	2(=)	0		0(=)	0		0(=)	0		8(=)	2	
	Juglans regia	3(=)	0		0(=)	0		0(=)	0		59(=)	48	
	Salix microstachya	0(=)	0		0(=)	0		0(=)	0		0(=)	0	
	No tree	86(=)	100		49(−)	91		92(=)	92		0(=)	0	
	Mixture	15(=)	0		43(+)	2		11(+)	1		0(=)	0	
TD	<2 tree/100 m²	114(=)	123	p < 0.05	51(−)	107	p < 0.01	96(=)	103	p < 0.01	75(+)	59	p < 0.01
	2~4 tree/100 m²	43(=)	46		10(−)	31		10(=)	18		31(=)	42	
	> 4 tree/100 m²	24(=)	12		85(+)	29		38(+)	17		21(=)	30	
TH	<4 m	148(−)	172	p < 0.01	72(−)	129	p < 0.01	113(=)	119	p > 0.05	116(=)	113	p < 0.01
	4~6 m	30(+)	7		53(+)	31		21(=)	14		13(=)	18	
	> 6 m	3(=)	2		21(+)	7		10(=)	5		2(=)	0	
DtT	<1 m	3(=)	1	p > 0.05	23(+)	2	p < 0.01	2(=)	0	p > 0.05	4(=)	9	p < 0.05
	1~3 m	64(=)	65		69(=)	70		37(=)	39		42(=)	50	
	> 3 m	114(=)	115		54(−)	95		105(=)	99		85(=)	72	
SD	<5 tree/100 m²	52(−)	81	p < 0.01	16(−)	73	p < 0.01	50(=)	49	p < 0.01	84(=)	75	p < 0.05
	5~10 tree/100 m²	52(−)	85		37(−)	81		25(=)	62		18(=)	31	
	> 10 tree/100 m²	77(+)	15		93(+)	13		69(+)	27		29(=)	25	
SH	<1.3 m	134(−)	172	p < 0.01	111(−)	161	p < 0.01	120(=)	138	p > 0.05	123(=)	126	p > 0.05
	1.3~1.7 m	21(+)	8		20(+)	4		11(=)	0		4(=)	3	
	> 1.7 m	26(+)	1		15(+)	2		11(=)	0		4(=)	2	
DtS	<0.5 m	11(+)	5	p < 0.01	28(+)	5	p < 0.01	14(=)	5	p < 0.01	10(=)	7	p < 0.01
	0.5~1.5 m	89(−)	106		75(=)	81		66(=)	71		22(−)	50	
	> 1.5 m	81(=)	70		43(−)	81		64(=)	62		99(=)	74	

Continues

Appendix 1. Continued.

Factors	Item	Spring			Summer			Autumn			Winter		
		Obs	Exp	χ²(Sig)	Obs	Exp	χ²(Sig)	Obs	Exp	χ²(Sig)	Obs	Exp	χ²(Sig)
HC	50%	48(–)	170	p < 0.01	16(–)	142	p < 0.01	9(–)	127	p < 0.01	33(–)	118	p < 0.01
	50~80%	73(+)	11		26(=)	19		16(=)	10		53(+)	11	
	> 80%	60(+)	0		104(+)	6		119(+)	1		45(+)	2	
SG	<10°	150(+)	16	p < 0.01	89(+)	10	p < 0.01	113(+)	2	p < 0.01	117(+)	27	p < 0.01
	10°~20°	24(–)	45		52(+)	29		31(=)	33		13(–)	45	
	> 20°	7(–)	120		5(–)	128		0(–)	103		1(–)	59	
SL	Upper slope	17(–)	53	p < 0.01	49(=)	65	p < 0.01	32(=)	45	p < 0.01	4(=)	7	p < 0.01
	Middle slope	25(–)	85		54(=)	75		76(=)	77		11(–)	68	
	Lower slope	139(+)	43		43(+)	27		36(+)	16		116(+)	56	
SA	Sunny slope	137(+)	55	p < 0.01	15(–)	118	p < 0.01	53(–)	93	p < 0.01	62(+)	39	p < 0.01
	Partial shade slope	35(–)	76		69(+)	34		75(+)	27		53(=)	49	
	Shady slope	9(–)	50		62(+)	15		16(=)	18		16(–)	43	
AL	< 1600 m	56(=)	52	p > 0.05	0(–)	44	p < 0.01	5(–)	27	p < 0.01	52(=)	48	p > 0.05
	1600~2000 m	52(=)	56		4(–)	29		6(–)	27		30(=)	34	
	2000~3000 m	62(=)	62		119(+)	74		94(+)	67		34(=)	31	
	> 3000 m	11(=)	11		23(=)	20		39(+)	17		15(=)	18	
DtW	< 500 m	101(=)	97	p > 0.05	42(–)	67	p < 0.01	44(=)	39	p > 0.05	54(=)	52	p > 0.05
	500~1 000 m	19(=)	22		19(=)	15		21(=)	15		19(=)	19	
	> 1 000 m	61(=)	62		85(=)	85		79(=)	84		58(=)	60	
DtH	< 500 m	34(=)	34	p > 0.05	2(=)	6	p > 0.05	0(=)	0	p > 0.05	8(=)	6	p > 0.05
	500~1 000 m	16(=)	16		4(=)	11		0(=)	5		18(=)	14	
	> 1 000 m	131(=)	131		140(=)	150		144(=)	133		105(=)	111	
DtR	< 50 m	80(–)	143	p < 0.01	14(–)	109	p < 0.01	7(–)	30	p < 0.01	87(–)	109	p < 0.01
	50~100 m	52(+)	18		44(+)	25		48(=)	53		30(+)	13	
	> 100 m	49(+)	20		88(+)	33		89(+)	55		14(=)	9	
HiC	> 25%	41(+)	2	p < 0.01	69(+)	0	p < 0.01	30(+)	0	p < 0.01	13(+)	0	p < 0.01
	25~50%	59(+)	7		35(+)	13		39(+)	4		17(+)	4	
	50~75%	46(=)	45		22(=)	21		33(+)	7		42(+)	12	
	> 75%	35(–)	127		20(–)	133		42(–)	127		59(–)	115	

The symbols (=), (+) and (–) located beside the observed values signify that those values were found to be (according to Bonferroni Intervals) in equal, greater, or lesser proportion than the respective expected values. (+) indicates that red deer preferred category i; (–) indicates that red deer avoided category i; (–) indicates that red deer showed no obvious selection according to category i.
Difference between used and comparison plots is statistically significant at p< 0.05 based on a chi–square goodness–of–fit test.

Molecular evidence for the polyphyly of *Bostryx* (Gastropoda: Bulimulidae) and genetic diversity of *Bostryx aguilari*

Jorge L. Ramírez[1, 2] & Rina Ramírez[1]

[1] *Departamento de Malacología y Carcinología, Museo de Historia Natural, Universidad Nacional Mayor de San Marcos, Apartado 14-0434, Lima-14, Perú.*
[2] *Coresponding author. E-mail: jolobio@hotmail.com*

ABSTRACT. *Bostryx* is largely distributed in Andean Valleys and Lomas formations along the coast of Peru and Chile. One species, *Bostryx aguilari*, is restricted to Lomas formations located in the Department of Lima (Peru). The use of genetic information has become essential in phylogenetic and population studies with conservation purposes. Considering the rapid degradation of desert ecosystems, which threatens the survival of vulnerable species, the aim of this study was, first, to resolve evolutionary relationships within *Bostryx* and to determine the position of *Bostryx* within the Bulimulidae, and second, to survey the genetic diversity of *Bostryx aguilari*, a species considered rare. Sequences of the mitochondrial 16S rRNA and nuclear rRNA regions were obtained for 12 and 11 species of Bulimulidae, respectively, including seven species of *Bostryx*. Sequences of the 16S rRNA gene were obtained for 14 individuals (from four different populations) of *Bostryx aguilari*. Phylogenetic reconstructions were carried out using Neighbor-Joining, Maximum Parsimony, Maximum Likelihood and Bayesian Inference methods. The monophyly of *Bostryx* was not supported. In our results, *B. solutus* (type species of *Bostryx*) grouped only with *B. aguilari*, *B. conspersus*, *B. modestus*, *B. scalariformis* and *B. sordidus*, forming a monophyletic group that is strongly supported in all analyses. In case the taxonomy of *Bostryx* is reviewed in the future, this group should keep the generic name. *Bostryx aguilari* was found to have both low genetic diversity and small population size. We recommend that conservation efforts should be increased in Lomas ecosystems to ensure the survival of *B. aguilari*, and a large number of other rare species restricted to Lomas.

KEY WORDS. Land snails; Lomas; molecular systematic; Orthalicoidea; rRNA.

Among Neotropical land snails, Bulimulidae is one of the most diverse (BREURE 1979, RAMÍREZ *et al.* 2003a). The phylogenetic relationships among its members, however, are still problematic. Genera such as *Bostryx* and *Scutalus* are distributed in desert ecosystems and are adapted to survive under some of the harshest climatic conditions (AGUILAR & ARRARTE 1974, RAMÍREZ *et al.* 2003b). *Bostryx* is found in Argentina, Bolivia, Chile, Peru, Ecuador, and possibly in Venezuela (BREURE 1979). It is spread throughout Peru, but is more prevalent in the Pacific coastal desert and the western Andean slopes (RAMÍREZ *et al.* 2003a). Among the *Bostryx* species, *B. aguilari* Weyrauch, 1967 (Fig. 2) is of particular interest, due to its vulnerable status and lack of information on the genetic diversity of its populations. It is a species associated with bushy Lomas and is found at elevations of 200 to 600 m. *Bostryx aguilari* was originally reported for the Lomas of Amancaes, Atocongo and Pachacamac, but there is also a record of an unknown locality near the city of Junín, in the Peruvian Andes (WEYRAUCH 1967). To date, this species has been reported for at least 12 Lomas in the Department of Lima, and is distributed from the Lomas of Lachay, in the north, to the Lomas of Pacta, in the south (R. Ramírez unpublished data). *Bostryx aguilari*, unlike other gastropod species of Lomas, is very

difficult to find, not only alive, but also as shell remnant. An exception is the Lomas of Atocongo, where *B. aguilari* can be found more easily. The Lomas formations are seasonal ecosystems occurring along the coast of Peru and Chile, between 8° and 30° SL (RUNDEL *et al.* 1990), where the main source of humidity are fogs brought from the Pacific Ocean during the winter months (DILLON *et al.* 2003). Periodically (every few years), the El Niño-Southern Oscillation (ENSO) alters the seasonality of the Lomas, causing summer drizzles that promote the development of out of season vegetation. The steady growth of cities is threatening the biodiversity in desert ecosystems, and particularly the Lomas, which are still poorly known and described. They are beginning to disappear at a fast pace, and with them, their endemic species. Our objectives are to resolve evolutionary relationships within *Bostryx* to clarify the position of the genus among the Bulimulidae, and to survey the genetic diversity of *B. aguilari*, a rare species threatened by loss of habitat and human pressure. Because the maintenance of genetic diversity is vital to the survival of populations and species, this information will be crucial to the establishment of guidelines for the conservation of *B. aguilari* and for the Lomas ecosystems they inhabit.

MATERIAL AND METHODS

Species of *Bostryx* were collected from several Peruvian localities comprising Lomas, Inter-Andean valleys and tropical forests (Table I). We also included species of *Scutalus*, *Drymaeus*, *Naesiotus* and *Neopetraeus* as outgroups (Table I and Figs 1-12). Samples were fixed in 96% ethanol and deposited in the collection at Department of Malacology and Carcinology, Museum of Natural History, San Marcos University. Individuals of *B. aguilari* were obtained from seven Lomas, all located in the Department of Lima in the central coast of Peru (Amancaes, Atocongo, Iguanil, Lúcumo, Manzano, Paraíso and Picapiedra), although live specimens were only found in three locations (Amancaes, Atocongo and Iguanil).

DNA was isolated using a modified CTAB method (DOYLE & DOYLE 1987) from 1-2 mm³ of tissue from the snail foot. The tissue sample was digested in 300 µL of extraction buffer (100 mM Tris/HCl, 1.4 M NaCl, 20 mM EDTA, 2% CTAB, 2% PVP and 0.2% of β-mercaptoethanol) with 0.05 mg Proteinase K and incubated at 60°C for approximately two hours. Proteins were removed twice with 310 µL of chloroform/isoamyl alcohol (24:1), centrifugation was at 13,000 rpm for 15 minutes before removal of the aqueous phase. The DNA was precipitated using 600 µL of cold absolute ethanol and 25 µL of 3M ammonium acetate and incubated at -20°C for at least 30 minutes, then centrifuged at 13,000 rpm for 15 min. The pellet obtained was washed twice in 1 mL of 70% ethanol and centri-fuged at 13,000 rpm for 15 min. Finally, the pellet was dried at room temperature for 24 hours, resuspended in 50 µL of double-distilled water at 37°C, and stored at -20°C.

Using total genomic DNA, we amplified and sequenced the 16S rRNA gene and the rRNA gene-cluster. Amplifications were carried out using the polymerase chain reaction (PCR) (SAIKI *et al.* 1988). For the amplification of the 16S rRNA gene, we used primers developed by (R. Ramírez unpublished data): 16SF-104 (5'-GACTGTGCTAAGGTAGCATAAT-3') and 16SR-472 (5'-TCGTAGTCCAACATCGAGGTCA-3'). To obtain the nuclear rRNA gene-cluster, including the 3'-end of the 5.8S rRNA gene, the complete internal transcribed spacer 2 (ITS-2) region, and the 5'-end of the large subunit (28S rRNA) gene, we used primers LSU1 and LSU3 developed for mollusks by WADE & MORDAN (2000).

For the 16S rRNA, PCR amplification were performed in a final volume of 30 µL, containing 1 U of *Taq* DNA polymerase (Fermentas Inc., Maryland, US), 1.5 mM MgCl$_2$, 0.2 mM dNTP and 0.2 iM of each primer, 1X buffer, and 3 µL of DNA template. Amplifications consisted of 35 cycles of denaturation at 94°C for 30s, annealing at 48°C for 30s, and extension at 72°C for 60s. PCR reagents used for the amplification of nuclear markers were the same as above; amplifications consisted of 35 cycles of 96°C for 60s, 50-55°C for 30s and 72°C for 60s. Amplicons were electrophoresed on 1% agarose gels to verify the amplification. PCR products were purified and sequenced for both strands using the commercial services at Macrogen USA.

Figures 1-12. Species of Bulimulidae analyzed in this work: (1) *Bostryx solutes*, MUSM 5515-82G1; (2) *B. aguilari* MUSM 5501-42A3; (3) *B. conspersus* MUSM 5505-23F1; (4) *B. modestus* MUSM 5507-74F1; (5) *B. sordidus* MUSM 5511-14A15; (6) *B. scalariformis* MUSM 5510-75.3; (7) *B. turritus* MUSM 5514-1F1; (8) *Scutalus proteus* MUSM 5519-35G1; (9) *S. versicolor* MUSM 5518-11.8; (10) *Drymaeus arcuatostriatus* MUSM 5516-59Eu; (11) *Neopetraeus tessellates* MUSM 4020-62E1; (12) *Naesiotus geophilus* MUSM 5517-18G1. Photographs: 1 and 8 by D. Maldonado; 2, 10 and 11 by J. Ramirez; 3-7 by A. Chumbe; and 12 by V. Borda. Escale bars: 1, 3, 4, 6, 7, 12 = 2 mm; 2, 5, 8-11 = 5 mm.

Table I. Voucher information and GenBank accession numbers for individuals included in the analyses. Sequences generated for this study are in bold. MUSM: Museum of Natural History, San Marcos University.

Species	Population	Voucher MUSM	GenBank accession 16S	GenBank accession LSU 1-3
Bostryx aguilari Weyrauch, 1967	Lima: Amancaes[1]	**MUSM 5501-42A3**	**HQ225813**	**HM116230**
	Lima: Amancaes[1]	**MUSM 5501-43A6**	**HQ225814**	
	Lima: Amancaes[1]	**MUSM 5500-53.10**	**HQ225815**	
	Lima: Amancaes[4]	MUSM 5041-Ama3	JQ669492	
	Lima: Iguanil[1]	**MUSM 5504-29A**	**HQ225820**	**JQ669461**
	Lima: Iguanil[1]	**MUSM 5504-31A**	**HQ225821**	
	Lima: Iguanil[1]	**MUSM 5504-26A**	**HQ225819**	
	Lima: Iguanil[1]	**MUSM 5504-32A**	**HQ225822**	
	Lima: Atocongo[1]	**MUSM 5503-25F1**	**HQ225816**	
	Lima: Atocongo[1]	**MUSM 5502-17F5**	**HQ225817**	
	Lima: Atocongo[1]	**MUSM 5502-19F8**	**HQ225818**	
	Lima: Atocongo[1]	**MUSM 5505-23F1**	**HM057172**	**JQ669462**
	Lima: Atocongo[4]	MUSM 5042-Atoc39	JQ669493	
	Lima: Lachay[4]	MUSM 5043-Lach.u	JQ669494	
Bostryx conspersus (Sowerby, 1833)	Lima: Atocongo[1]	**MUSM 5505-23F1**	**HM057173**	
	Lima: Iguanil[1]	**MUSM 5036-Ig5**		**JQ669463**
	Lima: Lachay[1]	**MUSM 5506-51G3**	**JQ669456**	**JQ669464**
Bostryx modestus (Broderip, in Broderip & Sowerby 1832)	Lima: Atocongo[1]	**MUSM 5507-74F1**	**HM057174**	
	Lima: Paraiso[1]	**MUSM 5508-6F1**	**JQ669457**	**JQ669465**
Bostryx scalariformis (Broderip, in Broderip & Sowerby 1832)	Lima: Pasamayo[1]	**MUSM 5510-75.3**	**HM057181.1**	**JQ669466**
	Lima: N Pan American Hwy Km 115[1]	MUSM 5509-83.a	FJ969796.1	
	Lima: N Pan American Hwy Km 115[1]	MUSM 5509-84.b		**JQ669467**
Bostryx solutus (Troschel, 1847)	Lima: Infiernillo[2]	**MUSM 5515-82G1**	**JQ669458**	**JQ669468**
	Lima: Infiernillo[2]	**MUSM 5515-80G6**	**HQ225824**	
Bostryx sordidus (Lesson, 1826)	Lima: Iguanil[1]	**MUSM 5511-14A15**	**HM057176.1**	
	Lima: Lupin[1]	MUSM 5512-62.12	FJ969797.1	
	Lima: Santa Eulalia[2]	**MUSM 5513-77E5**	**JQ669459**	**JQ669469**
Bostryx turritus (Broderip, in Broderip & Sowerby 1832)	Lima: Santa Eulalia[2]	**MUSM 5514-1F1**	**HM057175**	**JQ669470**
	Lima: Santa Eulalia[2]	**MUSM 5514-4F4**	**JQ669460**	**JQ669471**
Bostryx bilineatus (Sowerby, 1833)	Ecuador[5]			HM027501
Bostryx strobeli (Parodiz, 1956)	Argentina[5]			HM027498
Bulimulus guadalupensis (Bruguière, 1789)	Puerto Rico[6]			AY841298
Bulimulus tenuissimus (Férussac, 1832)	Brazil[5]			HM027507
Bulimulus sporadicus (d'Orbigny, 1835)	Brazil[6]			AY841299
Clessinia pagoda Hylton Scott, 1967	Argentina[5]			HM027497
Drymaeus discrepans (Sowerby, 1833)	Guatemala[6]			AY841300
Drymaeus inusitatus (Fulton, 1900)	Costa Rica[5]			HM027503
Dryamaues laticinctus (Guppy, 1868)	Dominica[5]			HM027492
Drymaeus serratus (Pfeiffer, 1855)	Peru[5]			HM027499
Drymaeus arcuatostriatus (Pfeiffer, 1855)	San Martin: Juan Guerra[3]	**MUSM 5516-59Eu**	**HM057178**	**JQ669472**
Naesiotus quitensis (Pfeiffer, 1848)	Ecuador[5]			HM027510
Naesiotus stenogyroides (Guppy, 1868)	Dominica[5]			HM027494
Naesiotus geophilus Weyrauch, 1967	San Martin: Juan Guerra[3]	**MUSM 5517-18G1**	**HM057180**	
Neopetraeus tessellatus (Shuttleworth, 1852)	Ancash: nr. Pontó[2]	**MUSM 4020-62E1**	**HM057179**	**JQ669473**
Plagiodontes multiplicatus Döring, 1874	Argentina[5]			HM027496
Scutalus proteus (Broderip, in Broderip & Sowerby 1832)	Lima: Santa Eulalia[2]	**MUSM 5519-35G1**	**HQ225823**	**JQ669474**
Scutalus versicolor (Broderip, in Broderip & Sowerby 1832)	Lima: Mongón[1]	**MUSM 5518-11.8**	**FJ969798**	**JQ669475**
Spixia popana Döring, 1876	Argentina[5]			HM027502
Placostylus bivaricosus (Gascoin, 1885)	Lord Howe Island[7]			AY165846
Placostylus bivaricosus (Gascoin, 1885)	Lord Howe Island[7]			AY165850

[1]Lomas, [2]Andean region, [3]Tropical forest, [4]R. Ramírez (unpublished data), [5]BREURE *et al.* (2010), [6]WADE *et al.* (2006); [7]PONDER *et al.* (2003).

Sequences of the mitochondrial 16S rRNA and nuclear rRNA regions were obtained for 12 and 11 species of Bulimulidae, respectively, including seven species of *Bostryx*. Eleven sequences of the partial 16S rRNA gene were obtained from different populations of *B. aguilari*; three samples were sequenced with the LSU1/LSU3 primer pair. Nineteen sequences were retrieved from GenBank. Voucher information and GenBank accession numbers are given in Table I.

Sequences were edited with Chromas (McCarthy 1996), assembled with CAP3WIN (Huang & Madan 1999), aligned with ClustalX 2.0 (Larkin *et al.* 2007) and adjusted manually in BioEdit v7.0.9 (Hall 1999). Gaps were treated as a fifth character. For the phylogenetic analyses we used, in addition to our data, seven sequences of the nuclear marker retrieved from GenBank (Table I). We were very careful when aligning the 16S rRNA marker, because it has a high mutation rate and indels are extremely common. In order to get a better hypothesis of homology, we used the secondary structure of the 16S rRNA of *Albinaria caerulea* (Lydeard *et al.* 2000, Ramírez & Ramírez 2010) as a template for the alignment.

Different phylogenetic analyses were performed. The cladogram for all taxa was constructed using Neighbor-Joining (NJ) (Saitou & Nei 1987) as implemented in PAUP* 4.0b10 (Swofford 2003). Tree searching was heuristic, with tree-bisection-reconnection branch swapping. Branch support was evaluated using bootstrap resampling (Felsenstein 1985) with 1,000 replicates. Maximum Parsimony (MP) was implemented using PAUP* 4.0b10 (Swofford 2003), initial heuristic searches were conducted with random stepwise addition, Tree-Bisection-Reconnection (TBR) branch swapping, and bootstrap with 1,000 replicates. Maximum Likelihood (ML) analyses were conducted using heuristic search, the initial tree was obtained by stepwise addition and TBR in PAUP* 4.0b10. Support for nodes was estimated with 1,000 bootstrap replicates. The nucleotide substitution model, base frequencies, proportion of invariant sites and shape parameter of the gamma distribution were estimated based on Akaike criterion using JModeltest (Posada 2008). Bayesian inference (BI) was performed using MrBayes 3.1.2 (Ronquist & Huelsenbeck 2003); four chains of a Markov Chain Monte Carlo algorithm were run simultaneously for 10 million generations, sampled every 1,000 generations, and burn-in of 9,000 generations. A consensus tree and final posterior probabilities were calculated using the remaining trees. The tree based on 16S rRNA was rooted using *Placostylus* (Placostylidae). For the nuclear rRNA gene-cluster, trees were rooted using species belonging Odontostomidae, which is sister to Bulimulidae according to Breure *et al.* (2010).

Sequences of *Bostryx aguilari* were evaluated in DAMBE v5.0.8 (Xia & Xie 2001). We calculated nucleotide frequencies, percentage of CpG islands, percentage of CG and the extent of saturation, by plotting pairwise genetic distances against the distribution of transitions and transversions. Values of genetic diversity, such as haplotype diversity (*h*) and nucleotide diver-

sity (π) were obtained using DnaSP v5.10 (Librado & Rozas 2009). Pairwise distances were obtained in MEGA v4.02 (Kumar *et al.* 2008) including all positions and using a Maximum Composite Likelihood method. Relationships among haplotypes of the 16S rRNA marker were evaluated using the Median Joining algorithm obtained in Network 4.5.1.0 (Bandelt *et al.* 1999). *Fst* statistics was calculated using Arlequin v3.11. In order to estimate the time to most recent common ancestor (TMRCA) for *B. aguilari*, we calibrated a Linearized NJ tree for a conservative rate for terrestrial mollusks (0.06 substitutions per site per million years) for the 16S rRNA, using MEGA (Excoffier *et al.* 2005).

RESULTS

Interspecific phylogeny

The alignment generated for the phylogenetic reconstruction of the partial 16S rRNA gene consisted of 26 sequences (only the four haplotypes of *B. aguilari* were used) corresponding to 13 species of Bulimulidae. This alignment had 382 positions, with 202 variable sites (of which 179 were informative), 170 conserved sites, and 22 singletons. The nucleotide substitution model selected was TPM1uf+G. For the nuclear rRNA, the alignment of 29 sequences resulted in 868 sites, 656 of which were conserved and 199 were variable sites (158 informative), and 41 were singletons. The nucleotide substitution model selected was GTR+G.

Phylogenetic reconstructions based on the partial 16S rRNA gene using NJ, MP, ML and BI resulted in trees with similar topologies (Fig. 13). The group of species known as the "*Bostryx modestus* species complex", which includes *B. modestus*, *B. sordidus* and *B. scalariformis* (R. Ramírez unpublished data), was strongly supported in our analyses. It grouped along with *B. solutus*, *B. aguilari*, and *B. conspersus* with weak to strong support. However, *B. turritus* did not cluster with any other species of *Bostryx*. Regarding the phylogenetic analyses using the nuclear rRNA marker, again the four phylogenetic methods used yielded trees with similar topologies (Fig. 14). Bulimulidae was supported by maximum values. The sequences of *B. modestus*, *B. scalariformis*, and *B. sordidus* (*B. modestus* species complex) grouped with strong support. The *B. modestus* species complex, along with *B. solutus*, *B. conspersus*, and *B. aguilari* grouped together with strong support. *Neopetraeus* and *Drymaeus* formed a monophyletic group with good to strong support. *Naesiotus quitensis* and *Bostryx strobeli* clustered with strong support and, in our data, formed a strongly supported monophyletic group with *Bulimulus*.

The trees obtained with the two markers have similar topologies. Both trees grouped *B. solutus* with *B. modestus* species complex, *B. aguilari* and *B. conspersus*.

Genetic diversity of *Bostryx aguilari*

The alignment of 14 sequences of the partial 16S rRNA gene of *B. aguilari* resulted in 345 sites without indels. There were three variable sites, which were informative. By compar-

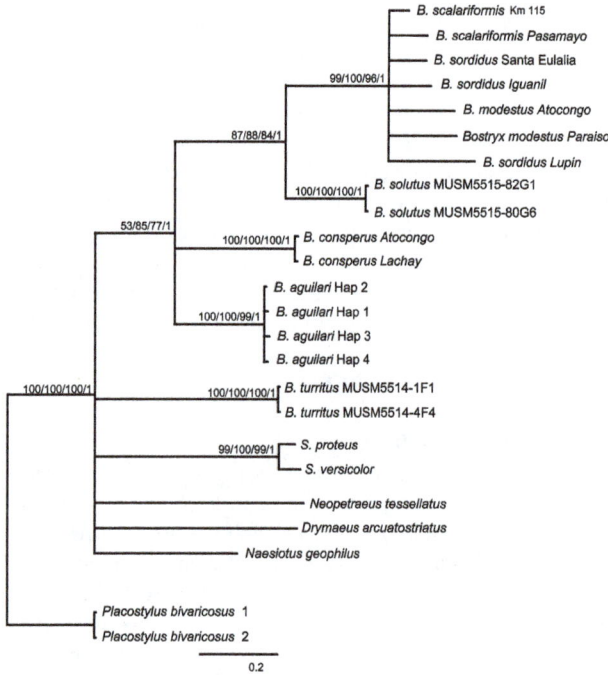

Figure 13. Phylogenetics relationships based on the 16S rRNA. Numbers correspond to bootstrap values for Neighbor-Joining, Maximum Parsimony and Maximum Likelihood, respectively, and posterior probabilities for Bayesian inference. Only nodes with bootstrap values greater than 50% and posterior probabilities of 0.9 are represented.

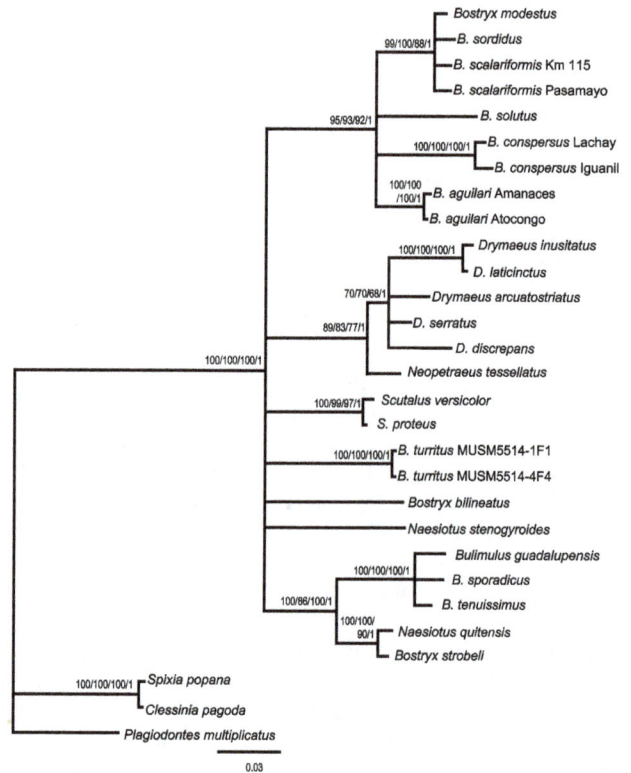

Figure 14. Phylogenetics relationships based on the nuclear rRNA gene cluster. Numbers represent bootstrap values for Neighbor-Joining, Maximum Parsimony and Maximum Likelihood, respectively, and posterior probabilities for Bayesian inference. Only nodes with bootstrap values greater than 50% and posterior probabilities of 0.9 are represented.

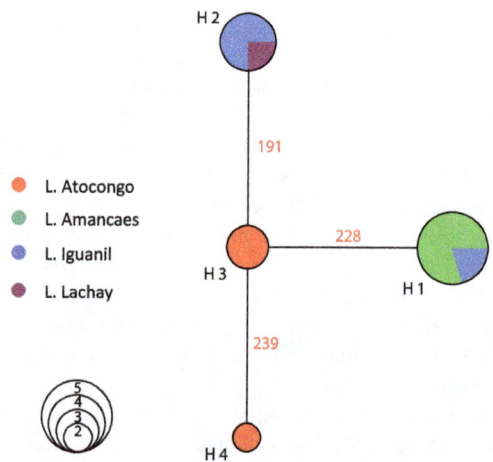

Figure 15. Haplotype Network based on 16S rRNA of *B. aguilari*. Circles are proportional to frequencies. Colors indicate locality of samples. There is only one mutation between haplotypes. Numbers indicate the position of mutation in the alignment.

ing these sequences with other Bulimulids, we were able to observe the presence of several indels up to 4 bp long. This region of the mitochondrial genome of *B. aguilari* is larger than that found in other *Bostryx* from Lomas, as well as in other genera of Bulimulidae evaluated so far (4 to 23 bp difference). The nucleotide composition showed a predominance of AT (71.66%) over GC (28.34%). Sequences obtained for the nuclear rRNA were 826 bp long. The three individuals had the same haplotype. The percentage of GC (55.83%) was slightly higher than that of AT.

The 14 16S rRNA sequences collapsed into four haplotypes. The haplotype diversity (h) was 0.7802 and π was 0.00347. By comparing these results with values found for other species of *Bostryx* from Lomas (R. Ramírez unpublished data), we observed that *B. aguilari* has the lowest values of haplotype diversity. The haplotype network in Figure 15 shows a correlation between haplotypes and the geographic distribution of *B. aguilari*, revealing the Atocongo population as the only one with unique haplotypes. The Amancaes population showed only one haplotype, which was shared with an individual from Iguanil, in spite of the geographic distance (70 km) and the apparent absence of intermediate populations between the two Lomas (no live individuals or shells recorded). The individual from Lachay shared its haplotype with individuals from Iguanil. These haplotypes are differentiated by a single mutational step between them. The *Fst* analyses showed significant values only between Atocongo and the remaining populations. The TMRCA for *B. aguilari* was estimated in 38,565 years.

DISCUSSION

The polyphyly of *Bostryx*

Breure (1979) conducted a study on evolutionary relationships and geographic distribution of the genera in Bulimulinae. More recently, molecular studies have shed new light on the diversity and the relationships within the land snails (Wade & Mordan 2000, Wade *et al.* 2001, 2006) and within Orthalicoidea (Ponder *et al.* 2003, Parent & Crespi 2006, Herbert & Mitchell 2009, Ramirez *et al.* 2009, Breure *et al.* 2010, Buckley *et al.* 2011). Breure *et al.* (2010), after revisiting the phylogeny of Orthalicoidea, found that Orthalicidae and Amphibulimulidae are the most basal families, whereas Placostylidae is basal to the clade consisting of Odontostomidae and Bulimulidae.

Bostryx belongs to Orthalicidae, after Bouchet & Rocroi (2005), which is composed of several subfamilies, including Bulimulinae. This subfamily had been considered as a separate family by several authors (Vaught 1989). In this study, we considered *Bostryx* as a member of Bulimulidae, following Breure *et al.* (2010). In our analyses based on the nuclear rRNA, we added more genera and species to the set of taxa analyzed by Breure *et al.* (2010), and confirmed that Bulimulidae is clearly a monophyletic group, and that *Bostryx* is a member of this family. Our results, obtained with both nuclear and mitochondrial markers, show that the *B. modestus* species complex, *B. aguilari* and *B. conspersus*, is related to *B. solutus*, a land snail that lives at 3300 m in the Western Andes. The position of *B. turritus*, a Peruvian species found in Inter-Andean valleys, was not resolved, showing low support for any relationships with the other *Bostryx* species analyzed. The monophyly of *Bostryx* was not supported by the different analyses. It is important to note that *Bostryx* was described using *Bostryx solutus* as the type species (Breure 1979). *Bostryx solutus* was recovered in a strongly supported group together with *B. aguilari*, *B. conspersus*, *B. modestus*, *B. scalariformis*, and *B. sordidus*. These results suggest that only this group should be considered as *Bostryx*. More studies are needed to establish the position of *B. turritus*, as well as the other two species of *Bostryx* (*B. bilineatus* and *B. strobeli*) included in the nuclear analyses.

Genetic diversity of *Bostryx aguilari*

Bostryx aguilari had the lowest value of haplotype diversity compared to other species of *Bostryx* from Lomas. This may be due to the small size of the populations (suggested by the extreme difficulty in locating live individuals), and their possible recent origin. To compare other values of genetic diversity such as π, we examined the work of P. Romero (unpublished data), where a value of 0.04028 for π was found for popula-

tions of *B. scalariformis*. This is about 10 times higher than what was observed for *B. aguilari*. P. Romero found intraspecific distance values up to 0.0608 for *B. scalariformis*, while the maximum value found for *B. aguilari* was 0.006.

The distribution of the 16S rRNA haplotypes of *B. aguilari* and *Fst* values are consistent with results reported by R. Ramírez (unpublished data) regarding the variation of shells, and revealed the population of Atocongo as the most differentiated. The fact that an individual from Amancaes shared its haplotype with individuals from Iguanil, two distant Lomas and without intermediate populations of *B. aguilari*, suggests a recent origin of these populations from a common ancestor. The possibility that this distribution is due to an event of recent geographic expansion after a genetic bottleneck (from refuge) cannot be discarded. The single individual of *B. aguilari* from Lachay shared the same haplotype with individuals from Iguanil. Coupled with the proximity of these Lomas (17 km), we propose a likely phenomenon for the historic gene flow between them. The occurrence of ENSO events could allow the establishment of corridors connecting Lomas that are considered islands of vegetation (Ramírez *et al.* 2003b).

Several ENSO events have left their marks on the genetic structure of populations of land snails from Lomas. ENSO events of greater magnitude have changed dramatically the landscape of the desert, generating larger Lomas and even connecting adjacent Lomas, whereas in dry periods and ENSO of low intensity, Lomas would become a refuge for these species (Ramírez *et al.* 2003b). Both Tudhope *et al.* (2001) and La Torre *et al.* (2002) reported a strong ENSO about 40 thousand years, which agrees with the estimated date for the geographical expansion of *B. aguilari*.

Implications for Conservation

The Lomas are unique ecosystems in the world. They harbor endemic species whose restricted distribution has been caused by different historical processes (drastic climatic changes, population expansion, bottlenecks, isolation of populations by physical barriers, etc.). Unfortunately, humans have started to invade and occupy different Lomas, threatening the local biodiversity. For instance, cities are an almost insurmountable physical barrier to desert species, generating a new type of isolation that cannot be overcome by periodic favorable conditions of the ENSO. In most localities where *B. aguilari* has been reported, live individuals could not be found, and in those where they were found alive, their numbers were low. Atocongo was an exception to this rule, as it had a larger number of individuals, greater variation in shells, and a more differentiated population with exclusive haplotypes. Major conservation efforts should be applied to this area, which is currently threatened by the expansion of shanty towns, and which has been temporarily put under the custody of a cement factory performing work in the area; the company has surrounded the place with a concrete fence to prevent imminent invasions by the surrounding shanty towns. A worrying situation is found in the Lomas of Amancaes, *B. aguilari* was originally reported for this Loma from 200 to 600 m, and at the present time the Loma is virtually occupied by urbanization up to the 400 meters, being restricted to a fraction of the original size. Due to the damage that these incursions cause, as well as the scarce conservation efforts, it is not difficult to imagine the immediate future of this ecosystem. A different picture is seen in Lachay and Iguanil; Lachay is a National Reserve of great extension (5070 ha.) and Iguanil is far from the city and surrounded by farming communities. Both Lomas guarantee the conservation of part of the low diversity of *B. aguilari*, whose populations are the most distinct besides Atocongo. *Bostryx aguilari* is considered a rare species that has low genetic diversity and small populations. Therefore, there is an urgent need to increase conservation efforts, which should focus on stopping the degradation of its habitat.

ACKNOWLEDGMENTS

We thank A. Chumbe, C. Congrains, D. Fernandez, J. Chirinos, N, Medina, P. Matos and P. Romero for assistance during laboratory and field work. We thank A. Chumbe, D. Maldonado and V. Borda for provided photographs of species. We also thank to M. Arakaki for helping to improve a former version of the manuscript. This study was sponsored by the Instituto de Investigación en Ciencias Biológicas Antonio Raimondi (ICBAR), and funded by Consejo Superior de Investigaciones (CSI) of Vicerrectorado Académico, UNMSM (061001071, 071001221).

LITERATURE CITED

Aguilar, P. & J. Arrarte. 1974. Moluscos de las lomas costeras del Perú. **Anales Científicos, Universidad Nacional Agraria, (Perú)** 12 (3-4): 93-98.

Bandelt, H.; P. Forster & A. Röhl. 1999. Median-joining networks for inferring intraspecific phylogenies. **Molecular Biology and Evolution** 16: 37-48.

Breure, A. 1979. Systematics, phylogeny and zoogeography of Bulimulinae (Mollusca). **Zoologische Verhandelingen 168:** 1-215.

Breure, A.; D. Groenenberg & M. Schilthuizen. 2010. New insights in the phylogenetic relations within the Orthalicoidea (Gastropoda, Stylommatophora) based on 28S sequence data. **Basteria 74** (1-3): 25-32.

Bouchet, P. & J. Rocroi. 2005. Classification and Nomenclator of Gastropod Families. **Malacologia 47:** 1-2.

Buckley, T.R.; I. Stringer; D. Gleeson; R. Howitt; D. Attanayake; R. Parrish; G. Sherley & M. Rohan. 2011. A revision of the New Zealand *Placostylus* land snails using mitochondrial DNA and shell morphometric analyses, with implications for conservation. **New Zealand Journal of Zoology 38** (1): 55-81.

Dillon, M.; M. Nakazawa & S. Leiva. 2003. The *Lomas* formations of coastal Peru: Composition and biogeographic history. **Fieldiana Botany 43:** 1-9.

Doyle, J.J. & J.L. Doyle. 1987. A rapid DNA isolation procedure for small amounts of fresh leaf tissue. **Phytochemical Bulletin 19:** 11-15.

Excoffier, L.; G. Laval & S. Schneider. 2005. Arlequin ver. 3.0: An integrated software package for population genetics data analysis. **Evolutionary Bioinformatics Online 1:** 47-50.

Felsenstein, J. 1985. Confidence limits on phylogenies: an approach using the bootstrap. **Evolution 39:** 783-791.

Hall, T. 1999. BioEdit: a user-friendly biological sequence alignment editor and analysis program for Windows 95/98/NT. **Nucleic Acids Symposium Series 41:** 95-98.

Herbert, D. & A. Mitchell. 2009. Phylogenetic relationships of the enigmatic land snail genus Prestonella – the missing African element in the Gondwanan superfamily Orthalicoidea (Mollusca: Stylommatophora). **Biological Journal of the Linnean Society 96:** 203-221.

Huang, X. & A. Madan. 1999. CAP3: A DNA Sequence Assembly Program. **Genome Research 9:** 868-877.

Kumar, S.; J. Dudley; M. Nei & K. Tamura. 2008. MEGA: A biologist-centric software for evolutionary analysis of DNA and protein sequences. **Briefings in Bioinformatics 9:** 299-306.

La Torre, C.; J. Betancourt; K. Rylander & J. Quade. 2002. Vegetation invasions into absolute desert: A 45 000 yr rodent midden record from the Calama-Salar de Atacama basins, northern Chile (lat 228-248S). **GSA Bulletin 114** (3): 349-366.

Larkin, M.; G. Blackshields; N. Brown; R. Chenna; P. Mcgettigan; H. Mcwilliam; F. Valentin; I. Wallace; A. Wilm; R. Lopez; J. Thompson; T. Gibson & D. Higgins. 2007. Clustal W and Clustal X version 2.0. **Bioinformatics 23:** 2947-2948.

Librado, P. & J. Rozas. 2009. DnaSP v5: A software for comprehensive analysis of DNA polymorphism data. **Bioinformatics 25:** 1451-1452.

Lydeard, C.; W. Holznagel; M. Schnare & R. Gutell. 2000. Phylogenetic analysis of molluscan mitochondrial LSU rDNA sequences and secondary structures. **Molecular Phylogenetics and Evolution 15:** 83-102.

McCarthy, C. 1996. **Chromas: version 1.3.** Brisbane, Griffith University.

Parent, C.E. & B.J. Crespi. 2006. Sequential colonization and diversification of Galápagos endemic land snail genus *Bulimulus* (Gastropoda, Stylommatophora). **Evolution 60:** 2311-2328.

Ponder, W.; D. Colgan; D. Gleeson & G. Sherley. 2003. The relationships of *Placostylus* from Lord Howe Island. **Molluscan Research 23:** 159-178.

Posada, D. 2008. jModelTest: phylogenetic model averaging. **Molecular Biology and Evolution 25:** 1253-1256.

Ramírez, J. & R. Ramírez, 2010. Utility of secondary structure of mitochondrial LSU rRNA in the phylogenetic reconstruction for land snails (Orthalicidae: Gastropoda). **Revista Peruana de Biología 17** (1): 53-57.

RAMIREZ, J.; R. RAMÍREZ; P. ROMERO; A. CHUMBE & P. RAMÍREZ. 2009. Posición evolutiva de caracoles terrestres peruanos (Orthalicidae) entre los Stylommatophora (Mollusca: Gastropoda). **Revista Peruana de Biología 16** (1): 51-56.

RAMÍREZ, R.; C. PAREDES & J. ARENAS. 2003a. Moluscos del Perú. **Revista de Biologia Tropical 51** (3): 225-284.

RAMÍREZ, R.; S. CÓRDOVA; K. CARO & J. DUÁREZ. 2003b. Response of a land snail species (*Bostryx conspersus*) in the Peruvian Central Coast *Lomas* Ecosystem to the 1982-1983 and 1997-1998 El Niño events. **Fieldiana, Botany 43**: 10-23.

RONQUIST, F. & J. HUELSENBECK. 2003. MrBayes 3: Bayesian phylogenetic inference under mixed models. **Bioinformatics 19** (12): 1572-1574.

RUNDEL, P.; M. DILLON; B. PALMA; H. MOONEY; S. GULMON & J.R. EHLERINGER. 1990. The Phytogeography and Ecology of the Coastal Atacama and Peruvian Deserts. **Aliso 13** (1): 1-50.

SAIKI, R.; D. GELFAND; S. STOFFEL; S. SCHARFJ; R. HIGUCHI; G. HORN; K. MULLIS & H. ERLICH. 1988. Primer-directed enzymatic amplification of DNA with a thermostable DNA polymerase. **Science 239**: 487-491.

SAITOU, N. & M. NEI. 1987. The neighbor-joining method: A new method for reconstructing phylogenetic trees.

Molecular Biology and Evolution 4: 406-425.

SWOFFORD, D. L. 2003. **PAUP*. Phylogenetic Analysis Using Parsimony (*and Other Methods).** Sunderland,Sinauer Associates, version 4.

TUDHOPE, A.; C. CHILCOTT; M. MCCULLOCH; E. COOK; J. CHAPPELL; R. ELLAM; D. LEA; J. LOUGH & G. SHIMMIELD. 2001. Variability in the El Niño-Southern Oscillation through a glacial-interglacial cycle. **Science 291**: 1511-1517.

VAUGHT, K. 1989. **A classification of the living Mollusca.** Melbourne, American Malacologist Inc.

XIA, X. & Z. XIE. 2001. DAMBE: Data analysis in molecular biology and evolution. **Journal of Heredity 92**: 371-373.

WADE, C. & P. MORDAN. 2000. Evolution within the gastropod mollusks; using the ribosomal RNA gene-cluster as an indicator of phylogenetic relationships. **Journal of Molluscan Studies 66**: 565-570.

WADE,C; P. MORDAN & B. CLARKE. 2001. A phylogeny of the land snails (Gastropoda: Pulmonata). **Proceedings of the Royal Society of London Series B 268**: 413-422.

WADE, C.; P. MORDAN & F. NAGGS. 2006. Evolutionary relationships among the Pulmonate land snails and slugs (Pulmonata, Stylommatophora). **Biological Journal of the Linnean Society 87**(4): 593-610.

WEYRAUCH, W. 1967. Treinta y Ocho Nuevos Gastropodos Terrestres de Perú. **Acta Zoológica Lilloana 21**: 349-351.

3

Habitat use and seasonal activity of insectivorous bats (Mammalia: Chiroptera) in the grasslands of southern Brazil

Marília A. S. Barros[1,3], Daniel M. A. Pessoa[1] & Ana Maria Rui[2]

[1] Departamento de Fisiologia, Centro de Biociências, Universidade Federal do Rio Grande do Norte. Campus Universitário Lagoa Nova, 59078-970 Natal, RN, Brazil.
[2] Departamento de Ecologia, Zoologia e Genética, Instituto de Biologia, Universidade Federal de Pelotas. Campus Universitário Capão do Leão, Caixa Postal 354, 96001-970 Pelotas, RS, Brazil.
[3] Corresponding author E-mail: barrosmas@gmail.com

ABSTRACT. In temperate zones, insectivorous bats use some types of habitat more frequently than others, and are more active in the warmest periods of the year. We assessed the spatial and seasonal activity patterns of bats in open areas of the southernmost region of Brazil. We tested the hypothesis that bat activity differs among habitat types, among seasons, and is influenced by weather variables. We monitored four 1,500-m transects monthly, from April 2009 to March 2010. Transects corresponded to the five habitat types that predominate in the region. In each sampling session, we detected and counted bat passes with an ultrasound detector (Pettersson D230) and measured climatic variables at the transects. We recorded 1,183 bat passes, and observed the highest activity at the edge of a eucalyptus stand (0.64 bat passes/min) and along an irrigation channel (0.54 bat passes/min). The second highest activity values (0.31 and 0.20 bat passes/min, respectively) were obtained at the edge of a riparian forest and at the margin of a wetland. The grasslands were used significantly less (0.05 bat passes/min). Bat activity was significantly lower in the winter (0.21 bat passes/min) and showed similar values in the autumn (0.33 bat passes/min), spring (0.26 bat passes/min), and summer (0.29 bat passes/min). Bat activity was correlated with temperature, but it was not correlated with wind speed and relative humidity of the air. Our data suggest that, in the study area, insectivorous bats are active throughout the year, and use mostly forest and watercourses areas. These habitat types should be considered prioritary for the conservation of bats in the southernmost region of Brazil.

KEY WORDS. Acoustic monitoring; activity patterns; Molossidae; South American Pampas; Vespertilionidae.

Insectivorous bats represent 70% of all bat species and are widely distributed (SIMMONS 2005). They play an important ecological role in the transfer of nutrients in ecosystems (PIERSON 1998) and in the control of insect populations, including agricultural pests (BOYLES et al. 2011). Insectivorous bats occupy high trophic levels, are indicators of habitat quality (JONES et al. 2009), and may undergo population decrease in response to environmental disturbances (TUTTLE 1979, GERELL & LUNDBERG 1993, O'DONNELL 2000).

In temperate zones, insectivorous bats use some types of habitat more frequently than others, and tend to respond positively to the presence of trees and water bodies (WALSH & HARRIS 1996). Bats are highly active in forests and forest fragments in rural areas (ERICKSON & WEST 2003, LUMSDEN & BENNETT 2005), mainly in hedgerows and forest edges (RUSS et al. 2003, PETTIT & WILKINS 2012), as well as around rivers, lakes, and lagoons (VAUGHAN et al. 1997, BROOKS 2009). Furthermore, in temperate regions the activity levels of bats vary seasonally in response to fluctuations in climatic conditions and their own energy

requirements throughout the year (CIECHANOWSKI et al. 2010, JOHNSON et al. 2011). Some studies highlighted the influence of abiotic factors on bat activity, such as temperature (HAYES 1997, RUSS et al. 2003), wind speed (AVERY 1985, JOHNSON et al. 2011), and relative humidity of the air (LACKI 1984, ADAM et al. 1994).

There is plenty of information available about habitat use and seasonal activity of insectivorous bats, which is mainly based on studies carried out in temperate regions of the northern hemisphere. In Brazil, studies focusing on habitat use by insectivorous bats are few, though there is one study carried out in an urban area in southeastern Brazil (ALMEIDA et al. 2007). One of the places where the Brazilian bat fauna is understudied is the state of Rio Grande do Sul, mainly its southern half, where the Pampa biome is located (BERNARD et al. 2011). The Pampa biome corresponds to the Brazilian part of the South American Pampas, which extends through Uruguay and Argentina and is characterized by plains covered by grasslands (IBGE 2004). Although the Pampa covers approximately 2% of the Brazilian territory (IBGE 2004), it occupies the third place among the six continental biomes of Brazil in terms of number of endangered species, surpassing even the Amazon, Caatinga, and Pantanal (Paglia *et al.* 2008). The main threats to the biodiversity of the Pampa are the expansion of agriculture, silviculture, and exotic grasses, which have been responsible for a considerable loss of natural grasslands in the past three decades (Pillar *et al.* 2009).

Information about patterns of habitat use is important for bat conservation (Fenton 1997) and, hence, for the conservation of ecological processes associated with bats. In the present study, we describe spatial and seasonal activity patterns of insectivorous bats in the southernmost region of Brazil, in a Pampa area. Our objectives were: a) to compare the main characteristic habitats of this region in terms of bat activity; b) to test for seasonal variations in bat activity; and c) to test whether bat activity is influenced by abiotic factors, such as temperature, wind speed, and relative humidity of the air. We hypothesized that bat activity is higher in habitats with trees and/or water, and also that bats decrease their activity in the colder months of the year.

MATERIAL AND METHODS

We carried out the present study in private properties of the rural area of Santa Vitória do Palmar, in the southernmost region of the state of Rio Grande do Sul, Brazil. The study area is located in the geomorphologic region of Planície Costeira (a coastal lowland), in the Pampa biome, classified as steppe according to the international phytogeographic system of world vegetation (IBGE 2004).

The landscape is wide lowland with sandy grasslands, located between Mirim Lagoon and the Atlantic Ocean. The predominant vegetation in the region is grassland, which is mainly formed by grasses, sedges, and other herbs and subshrubs (Rambo 2000). In areas under freshwater influence there is paludal and shrubby vegetation, forming small riparian forests along watercourses (Rambo 2000). In addition, there are lines and stands of introduced eucalyptus isolated in the grassland, planted to serve as windbreaks and shelter for farm livestock.

According to the Köppen system, the climate of the region is classified as Cfa, i.e., temperate without dry season and with hot summer (Peel *et al.* 2007). According to data obtained between 1990 and 2010 at the meteorological station of Santa Vitória do Palmar (Instituto Nacional de Meteorologia/INMET), the average annual temperature is 17°C. The coldest months are June and July, with an average temperature of 12°C, and the warmest months are January and February, with an average temperature of 22°C. The average monthly rainfall is 109 mm, and the annual total rainfall was 1,153 mm in 2009 and 1,176 in 2010.

The term "activity" has broad meaning; in our study we are referring to the flight/foraging activity of bats, outside day roosts. Monitoring of bat activity was carried out in four 1,500-m long transects, located in the different types of habitats that

predominate in the region (Table I, Figs 1-6. 1). We marked 30 fixed points at each transect, 50 m away from each other. We marked the transects TR II and TR IV in structurally homogeneous environments in terms of vegetation, and considered them to contain the same type of habitat. Transects TR I and TR III comprised two types of habitat each. Each point at the transects comprised only one type of habitat, considering the type of vegetation in a radius of 50 m around the point.

Every month from April 2009 to March 2010 we monitored one transect each night and, whenever possible, during four consecutive nights. We used the ultrasound detector Pettersson Elektronik AB D230 (frequency range: 10-120 kHz, bandwidth: 8 kHz ± 4 kHz), in the mode heterodyne, which artificially reduces the ultrasound frequency and makes it audible to humans in real time (Parsons & Szewczak 2009). The transects were always monitored by the same observer, and the ultrasound detector was used for three minutes in each one of the 30 points. Transect monitoring began 10 minutes after sunset and was carried out at night when it was not raining. At each point, the observer kept the ultrasound detector at approximately 1 m above ground, at 45° of inclination, and turned it 360° covering all directions. For each point, we recorded the number of bat passes, which are defined as a sequence of a bat echolocation calls on the detector from beginning to end (following Kuenzi & Morrison 2003). To increase the chance of recording the highest possible number of species, we constantly altered the frequency of the bat detector between 10 and 120 kHz (following Celuch & Kropil 2008).

We considered the activity of the bat assemblage as a whole. Spatial and seasonal activity patterns may vary among species (Brooks & Ford 2005, Ciechanowski *et al.* 2010), but information on general bat activity is useful for the identification of general tendencies of habitat use and priority areas for bat conservation (Estrada *et al.* 2004, Walsh & Harris 1996). In the study area, the fauna of insectivorous bats corresponds to molossid and vespertilionid species, and sampling with mist nets confirmed the presence of *Tadarida brasiliensis* (I. Geoffroy, 1824), *Molossus molossus* (Pallas, 1766), *Molossus rufus* É. Geoffroy, 1805, and *Eptesicus brasiliensis* (Desmarest, 1819) in the region (M.A.S. Barros, unpublished data) – although probably more species occur in the area.

The sampling effort was 72 hours on 48 sampling nights. The average time to completely monitor transects was 132.4 ± 9.5 minutes (time with the bat detector switched on plus movement time on foot between points). According to information from temperate zones of North America (Kunz 1973, Hayes 1997), and also to data from other parts of the region of Planície Costeira in Rio Grande do Sul (Rui & Barros, unpublished data), our survey was carried out in the typical activity peak of insectivorous bats in the first two-three hours after sunset. Because of this, we assume that the timing of activity monitoring did not bias our results, and transects were always surveyed in the same direction (from point 1 to 30).

Figures 1-6. Habitat types monitored for bat activity with acoustic surveys, from April 2009 to March 2010, in the grasslands of southernmost Brazil: (1) Eucalyptus stand (TR I); (2) Grassland (TR I); (3) Channel (TR II); (4) Riparian forest (TR III); (5) Wetland (TR III); (6) Grassland (TR IV).

Table I. Transects and habitat types monitored for bat activity with acoustic surveys, from April 2009 to March 2010, in the grasslands of southernmost Brazil.

Transect	Sites	Coordinates (UTM)	Habitat	Description
TR I	01 to 10	288369/6282442	Eucalyptus stand	Edge of eucalyptus stand, with an area of 600 m^2 and maximum height of 15 m.
TR I	11 to 30	288914/6281884	Grassland	Grassland with only underbrush vegetation, used as pasture for bovine livestock.
TR II	01 to 30	293353/6284182	Channel	Margin of an artificial managed channel of 15 m in width and 3 m in depth, used for water extraction for rice irrigation.
TR III	01 to 15	293208/6280102	Riparian forest	Edge of dense arboreal-shrubby riparian forest fragment, with maximum height of 5 m, located on the margin of a native wetland.
TR III	16 to 30	293107/6280711	Wetland	Margin of native wetland, composed of a flooded field covered by herbaceous and shrubby vegetation.
TR IV	01 to 30	289340/6287498	Grassland	Grassland with only underbrush vegetation, used as pasture for bovine livestock.

We recorded the following abiotic parameters: temperature, relative humidity of the air, and wind velocity. We collected these data four times while monitoring transects: at point 01 (beginning of the transect), at point 10 (500 m), at point 20 (1,000 m), and at point 30 (end of the transect).

The statistical analysis was carried out in the program PASW Statistics 18 (Statistical Package for the Social Sciences/SPSS Inc.). Since our data were not normally distributed (Kolmogorov-Smirnov test, $\alpha = 0.01$), we used non-parametric tests.

We compared the number of bat passes/3 min among the five habitat types, pooling the 12 sampling months, and compared bat activity between habitats for each season. In both analyses, we used Kruskal-Wallis tests ($\alpha = 0.05$) and Mann-Whitney *post hoc* tests ($\alpha = 0.005$, with Bonferroni correction).

For the assessment of seasonal variations in bat activity, we compared the number of bat passes/3 min among seasons, pooling data of the five types of habitats. We also compared bat activity among seasons for each habitat alone, carrying out five additional analyses, one for each type of habitat. For these analyses, we used the Friedman ANOVA ($\alpha = 0.05$) and the Wilcoxon *post hoc* test ($\alpha = 0.008$ with Bonferroni correction).

To assess the influence of abiotic factors on bat activity, we calculated correlations between the number of bat passes and temperature (°C), relative humidity of the air (%), and wind speed (m/s), with the Spearman coefficient ($\alpha = 0.05$). Since the abiotic factors were measured only four times when we monitored transects, and did not vary much from one point to the next, we extrapolated the values measured in one point to its neighbors. For the analyses, we extrapolated the abiotic data obtained at point 01 to points 02 to 05, the data of point 10 to points 6 to 15, the data of point 20 to points 16 to 25, and the data of point 30 to points 26 to 29.

RESULTS

Use of habitat by bats

We recorded 1,183 bat passes during one year. Bat activity differed significantly among habitat types (H = 311.38, df = 4, p < 0.001). We observed the highest activity in the eucalyptus stand and in the channel, followed by the riparian forest and the wetland (Fig. 7). The grassland was the least used habitat (Fig. 7). All habitats differed from each other (p < 0.001), except for the comparisons "eucalyptus vs. channel" (p = 0.038) and "riparian forest vs. wetland" (p = 0.162) (Fig. 8).

Bat activity differed also among habitat types in the different seasons (Fig. 9). The difference in activity among habitats was significant in the autumn (H = 190.56, df = 4, p < 0.001), the winter (H = 81.65, df = 4, p < 0.001), the spring (H = 71.48, df = 4, p < 0.001), and the summer (H = 90.54, df = 4, p < 0.001). In the autumn, the highest activity levels were recorded in the eucalyptus stand and in the channel; all habitats differed from each other (p ≤ 0.002), except for the com-

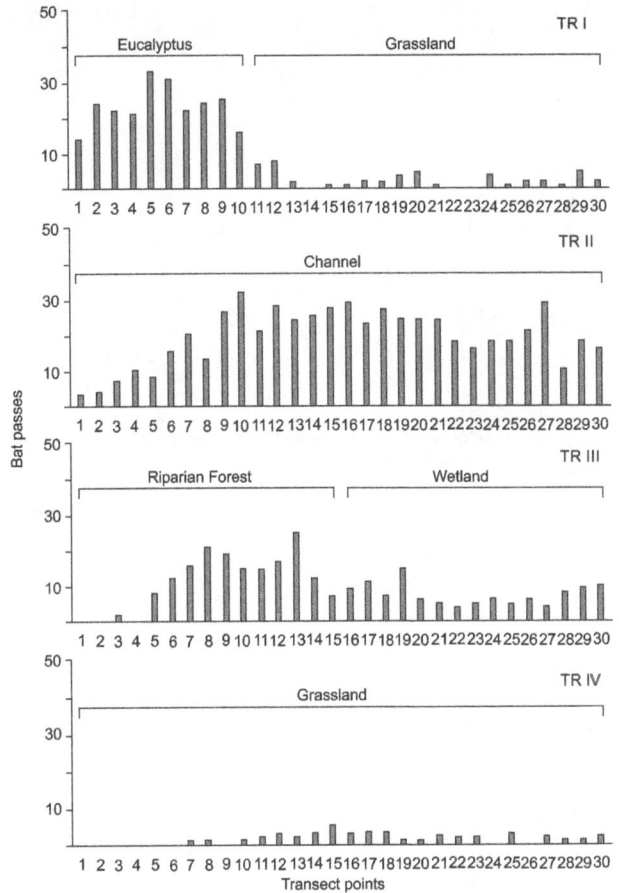

Figure 7. Total number of bat passes recorded with a bat detector in each fixed point on 1,500-m transects monitored for bat activity, from April 2009 to March 2010, in the grasslands of southernmost Brazil.

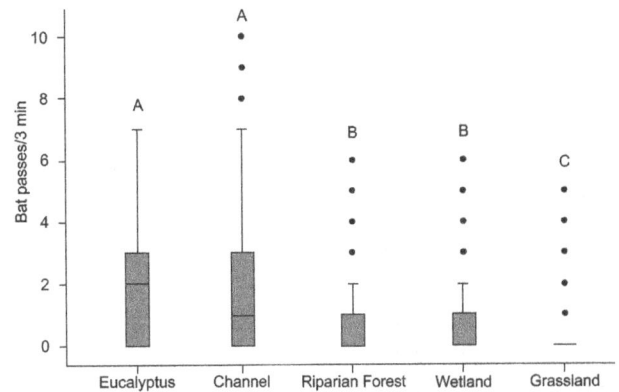

Figure 8. Box plot of the number of bat passes/3 min recorded with a bat detector in each habitat type, from April 2009 to March 2010, in the grasslands of southernmost Brazil. Different letters indicate significant differences at the 0.005 probability level (with Bonferroni correction).

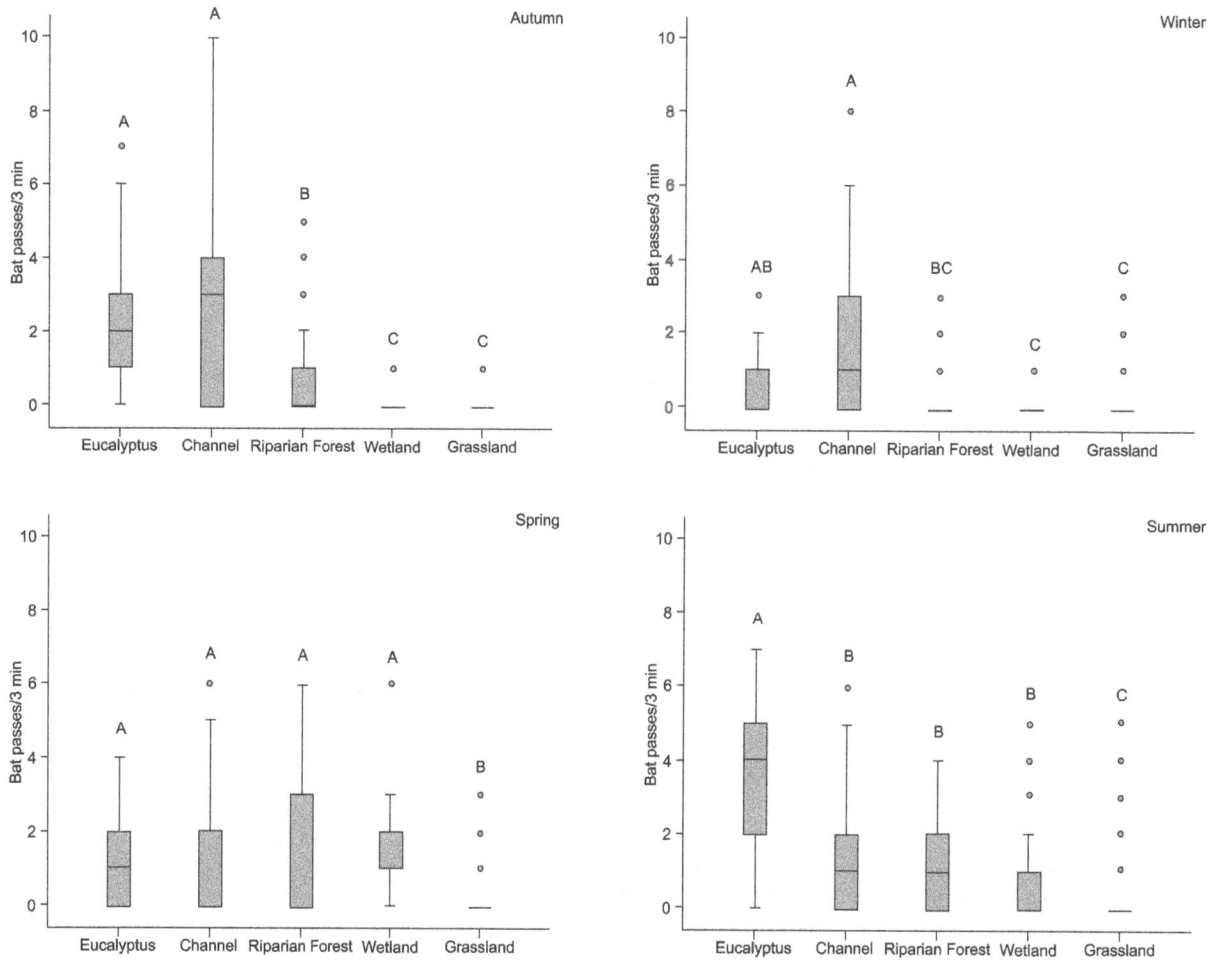

Figure 9. Box plot of the number of bat passes/3 min recorded with a bat detector in each habitat type in different seasons, from April 2009 to March 2010, in the grasslands of southernmost Brazil. Different letters indicate significant differences at the 0.005 probability level (with Bonferroni correction).

parisons "eucalyptus vs. channel" (p = 0.834) and "grasslands vs. wetland" (p = 0.113). In the winter, activity was higher in the channel than in the other habitats (p < 0.001), except for the eucalyptus stand (p = 0.018). In the spring, bat activity was lower in the grassland than in the other habitats (p < 0.001), which did not differ from each other (p ⩾ 0.010). In the summer, the eucalyptus stand showed the highest activity, differing from all other types of habitats (p < 0.001).

Seasonal variations in bat activity

Bat activity varied among seasons (F_r = 10.34, df = 3, p = 0.016). There was no significant difference in the number of bat passes between the autumn (356 bat passes in total), summer (316), and spring (282) (p ⩾ 0.116). The number of bat passes in the winter (229) was significantly smaller than in the autumn and summer (p ⩽ 0.004), and statistically similar to the number of bat passes in the spring (p = 0.106).

We also observed seasonal variations in bat activity for each habitat alone, in the eucalyptus stand (F_r = 36.42, df = 3, p < 0.001), the channel (F_r = 32.46, df = 3, p < 0.001), the riparian forest (F_r = 12.12, df = 3, p = 0.007), the wetland (F_r = 61.94, df = 3, p < 0.001), and the grassland (F_r = 28.14, df = 3, p < 0.001). In the eucalyptus stand, riparian forest, wetland and grassland, the highest bat activity was observed in the spring and summer (Fig. 10). However, in the channel, bat activity was higher in the autumn and winter; bat activity in the autumn was significantly higher than in the other seasons (p ⩽ 0.006) (Fig. 10).

Influence of climatic factors in bat activity

In the winter we recorded the lowest average temperature, and also the highest humidity and highest wind speed (Table II). We observed the highest bat activity (⩾ 7 bat passes/3 min) in the points where the temperature was higher than 15°C, and we observed null or very low values (⩽ 1 bat pass/

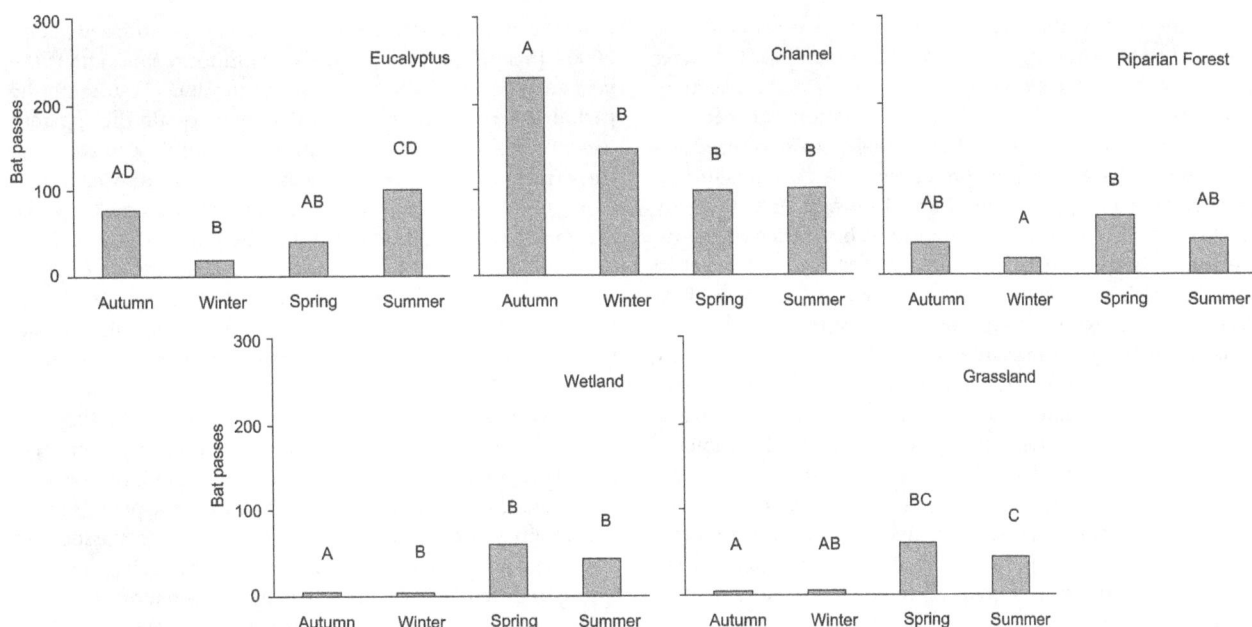

Figure 10. Total number of bat passes recorded with a bat detector in each season for the five habitat types, from April 2009 to March 2010, in the grasslands of southernmost Brazil. Different letters indicate significant differences at the 0.008 probability level (with Bonferroni correction).

Table II. Average values and standard deviation of the climatic factors measured in the transects monitored for bat activity, from April 2009 to March 2010, in the grasslands of southernmost Brazil.

Season	Temperature (°C)	Humidity (%)	Wind speed (m/s)
Autumn	15.0 ± 4.7	80.5 ± 9.5	1.5 ± 1.6
Winter	11.9 ± 2.5	92.7 ± 5.3	2.6 ± 1.5
Spring	14.7 ± 4.5	88.7 ± 8.9	2.1 ± 1.3
Summer	17.8 ± 2.3	85.2 ± 11.8	1.2 ± 0.9

3 min) when temperatures were between 5°C and 10°C. The number of bat passes was correlated with temperature (r_s = 0.138, p < 0.001). We recorded high levels of activity (≥ 7 bat passes/3 min) in nights with different values of wind speed and relative humidity of the air. Bat passes were not significantly correlated with relative humidity of the air (r_s = 0.009, p = 0.722) or wind speed (r_s = -0.030, p = 0.258).

DISCUSSION

Use of habitat

The hypothesis that bats mostly use areas with tall trees and water bodies was corroborated. The association of insectivorous bats with forest edges has been broadly documented for other regions (Lumsden & Bennett 2005, Kofoky et al. 2007, Morris et al. 2010). This is also true for aquatic habitats (Lunde & Harestad 1986, Vaughan et al. 1997, Brooks 2009). The use of the

eucalyptus stand and the channel were probably due to the availability of aerial insects, which are frequently abundant in forest edges (Lewis 1970) and water bodies (Barclay 1991).

The canopy of the eucalyptus stand is approximately three times higher than the canopy of the riparian forest. Since linear vegetation edges serve as landmarks and also offer protection against the wind and predators (Limpens & Kapteyn 1991), eucalyptus probably are more efficient for spatial orientation and protection than riparian forest. In addition, the eucalyptus trees have a large diameter, a reasonable space between each other, and trunks with cavities, cracks and loose bark, which favors their use as diurnal roost by bats, and also for flights inside the stand. The vegetation of the riparian forest, on the contrary, harbors trees with smaller diameter and high density, a factor that is negatively correlated with bat activity (Erickson & West 2003).

In southern Brazil, there are records of association between insectivorous bats and tree species introduced for commercial use (*Pinus* spp.), which are used by vespertilionid bats to roost during the day (Barros & Rui 2011). Exotic trees tends to acquire particular significance when the original vegetation has been altered and the availability of tall native trees is reduced. In the study area, the natural vegetation is known as *Butiazal*, which is composed of jelly palm (*Butia capitata*) clusters that occur along the coastal lowland of southern Brazil (Rambo 2000). These formations are currently rare in the state of Rio Grande do Sul, and *Butia capitata* is endangered in the state (Rio Grande do Sul 2002).

The channel may represent an important water source, since frequent water ingestion is a critical factor for the water balance of insectivorous bats (Neuweiler 2000). Since insectivorous bats use watercourses for spatial orientation (Racey & Swift 1985, Serra-Cobo et al. 2000), the channel can also be used as a flying route. Apparently, the presence of open water is important for bat activity, since the channel showed higher activity levels than the wetland, where portions of flooded grassland are completely covered by vegetation. A study carried out in temporary lakes showed that the activity of insectivorous bats decreases as the exposed water area decreases (Francl et al. 2008). Insect availability in riparian areas may be high (Racey & Swift 1985). However, the activity of bats that forage over water may be negatively correlated with the presence of floating vegetation, which hinders the detection of prey through echolocation (Ciechanowski et al. 2007, Siemers et al. 2001).

The low bat activity in the grassland was expected, since insectivorous bats, in general, use field and pasture areas less frequently (Fenton 1970, Estrada et al. 2004). The activity recorded for grasslands is probably a result of the movement of individuals between roosts and feeding areas, or between feeding areas.

In all seasons, the eucalyptus stand and the channel were the habitats with highest activity, except in the spring. During this season, strong rains caused a more intense flood in the riparian forest and the wetland, which attracted more bats in relation to other seasons of the year, probably due to an increase in insect abundance.

Our data suggest that watercourses and edges of tree stands are priority habitats for bat conservation in the region. Areas of native wetland, with and without riparian forest, were less used, but also presented expressive levels of bat activity. The association of these types of habitat with insectivorous bats indicates that alterations in aquatic habitats or removal of trees tend to negatively affect bat activity in the region.

Seasonal variations in bat activity and influence of climatic factors

As expected, bat activity varied seasonally, and activity in the winter was low. This result is consistent with the common pattern observed for insectivorous bats (Hays et al. 1992, Carmel & Safriel 1997, Hayes 1997). The winter is the most critical period for the energy balance of insectivorous bats because low temperatures reduce insect abundance and activity (Taylor 1963, Wolda 1988) and increase heat loss (Ransome 1990), making foraging less profitable. In the study area, bat activity was positively correlated with temperature, as in other temperate regions (Avery 1985, Brooks 2009, Broders et al. 2006), and the low activity levels in winter was a response of bats to low temperatures.

Like other authors (e.g., Johnson et al. 2011, Verboom & Spoelstra 1999), we did not detect a significant relationship between bat activity and humidity or wind speed. In the study area, the climate frequently varies more expressively in terms

of temperature than in terms of humidity or wind speed (Rambo 2000). In southern Brazil, rainfall is uniform throughout the year and the relative humidity of the air during our sampling period was never below 60%, which may explain the apparent lack of response of bats to variations in humidity. In addition, the wind may affect bat activity only when wind speed is high (> 4 m/s) (O'Farrell et al. 1967, O'Farrell & Bradley 1970), which was rarely observed in our sampling period. The lack of a correlation between bat activity and climatic factors may also result from the pooled analysis of the bat assemblage, since some species at an area may be influenced by specific climatic factors, such as wind speed, whereas others may not (Russo & Jones 2003, Russ et al. 2003).

We observed a tendency for high bat activity in the summer and spring and low activity in the winter in the eucalyptus stand, grassland, riparian forest and wetland. By contrast, in the channel, the spring and summer were the periods when bat activity was the lowest. This pattern is probably associated with the cultivation of rice in the area. From the beginning of spring to the end of summer, the grassland is removed, the soil is plowed and flooded, and the crop is treated with pesticides. It is likely that these actions may have negatively influenced bat activity in the channel, since bat activity is lower in areas with intensively managed crops (Walsh & Harris 1996), probably due to a reduction in insect availability due to pesticides (Carmel & Safriel 1998, Wickramasinghe et al. 2003).

Our study first demonstrated that insectivorous bats show seasonal variation in activity in the grasslands of southern Brazil (Pampa biome). Their activity patterns are markedly reduced in the winter, and are influenced by temperature.

ACKNOWLEDGMENTS

Rodrigo G. de Magalhães, Leandro V. Umann, Marcelo D. Freire, Estevão J.Comitti, Cristian M. Joenck, and Jeferson Bugoni provided us with logistic support in the study area. We thank Edison C. de Souza for the great help and friendship during fieldwork. Felipe N. Castro helped us in the statistical analysis. The company Maia Meio Ambiente LTDA granted us financial support. Coordenação de Aperfeiçoamento Pessoal de Nível Superior (CAPES) granted M.A.S. Barros a Master's scholarship.

LITERATURE CITED

Adam, M.D.; M.J. Lacki & L.G. Shoemaker. 1994. Influence of environmental conditions on flight activity of *Plecotus townsendii virginianus* (Chiroptera: Vespertilionidae). **Brimleyana** 21: 77-85.

Almeida, M.H.; A.D. Ditchifield & R.S. Tokumaru. 2007. Atividade de morcegos e preferência por hábitat na zona urbana da Grande Vitória, ES, Brasil. **Revista Brasileira de Zoociências** 9 (1): 13-18.

AVERY, M.I. 1985. Winter activity by pipistrelle bats. **Journal of Animal Ecology 54** (3): 721-738.

BARCLAY, R.M.R. 1991. Population Structure of Temperate Zone Insectivorous Bats in Relation to Foraging Behaviour and Energy Demand. **Journal of Animal Ecology 60** (1): 165-178.

BARROS, M.A.S. & A.M. RUI. 2011. Occurrence and Mortality of *Lasiurus ega* (Chiroptera, Vespertilionidae) in Monocultures of *Pinus* sp. in Rio Grande do Sul, Southern Brazil. **Chiroptera Neotropical 17** (2): 997-1002.

BERNARD, E.; L.M.S. AGUIAR & R.B. MACHADO. 2011. Discovering the Brazilian bat fauna: a task for two centuries? **Mammal Review 41** (1): 23-39. doi: 10.1111/j.1365-2907.2010.00164.x.

BOYLES, J.G.; P.M. CRYAN; G.F. MCCRACKEN & T.H. KUNZ. 2011. Economic Importance of Bats in Agriculture. **Science 332** (6025): 41-42. doi: 10.1126/science.1201366.

BRODERS, H.G.; G.J. FORBES; S. WOODLEY & I.D. THOMPSON. 2006. Range extent and stand selection for roosting and foraging in forest-dwelling northern long-eared bats and little brown bats in the Greater Fundy ecosystem New Brunswick. **Journal of Wildlife Management 70** (5): 1174-1184. doi: 10.2193/0022-541X(2006)70[1174:REASSF]2.0.CO;2.

BROOKS, R.T. 2009. Habitat-associated and temporal patterns of a bat activity in a diverse forest landscape of southern New England, USA. **Biodiversity and Conservation 18** (3): 529-545. doi: 10.1007/s10531-008-9518-x.

BROOKS, R.T. & W.M. FORD. 2005. Bat activity in a forest landscape of Central Massachusetts. **Northeastern Naturalist 12** (4): 447-462. doi: 10.1656/1092-6194(2005)012[0447:BAIAFL]2.0.CO;2.

CARMEL, Y. & U. SAFRIEL. 1998. Habitat use by bats in a Mediterranean ecosystem in Israel – conservation implications. **Biological Conservation 84** (3): 245-250. doi: 10.1016/S0006-3207(97)00131-6.

CELUCH, M. & R. KROPIL. 2008. Bats in a Carpathian beech-oak forest (Central Europe): habitat use, foraging assemblages and activity patterns. **Folia Zoologica 57** (4): 358-372.

CIECHANOWSKI, M.; T. ZAJ¹C; A. BITAS & R. DUNAJSKI. 2007. Spatiotemporal variation in activity of bat species differing in hunting tactics: effects of weather, moonlight, food abundance, and structural clutter. **Canadian Journal of Zoology 85** (12): 1249-1263. doi: 10.1139/Z07-090.

CIECHANOWSKI, M.; T. ZAJ¹C; A. ZIELIŃSKA & R. DUNAJSKI. 2010. Seasonal activity patterns of seven vespertilionid bat species in Polish lowlands. **Acta Theriologica 55** (4): 301-314. doi: 10.4098/j.at.0001-7051.093.2009.

ERICKSON, J.L. & S.D. WEST. 2003. Associations of bats with local structure and landscape features of forested stands in western Oregon and Washington. **Biological Conservation 109** (1): 95-102. doi: 10.1016/S0006-3207(02)00141-6.

ESTRADA, A.; C. JIMÉNEZ; A. RIVERA & E. FUENTES. 2004. General bat activity measured with an ultrasound detector in a fragmented tropical landscape in Los Tuxtlas, Mexico. **Animal Biodiversity and Conservation 27** (2): 1-9.

FENTON, M.B. 1970. A technique for monitoring bat activity with results obtained from different environments in southern Ontario. **Canadian Journal of Zoology 48** (4): 847-851. doi: 10.1139/z70-148.

FENTON, M.B. 1997. Science and the Conservation of Bats. **Journal of Mammalogy 78** (1): 1-14.

FRANCL, K.E. 2008. Summer bat activity at woodland seasonal pools in the Northern Great Lakes Region. **Wetlands 28** (1): 117-124. doi: 10.1672/07-104.1.

GERELL, R. & K.G. LUNDBERG. 1993. Decline of a bat *Pipistrellus pipistrellus* population in an industrialized area in south Sweden. **Biological Conservation 65** (2): 153-157. doi: 10.1016/0006-3207(93)90444-6.

HAYES, J.P. 1997. Temporal variation in activity of bats and the design of echolocation-monitoring studies. **Journal of Mammalogy 78** (2): 514-524.

HAYS, G.C.; J.R. SPEAKMAN & P.I. WEBB. 1992. Why do brown long-eared bats (*Plecotus auritus*) fly in winter? **Physiological Zoology 65** (3): 554-567.

IBGE. 2004. **Mapa de Biomas do Brasil.** Ministério do Meio ambiente, Instituto Brasileiro de Geografia e Estatística. Available online at: http://www.ibge.gov.br/home/presidencia/noticias/21052004biomas.shtm [Accessed: 01/VI/2013].

JOHNSON, J.B.; J.E. GATES & N.P. ZEGRE. 2011. Monitoring seasonal bat activity on a coastal barrier island in Maryland, USA. **Environmental Monitoring and Assessment 173** (1): 685-699. doi: 10.1007/s10661-010-1415-6.

JONES, G.; D.S. JACOBS; T.H. KUNZ; M.R. WILLIG & P.A. RACEY. 2009. Carpe noctem: the importance of bats as bioindicators. **Endangered Species Research 8** (1-2): 93-115. doi: 10.3354/esr00182.

KOFOKY, A.; D. ANDRIAFIDISON; F. RATRIMOMANARIVO; H. J. RAZAFIMANAHAKA; D. RAKOTONDRAVONY; P.A. RACEY & R.K.B. JENKINS. 2007. Habitat use, roost selection and conservation of bats in Tsingy de Bemaraha National Park, Madagascar. **Biodiversity and Conservation 16** (4): 1039-1053. doi: 10.1007/s10531-006-9059-0.

KUENZI, A.J. & M.L. MORRISON. 2003. Temporal Patterns of Bat Activity in Southern Arizona. **Journal of Wildlife Management 67** (1): 52-64.

KUNZ, T.H. 1973. Resource utilization: temporal and spatial components of bat activity in central Iowa. **Journal of Mammalogy 54** (1): 14-32.

LACKI, M.J. 1984. Temperature and humidity-induced shifts in the flight activity of little brown bats. **The Ohio Journal of Science 84** (5): 264-266.

LEWIS, T.S. 1970. Patterns of distribution of insects near a windbreak of tall trees. **Annals of Applied Biology 65** (2): 213-220. doi: 10.1111/j.1744-7348.1970.tb04581.x.

LIMPENS, H.J.G.A. & K. KAPTEYN. 1991. Bats, their behaviour and linear landscape elements. **Myotis 29**: 63-71.

LUMSDEN, L.F. & A.F. BENNETT. 2005. Scattered trees in rural landscapes: foraging habitat for insectivorous bats in southeastern Australia. **Biological Conservation 122** (2): 205-222. doi:10.1016/j.biocon.2004.07.006.

LUNDE, R.E. & A.S. HARESTAD. 1986. Activity of Little Brown Bats in Coastal Forest. **Northwest Science 60** (4): 206-209.

MORRIS, A.D.; D.A. MILLER & M.C. KALCOUNIS-RUEPPELL. 2010. Use of forest edges by bats in a managed pine forest landscape. **Journal of Wildlife Management 74** (1): 26-34. doi: 10.2193/2008-471.

NEUWEILER, G. 2000. **The biology of bats.** Oxford, Oxford University Press, VI+310p.

O'DONNELL, C.F.J. 2000. Conservation status and causes of decline of a threatened New Zealand Long-tailed bat *Chalinolobus tuberculatus* (Chiroptera: Vespertilionidae). **Mammal Review 30** (2): 89-106. doi: 10.1046/j.1365-2907.2000.00059.x.

O'FARRELL, M.J. & BRADLEY W.G. 1970. Activity Patterns of Bats over a Desert Spring. **Journal of Mammalogy 51** (1): 18-26.

O'FARRELL, M.J.; W.G. BRADLEY & G.W. JONES. 1967. Fall and winter bat activity at a desert spring in Southern Nevada. **The Southwestern Naturalist 12** (2): 163-171.

PAGLIA, A.P.; G.A.B. FONSECA & J.M.C. SILVA. 2008. A Fauna Brasileira Ameaçada de Extinção: Síntese Taxonômica e Geográfica, p. 63-70. *In*: A.B.M. MACHADO; G.M. DRUMMOND & A.P. PAGLIA (Eds). **Livro Vermelho da Fauna Brasileira Ameaçada de Extinção.** Belo Horizonte, Fundação Biodiversitas, vol. 2, 908p.

PARSONS, S. & J.M. SZEWCZAK. 2009. Detecting, Recording, and Analyzing the Vocalizations of Bats, p. 91-111. *In*: T.H. KUNZ & S. PARSONS (Eds). **Ecological and behavioral methods for the study of bats.** Baltimore, The Johns Hopkins University Press, XVII+901p.

PEEL, M.C.; B.L. FINLAYSON & T.A. McMAHON. 2007. Updated world map of the Köppen-Geiger climate classification. **Hydrology and Earth System Sciences 11** (5): 1633-1644. doi: 10.5194/hess-11-1633-2007.

PETTIT, T.W. & K.T. WILKINS. 2012. Canopy and edge activity of bats in a quaking aspen (*Populus tremuloides*) forest. **Canadian Journal of Zoology 90** (7): 798-807. doi:10.1139/Z2012-049.

PIERSON, E.D. 1998. Tall Trees, Deep Holes, and Scarred Landscapes – Conservation Biology of North American Bats, p. 309-325. *In*: T.H. KUNZ & P.A. RACEY (Eds). **Bat biology and conservation.** Washington, D.C., Smithsonian Institution Press, XIV+365p.

PILLAR, V.P.; S.C. MÜLLER; Z.M.S. CASTILHOS & A.V.A. JACQUES. 2009. **Campos Sulinos – Conservação e Uso Sustentável da Biodiversidade.** Brasília, Ministério do Meio Ambiente, 403p.

RACEY, P.A. & S.M. SWIFT. 1985. Feeding ecology of *Pipistrellus pipistrellus* (Chiropteran: Vespertilionidae) during pregnancy and lactation – I. Foraging behaviour. **Journal of Animal Ecology 54** (1): 205-215.

RAMBO, B. 2000. **A fisionomia do Rio Grande do Sul: ensaio de monografia natural.** São Leopoldo, Editora Unisinos, 3rd ed., XXVII+473p.

RANSOME, R.D. 1990. **The Natural History of Hibernating Bats.** London, Christopher Helm, XXI+235p.

RIO GRANDE DO SUL. 2002. **Decreto Estadual Nº 42.099, de 31 de dezembro de 2002.** Declara as espécies da flora nativa ameaçadas de extinção no estado do Rio Grande do Sul e dá outras providências. Porto Alegre, Diário Oficial do estado do Rio Grande do Sul de 01/01/2003.

RUSS, J.M.; M. BRIFFA & W.I. MONTGOMERY. 2003. Seasonal patterns in activity and habitat use by bats (*Pipistrellus* spp. and *Nyctalus leisleri*) in Northern Ireland, determined using a driven transect. **Journal of Zoology 259** (3): 289-299. doi: 10.1017/S0952836902003254.

RUSSO, D. & G. JONES. 2003. Use of foraging habitats by bats in a Mediterranean area determined by acoustic surveys: conseration implications. **Ecography 26** (2): 197-209. doi: 10.1034/j.1600-0587.2003.03422.x.

SERRA-COBO, J.; M. LOPEZ-ROIG; T. MARQUES-BONET & E. LAHUERTA. 2000. Rivers as possible landmarks in the orientation flight of *Miniopterus schreibersii*. **Acta Theriologica 45** (3): 347-352.

SIEMERS, B.M.; P. STILZ & H. SCHNITZLER. 2001. The acoustic advantage of hunting at low heights above water: behavioural experiments on the European 'trawling' bats *Myotis capaccinii*, *M. dasycneme* and *M. daubentonii*. **The Journal of Experimental Biology 204** (22): 3843-3854.

SIMMONS, N.B. 2005. Order Chiroptera, p. 312-529. *In*: D.E. WILSON & D.M. REEDER (Eds). **Mammal species of the world: a taxonomic and geographic reference.** Baltimore, The Johns Hopkins University Press, vol. 1, XXXV+743p.

TAYLOR, L.R. 1963. Analysis of the Effect of Temperature on Insects in Flight. **Journal of Animal Ecology 32** (1): 99-117.

TUTTLE, M.D. 1979. Status, Causes of Decline, and Management of Endangered Gray Bats. **The Journal of Wildlife Management 43** (1): 1-17.

VAUGHAN, N.; G. JONES & S. HARRIS. 1997. Habitat use by bats (Chiroptera) assessed by means of a broad-band acoustic method. **Journal of Applied Ecology 34** (3): 716-730.

VERBOOM, B. & K. SPOELSTRA. 1999. Effects of food abundance and wind on the use of tree lines by an insectivorous bat, *Pipistrellus pipistrellus*. **Canadian Journal of Zoology 77** (9): 1393-1401. doi: 10.1139/z99-116.

WALSH, A.L. & S. HARRIS. 1996. Foraging habitat preferences of vespertilionid bats in Britain. **Journal of Applied Ecology 33** (3): 508-518.

WICKRAMASINGHE, L.P.; S. HARRIS; G. JONES & N. VAUGHAN. 2003. Bat activity and species richness on organic and conventional farms: impact of agricultural intensification. **Journal of Applied Ecology 40** (6): 984-993. doi: 10.1111/j.1365-2664.2003.00856.x.

WOLDA, H. 1988. Insect Seasonality: Why? **Annual Review of Ecology and Systematics 19**: 1-18. doi: 10.1146/annurev.es.19.110188.000245.

4

Population biology of *Aegla platensis* (Decapoda: Anomura: Aeglidae) in a tributary of the Uruguay River, state of Rio Grande do Sul, Brazil

Marcelo M. Dalosto[1], Alexandre V. Palaoro[1], Davi de Oliveira[1],
Évelin Samuelsson[2] & Sandro Santos[1,3]

[1] *Núcleo de Estudos em Biodiversidade Aquática, Programa de Pós-Graduação em Biodiversidade Animal, Centro de Ciências Naturais e Exatas, Universidade Federal de Santa Maria. Avenida Roraima 1000, 97105-900 Santa Maria, RS, Brazil.*
[2] *Programa de Pós-Graduação em Ecologia, Departamento de Ciências Biológicas, Universidade Regional Integrada do Alto Uruguai e das Missões. Avenida Sete de Setembro 1621, 99700-000 Erechim, RS, Brazil.*
[3]*Corresponding author. E-mail: sandro.santos30@gmail.com*

ABSTRACT. Aeglids are freshwater anomurans that are endemic from southern South America. While their population biology at the species-level is relatively well understood, intraspecific variation within populations has been poorly investigated. Our goal was to investigate the population biology of *Aegla platensis* Schmitt, 1942 from the Uruguay River Basin, and compare our data with data from other populations. We estimated biometric data, sex ratio, population density and size-class frequencies, and frequencies of ovigerous females and juveniles, from the austral spring of 2007 until autumn 2008. Sexual dimorphism was present in adults, with males being larger than females. Furthermore, males and females were significantly larger than previously recorded for the species. The overall sex ratio was 1.33:1 (male:female), and population density ranged from 1.8 (spring) to 3.83 ind.m^{-2} (winter). Data from this population differ from published information about *A. platensis* in almost all parameters quantified except for the reproductive period, which happens in the coldest months, and a population structure with two distinct cohorts. Difference among studies, however, may be in part due to methodological differences and should be further investigated in order to determine their cause. In addition to different methodologies, they may result from ecological plasticity or from the fact that the different populations actually correspond to more than one species.

KEY WORDS. Aeglids; ecological plasticity; intraspecific variation; population density; sampling methods; sex ratio.

Aeglids are freshwater anomuran crustaceans with benthonic habits, whose distribution is restricted to temperate and subtropical regions of South America (BUCKUP & BOND-BUCKUP 1999, BOND-BUCKUP 2003). These crustaceans occur in river basins in Southern Brazil, Uruguay, Argentina, Southern Bolivia, Paraguay and South-central Chile, with 70 species currently described (SANTOS *et al.* 2013). Albeit ubiquitous in well-oxygenated running waters in these regions (DALOSTO & SANTOS 2011), several species have a very restricted distribution (BOND-BUCKUP *et al.* 2008).

Understanding the basic traits of an organism's biology is important because it provides basic information for a wide array of studies. In the case of aeglids, these range from conservation efforts (PÉREZ-LOSADA *et al.* 2009) to the use of model species in laboratory studies (PALAORO *et al.* 2013, SIQUEIRA *et al.* 2013). There is a considerable amount of studies on the basic biology of aeglids, such as population structure and dynamics (e.g., BUENO & BOND-BUCKUP 2000, FRANSOZO *et al.* 2003, COHEN *et al.* 2011, GRABOWSKI *et al.* 2013). However, previous studies on aeglids focused solely on the population dynamics of one species in a single location (e.g., BUENO & BOND-BUCKUP 2000, FRANSOZO *et al.* 2003, COHEN *et al.* 2011, GRABOWSKI *et al.* 2013). Ecological plasticity and variation in population biology parameters have been documented for other freshwater organisms, such as crayfish (HONAN & MITCHELL 1995, AUSTIN 1998, BEATTY *et al.* 2004, 2011), which share ecological similarities to aeglids (BOND-BUCKUP & BUCKUP 1994, NYSTRÖM 2002, AYRES-PERES *et al.* 2011, BURRESS *et al.* 2013, COGO & SANTOS 2013).

Unlike other aeglids, *Aegla platensis* Schmitt, 1942 has broad distribution and relatively large populations. This species is recorded for Paraguay, Uruguay, Argentina and Brazil, where it occurs in the states of Rio Grande do Sul and Santa Catarina (BOND-BUCKUP 2003). The population dynamics and growth of *A. platensis* have been studied for a population in the Guaíba Basin in the state of Rio Grande do Sul (BUENO & BOND-BUCKUP 2000 and BUENO *et al.* 2000, respectively). More recently, OLIVEIRA & SANTOS (2011) investigated the morphological sexual maturity of another population that inhabits the Uruguay River Basin, obtaining markedly different results from those reported by BUENO & BOND-BUCKUP (2000) and BUENO *et al.* (2000).

Our goal was to investigate several characteristics of the population biology of *A. platensis*, such as sex ratio, population structure, reproductive/recruitment seasons and population density. Also, we compare our results with data already available for this species, and with information available for other aeglids. Lastly, we discuss the variations in the population biology of this group of crustaceans and whether or not it is productive to compare among data obtained using different methods.

MATERIAL AND METHODS

The Lajeado Bonito stream (27°25'27"S; 53°24'39"W) is located in the municipality of Frederico Westphalen, state of Rio Grande do Sul. The dominant vegetation in the area is the Atlantic Forest and the climate is subtropical. The stream is a first order tributary of the Várzea River, in the Uruguay River Basin. The study site is located 470 m above sea level. Even though agricultural and livestock activities happen in the areas located upstream of the collection sites, the studied area harbors riparian vegetation on both margins of the stream. The streambed is composed of rocks of various sizes, sand, and bedrock.

Monthly collections of *A. platensis* were performed in a 160 m section of the stream from July 2007 to June 2008. This section was divided into 16 subunits. Aeglids were captured with traps (N = 16, one per subunit) placed before the dusk and revised in the morning of the following day. In order to sample the population more thoroughly, a 30 x 50 cm hand net with a 60 cm deep mouth and 1 mm mesh was also employed. The sampling effort, performed by two people, lasted approximately five minutes per subunit. Environmental variables (water temperature, dissolved oxygen, pH, flow speed, stream depth, stream width and conductivity) were measured monthly in three predetermined locations of the stream (Table I).

Table I. Enviromental parameters recorded for the Lajeado Bonito stream, Uruguay Basin, Rio Grande do Sul state, Brazil.

Parameters	Spring	Summer	Autumn	Winter
Temperature (°C)	19.830	19.830	16.270	15.200
Dissolved oxygen (mg/L)	6.250	6.280	8.040	7.990
Flow speed (m/s)	0.420	0.340	0.620	0.240
Conductivity (μS/cm)	68.680	88.510	71.990	76.960
pH	7.520	7.660	7.760	7.320
Stream depth (cm)	17.780	6.480	18.670	14.890
Stream width (m)	1.790	1.210	2.160	1.790
Discharge (m³/s)	0.277	0.024	0.075	0.096
Rainfall (mm/month)	246.000	97.000	198.000	147.330

After being separated from other animals captured accidentally, aeglids were identified, sexed, and females were checked for the presence of eggs. Sexing was based on morphological traits, such as the presence of pleopods in adult females, and their absence in adult males. Since younger individuals have inconspicuous pleopods, their sex was determined by observing the genital pores at the base of the third pereiopods (females) or their absence (males) (BOND-BUCKUP 2003). Biometric measurements were then taken with a digital caliper (0.01 mm accuracy), including carapace length (CL – from the tip of the rostrum to the posterior edge of the carapace), carapace width (CW – taken on the height of the upper suture of the gastric region), abdomen width (AW – measured on the second abdominal segment), length of the propodus of the left (LPL) and right (RPL) chelipeds (measured from the posterior proximal margin of the propodus to the tip of the fixed finger) and height of the chelar propodus (HCP – measured perpendicularly to the propodus length). Aeglids were classified according to their CL following OLIVEIRA & SANTOS (2011): males larger than 19.15 mm were considered adults, and females larger than 16.5 mm were determined adults. Aeglids smaller than 8 mm CL were measured with the help of a stereomicroscope taken to the field site. In order to minimize impact on the studied population, most animals were released back at their capture sites after data recording, with the exception of a few large males from the first collections, which were preserved as vouchers in the scientific collection of the Núcleo de Estudos em Biodiversidade Aquática, Universidade Federal de Santa Maria (voucher number UFSM-C 298). To test for differences in the body measurements of males and females, a Mann-Whitney test was used due to heterocedasticity and non-normality of the data (ZAR 2010). The test was performed in two different configurations: 1) using all captured individuals, and 2) using only adult individuals. The exception was the AW in the all-animals sample and the CL and CW in the adults-only, for which normality and homocedasticity could be attained through a log10 transformation, and for which a Welch two-sample t-test was used.

Sex ratio (male/female) was calculated for each season separately. A chi-square with Yates' correction for small samples was performed to test whether the sex ratio differed from the expected proportion of 1:1 within each season (ZAR 2010). Additionally, data obtained from traps and from handnets were plotted separately to check for possible influences of the sampling method on the sex ratio. The chi-square test performs poorly with small sampling numbers, which can generate spurious results (CRAWLEY 2012). Thus, the test was performed with pooled data from captures using traps and hand nets because of the low capture rates in certain seasons. The reproductive and recruitment seasons were estimated qualitatively through the frequency of ovigerous females and unsexed juveniles. Afterwards, we tested if the proportion of captures of ovigerous females and juveniles differed from the expected equal proportion of captures among the seasons using a binomial proportion test (ZAR 2010).

To estimate population density for each season four field samplings (August and November 2007, February and May

2008) were performed differently. Traps were set on a given day and revisited the morning of the following day, as usual, and then all captured aeglids were marked with a plastic tag placed in their dorsal region. This tag indicated the initial capture site and month of capture. The aeglids were then released back in the stream. The writing on the tags was made with Nanking ink, and the tags were fixed on the aeglid's carapace with cianoacrilate glue. Differently from regular collections, traps were then put back on the stream and the sampling procedure was repeated the following day. The amount of recaptured tagged individuals was recorded. Peterson's estimate (BEGON 1979) was applied to estimate population size: N = r*n/m, where: (N) estimate of the population size, (r) number of animals marked in the first day, (n) number of animals collected in the second day, and (m) number of tagged animals recaptured in the second day.

All data were tested for normality and heterocedasticity with the tests of Shapiro-Wilk and Levene, respectively. All tests were performed in the BioEstat 5.0 software (ZAR 2010, AYRES et al. 2007), except for the Welch two-sample t-tests, which were performed in the R environment (R CORE TEAM 2013).

RESULTS

A total of 957 individuals were collected, of which 76 were non-sexed juveniles, 503 males (323 juveniles and 180 adults), 378 females (187 juveniles, 169 adults and 22 ovigerous) (Table II). The CL ranged from 6 to 31.75 mm for males (median ± SD: 15.09 ± 7.35 mm), and from 6.08 to 27.92 for females (median ± SD: 16.11 ± 5.95 mm). There were significant differences only for AW (t = 2.215, p = 0.027) between males and females when all aeglids were considered. However, when only adults were considered there was a significant difference between males and females in all dimensions compared (U = 3.793, 9.781, 11.150, 12.452, p < 0.001; for AW, LPL, RPL and HCP, respectively; and t = -12.226, -11.392, and p < 0.001 for CL and CW, respectively) (Table III). The frequency distribution of size-classes of males and females of *A. platensis* for each season presented a bimodal distribution in all seasons (Figs 1-4).

Table II. Number of individuals of *Aegla platensis* collected during the four seasons in the Lajeado Bonito stream, Uruguay Basin, Rio Grande do Sul state. (JM) Juvenile males, (AM) adult males, (JF) juvenile females, (AF) adult females, (OF) ovigerous females, (NS) non-sexed juveniles.

Seasons	JM	AM	JF	AF	OF	NS	Total
Spring	81	43	33	44	4	50	255
Summer	102	22	63	18	7	7	219
Autumn	89	51	61	68	1	13	283
Winter	51	64	30	39	10	6	200
Total	323	180	187	169	22	76	957

Table III. Medians of the biometric measurements (mm) of the adult individuals of *Aegla platensis* captured in the Lajeado Bonito stream, Uruguay Basin, Rio Grande do Sul state, Brazil. Different letters in the column indicate statistically significant differences (p < 0.05) in the Mann-Whitney (a,b) or t-tests (c,d).

	CL (mm)	CW (mm)	AW (mm)	RPL (mm)	LPL (mm)	ACP (mm)
Males	24.87[c]	14.67[c]	17.67[a]	14.12[a]	15.91[a]	9.55[a]
Females	21.29[d]	12.55[d]	16.48[b]	10.28[b]	10.74[b]	6.34[b]

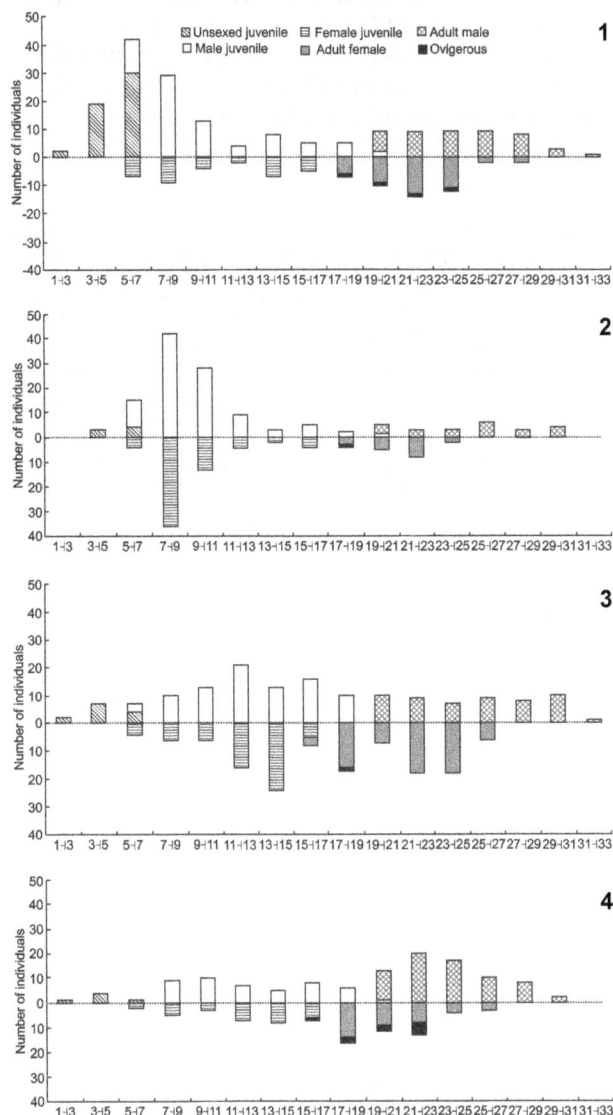

Figures 1-4. Absolute frequencies distribution of cephalothoracic length (CL) classes of individuals of *Aegla platensis* collected in the Lajeado Bonito stream, Uruguay Basin, Rio Grande do Sul state. Different letters indicate seasons: (1) spring; (2) summer; (3) autumn and (4) winter.

Males were more common in the spring and winter, with no difference for the other seasons, when all captures were considered (spring: $\chi^2 = 8.605$, df = 1, p = 0.003; summer: $\chi^2 = 2.151$, df = 1, p = 0.142; autumn: $\chi^2 = 0.034$, df = 1, p = 0.853; winter: $\chi^2 = 6.314$, df = 1, p = 0.012). The highest proportion of males was 60.48%, in the spring (Fig. 5). The number of ovigerous females caught (22) represented 5.82% of all the females. These were caught in all seasons, with a higher frequency in winter and summer, and the lowest frequency in the autumn. There was significant difference among the seasons, probably due to the small number of ovigerous females caught in the autumn (only 1; $\chi^2 = 24.735$, df = 3, p < 0.001; Fig. 6). Juveniles were caught throughout the sampling period, with a higher frequency during the spring ($\chi^2 = 63.104$, df = 3, p < 0.001; Fig. 6). Density ranged from 1.80 to 3.83 ind.m^{-2}, with the highest values in the winter (Table IV).

Table IV. Petersen's estimate of population size of *Aegla platensis* in the Lajeado Bonito stream, Uruguay Basin, Rio Grande do Sul state, Brazil.

Season	Marked in the 1st day	Captured 2nd day	Marked and recaptured	Population estimate	Individuals/m^2
Winter	25	72	2	900	3.83
Spring	29	73	5	423	1.80
Summer	18	86	2	772	3.29
Autumn	23	95	4	547	2.33

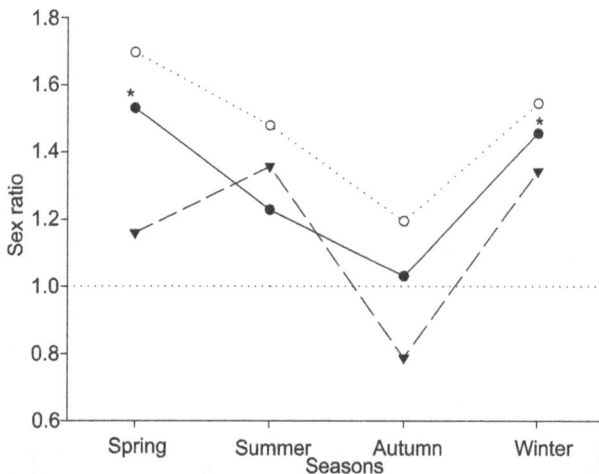

Figure 5. Relative frequency of males and females of *Aegla platensis* in the four seasons during the sampling period in the Lajeado Bonito stream, Uruguay Basin, Rio Grande do Sul state. Trap captures, handnet captures, and global (traps + handnet captures) are plotted separately. The asterisk (*) denotes statistical difference between the number of males and females by the Chi-square test (p < 0.05) for the global dataset. (●) Global, (○) hand net, (▼) trap.

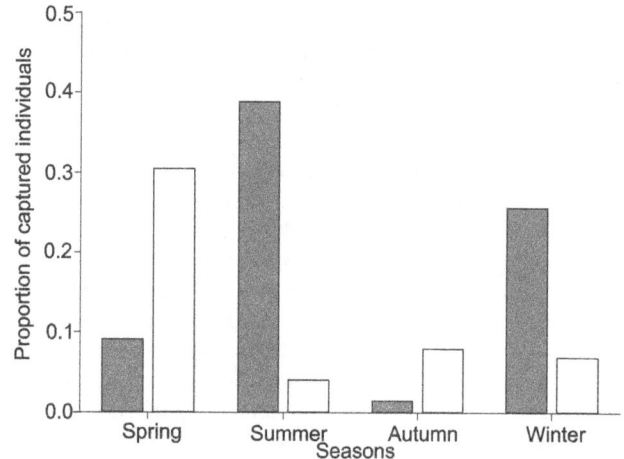

Figure 6. Number of ovigerous females and unsexed juveniles of *Aegla platensis* in the Lajeado Bonito stream, Uruguay Basin, Rio Grande do Sul state. Individuals were captured during four seasons. Frequencies of capture were distinct from the expected among the four seasons (Chi-square test, p < 0.05 for ovigerous females and juveniles). (■) Ovigerous, (□) unsexed juveniles.

DISCUSSION

In the population studied here (Uruguay River Basin, henceforth, UB), adult males had larger body dimensions than their female counterparts (except for AW, which is larger in females, even considering the juveniles). These data agree with the pattern described for most aeglids studied so far, where there is clear sexual dimorphism, with males prevailing in the larger size-classes (Colpo *et al.* 2005, Trevisan & Santos 2012, Trevisan *et al.* 2012), and females possessing broader abdomens for egg-incubation (López-Greco *et al.* 2004). Interestingly enough, one of the few studies that does not fit this pattern is the other previously studied population of *A. platensis* in the Guaíba Basin (henceforth, GB), where females were larger than males (Bueno & Bond-Buckup 2000). Since sexual dimorphism with larger males is a characteristic of the group (Barría *et al.* 2014, Bond-Buckup & Buckup 1994), a possible explanation for these opposite results might be the sampling methods adopted (see below for a more detailed discussion).

The maximum size of the animals also differed between UB and GB: maximum CL for males and females were of 17.39 and 19.12 mm for GB, and of 31.75 and 27.92 mm for UB, respectively (Bueno & Bond-Buckup 2000, Table V). Animals from UB were 50% larger than those registered for the GB. The different sampling and measuring methods adopted could account for this. In GB, the CL measurements did not include the rostrum, while our data include it. However, the rostrum would increase the CL size of the aeglids in approximately 10%, which is certainly not enough to compensate for a size difference of 66% in relation to animals from the UB. Additionally, Bueno &

BOND-BUCKUP (2000) only employed manual search, while we employed a combination of traps and handnet. The use of traps could affect the mean and maximum size of the animals captured. We found differences in the mean size of individuals captured by traps (21.83 ± 4.84 mm) and handnets (12.76 ± 6.34 mm), similar to what has been found for other decapod crustaceans. However, this was not the case for the maximum size, since the largest aeglid caught by traps had 31.75 mm CL, while the largest caught by hand-net had 30.65 mm CL. This shows that despite the fact that traps tend to capture larger individuals, handnet is still able to sample large specimens, and is thus considered an appropriate collecting gear to estimate size range. We think that it is safe to conclude that animals from the UB are considerably larger than GB.

Table V. Sampling methods and population parameters of *Aegla platensis* evaluated for the populations of the Uruguay River (this study) and Guaíba Basins (BUENO & BOND-BUCKUP 2000), respectively.

Parameters	Uruguay River Basin	Guaíba Basin
Sex ratio (M:F)	1.33:1	1.08:1
Population density (ind/m²)	1.8-3.83	8.7-19
Sexual dimorphism	Present, males larger	Present, females larger
Size of the largest male (mm)	31.75	17.39
Size of the largest female (mm)	27.92	19.12
Population structure	Bimodal	Bimodal
Reproductive period	Year-round, peak in coldest months	Year-round, peak in coldest months
Sampling technique	Handnet + traps	Handnet

The sex ratio also differed between GB and UB. In GB, it did not differ significantly from 1:1, while the opposite was found in UB, where it differed from the 1:1 expected proportion in the spring and winter, with an overall sex ratio of 1.33:1. Both results fit the pattern recorded for aeglids, in which the sex ratio ranges from 1:1 to values skewed towards males (Table VI). Once again, the effect of the sampling method makes it difficult to distinguish between actual differences between the species/populations, and the effects of the different sampling methods chosen by each author. The tendency to capture more large adult males using traps had already been demonstrated in crayfish surveys (BEATTY et al. 2004, 2011), and it is relatively safe to infer that the same is true for aeglids (BUENO et al. 2007, TEODÓSIO & MASUNARI 2009, GRABOWSKI et al. 2013). In this study, however, when analyzing the sex ratio of aeglids captured by hand-net and by traps (Fig. 5), we can see that the ratio was more skewed towards males in the hand-net captures than in the traps. Thus, we conclude that the difference regarding the sex ratio between GB and UB is not an effect of the methods chosen, but that it reflects an actual difference between these populations.

Table VI. Sex ratio and population density parameters of published studies on Brazilian species of *Aegla*.

Species	Sex ratio (M:F)	Density (ind/m²)	Authors
A. castro	1:1	–	SWIECH-AYOUB & MASUNARI (2001)
A. castro	1.08:1	–	FRANSOZO et al. (2003)
A. franca	–	2.2-2.7	BUENO et al. (2007)
A. franciscana	1:1	–	GONÇALVES et al. (2006)
A. leptodactyla	1.19:1	–	NORO & BUCKUP (2002)
A. longirostri	1:1	–	COLPO et al. (2005)
A. parana	2:1*	–	GRABOWSKI et al. (2013)
A. paulensis	1.66:1*	–	COHEN et al. (2011)
A. platensis	1.08:1	8.7-19	BUENO & BOND-BUCKUP (2000)
A. platensis	1.33:1*	1.8-3.83	Current study
A. schmitti	2:1*	–	TEODÓSIO & MASUNARI (2009)

* Sex ratio statistically different from an 1:1 expected proportion.

Ovigerous females were captured year-round, with a peak in the colder months (late winter and early spring). Thus, our data agree with the pattern known for other Brazilian species of *Aegla*, where the reproduction is either year-round with peaks in the colder months (BUENO & BOND-BUCKUP 2000, COLPO et al. 2005), or just concentrated in the colder months (TEODÓSIO & MASUNARI 2009, GRABOWSKI et al. 2013), including GB. The frequency of juveniles also follows a similar pattern: juveniles were captured year-round, being more abundant in the spring following the peak of the reproductive season (BUENO & BOND-BUCKUP 2000) (Fig. 6). The population structure was bimodal, with two age groups easily distinguishable in the size-class frequency distribution (Figs 1-4). This is in agreement with information for other aeglids, which also show two distinct cohorts in the population (e.g., BUENO & BOND-BUCKUP 2000, FRANSOZO et al. 2003).

The density also differed markedly between localities, being much lower in UB than in GB. In fact, the density of A. platensis in UB was much more similar to the density of another species, A. franca Schmitt, 1942, in the Barro Preto stream (Minas Gerais state, Brazil), than to the density of its conspecific in GB (Table V). Even though the capture methods used by BUENO & BOND-BUCKUP (2000) differ from ours, and may have affected our density results, the difference between both estimates is over 120%. Considering the markedly larger size of the aeglids in UB, one can expect that their populations will exhibit lower densities. This is even more likely if we consider the aggressive nature of aeglids (AYRES-PERES et al. 2011, PALAORO et al. 2013). In crayfish, spatial patterns investigated in natural environments show that dominant animals (i.e., the largest) are more spaced from other crayfish than smaller individuals (FERO & MOORE 2008). Although there are no such studies for aeglids, their ecological (BURRESS et al. 2013) and behavioral (MOORE 2007, AYRES-PERES et al. 2011) similarities with crayfish, along with our results, support the idea of a negative relationship between body size and density in *A. platensis*.

In general terms, the population structure of UB agrees with the known pattern for aeglids, presenting sexual dimorphism with larger males, a bimodal distribution of the size-class frequencies, reproduction concentrated in the coldest months of the year and release of juveniles in the following season (ROCHA *et al.* 2010, COHEN *et al.* 2011). When compared to GB, however, some differences can be highlighted: aeglids were much larger in UB than GB; the larger aeglids were males in UB and females in GB; the sex ratio was skewed towards males in UB, and similar to 1:1 in GB; and population density values were at least two times higher in GB than in UB. Differences in these population biology characteristics can also be a result of different environmental pressures. BÜCKER *et al.* (2008) have shown that the spatial micro distribution of *A. platensis* and *A. itacolomiensis* Bond-Buckup & Buckup, 1994 are correlated with the availability of coarse organic matter: the distribution of *A. platensis* was explained by the availability of twigs, followed by fragmented leaves, while the distribution of *A. itacolomiensis* was explained by fragmented leaves, followed by twigs. However, BUENO & BOND-BUCKUP (2000) do not present any environmental variable other than temperature. Furthermore, BÜCKER *et al.* (2008) make a much more detailed surveillance of environmental variables, but do not provide any data regarding the size, density, or other population biology parameters.

The only similarities between UB and GB were the bimodal population structure and the reproduction peak on the colder months (Table V). Nevertheless, these characteristics are shared by most *Aegla* species studied so far (e.g., FRANSOZO *et al.* 2003, GONÇALVES *et al.* 2006, TEODÓSIO & MASUNARI 2009), and thus, cannot be considered a species-specific characteristic. Conversely, the maximum size, sexual dimorphism, sex ratio and population density clearly differed between the two populations. Albeit variation in population parameters is expected, and sampling methods can bias the results, the differences between UB and GB are very marked, eventually presenting differences of over 100% in certain values. If we consider the geographical isolation between the two river basins (SCHWARZBOLD 2010), alongside the evidences for ecological differences, it becomes clear that molecular studies might be the best choice to elucidate if this is a case of ecological differences between populations, or if this a case of cryptic species (MARCHIORI *et al.* 2014).

The variety of methods employed by researchers is by far the greatest obstacle to reliable comparisons between population studies on *Aegla*. More specifically, the choice of the capture method (baited traps, manual search, handnet, Surber sampler, or any combination of these) seems to bias the results. A clear example can be seen in Table VI. The four studies of *Aegla* where the sex ratio differed significantly from 1:1 were the ones that employed traps, with the three with the more skewed sex ratios being those that relied solely on traps as the sampling method. This issue has already been addressed by previous authors (e.g., BUENO *et al.* 2007, GRABOWSKI *et al.* 2013). Despite this, there is still no consensus among researchers on the best methods.

In conclusion, *A. platensis* presented marked differences from one population to another. These differences can be attributed partially to the different sampling methods used by different authors. These differences, along with isolation between the two river basins, suggest that molecular studies are needed to elucidate the taxonomic status of the populations of this species. The only similarities between the populations were common to many *Aegla* species, which highlights the need of a standardized technique to perform population studies in these anomurans, so that more reliable and less speculative comparisons can be made.

ACKNOWLEDGMENTS

We would like to thank CAPES for the scholarships for AVP and DO; CAPES/FAPERGS for the scholarship for MMD, and CNPq for the productivity grant for SS (308598/2011-3). We would also like to thank our colleagues at the Núcleo de Estudos em Biodiversidade Aquática for their help in the field work, T.M. Dias, two anonymous reviewers, and A.S. Melo for the helpful comments and suggestions that certainly improved the manuscript.

LITERATURE CITED

AUSTIN, C.M. 1998. Intra-specific variation in clutch and brood size and rate of development in the yabby, *Cherax destructor* (Decapoda: Parastacidae). **Aquaculture 167**: 147-159. doi: 10.1016/S0044-8486(98)00306-8.

AYRES, M.; M. AYRES JR; D.L. AYRES & A.S. SANTOS. 2007. **Bioestat 5.0: aplicações estatísticas nas áreas das Ciências Biológicas e Médicas.** Belém, Sociedade Civil Mamirauá.

AYRES-PERES, L.; P.B. ARAUJO & S. SANTOS. 2011. Description of the agonistic behavior of *Aegla longirostri* (Decapoda: Aeglidae). **Journal of Crustacean Biology 31** (3): 379-388. doi: 10.1651/10-3422.1.

BARRÍA, E.M.; S. SANTOS; C.G. JARA & C.J. BUTLER. 2014. Sexual dimorphism in the cephalothorax of freshwater crabs of the genus *Aegla* Leach from Chile (Decapoda, Anomura, Aeglidae): an interspecific approach based on distance variables. **Zoomorphology**. doi: 10.1007/s00435-014-0231-x.

BEATTY S.J.; D.L. MORGAN & H.S. GILL. 2004. Biology of a translocated population of *Cherax cainii* Austin & Ryan, 2002 in a western Australian river. **Crustaceana 77** (11): 1329-1351. doi: 10.1163/1568540043166010.

BEATTY S.J.: M. DE GRAAF; B. MOLONY; V. NGUYEN & K. POLLOCK. 2011. Plasticity in population biology of *Cherax cainii* (Decapoda: Parastacidae) inhabiting lentic and lotic environments in south-western Australia: Implications for the sustainable management of the recreational ûshery. **Fisheries Research 110**: 312-324. doi: 10.1016/j.fishres.2011.04.021.

BEGON, M. 1979. **Investigating animal abundance: capture-recapture techniques for biologists.** London, Edward Arnold.

BOND-BUCKUP, G. 2003. A Família Aeglidae, p. 21-116. *In*: G.A.S. MELO (Ed.). **Manual de identificação dos Crustacea Decapoda de água doce do Brasil.** São Paulo, Editora Loyola.

BOND-BUCKUP, G. & L. BUCKUP. 1994. A Família Aeglidae (Crustacea, Decapoda, Anomura). **Arquivos de Zoologia 32** (4): 159-347.

BOND-BUCKUP, G.; C.G. JARA; M. PÉREZ-LOSADA; L. BUCKUP & K.A. CRANDALL. 2008. Global diversity of crabs (Aeglidae: Anomura: Decapoda) in freshwater. **Hydrobiologia 595**: 267-273. doi: 10.1007/s10750-007-9022-4.

BÜCKER, F; R. GONÇALVES; G. BOND-BUCKUP & A.S. MELO. 2008. Effect of the environmental variables on the distribution of two freshwater crabs (Anomura: Aeglidae). **Journal of Crustacean Biology 28** (2): 248-251. doi: 10.1651/0278-0372(2008)028[0248:EOEVOT]2.0.CO;2.

BUCKUP, L. & G. BOND-BUCKUP. 1999. **Os crustáceos do Rio Grande do Sul.** Porto Alegre, Editora UFRGS.

BUENO, A.A.P. & G. BOND-BUCKUP. 2000. Dinâmica populacional de *Aegla platensis* Schmitt (Crustacea, Decapoda, Aeglidae). **Revista Brasileira de Zoologia 17** (1): 43-49. doi: 10.1590/S0101-81752000000100005.

BUENO, A.A.P.; G. BOND-BUCKUP & L. BUCKUP. 2000. Crescimento de *Aegla platensis* Schmitt em ambiente natural (Crustacea, Decapoda, Aeglidae). **Revista Brasileira de Zoologia 17** (1): 51-60. doi: 10.1590/S0101-81752000000100006.

BUENO, S.L.S.; R.M. SHIMIZU & S.S. DA ROCHA. 2007. Estimating the population size of *Aegla franca* (Decapoda: Anomura: Aeglidae) by mark-recapture technique from an isolated section of Barro Preto stream, county of Claraval, state of Minas Gerais, southeastern Brazil. **Journal of Crustacean Biology 27** (4): 553-559. doi: 10.1651/S-2762.1.

BURRESS, E.D.; M.M. GANGLOFF & L. SIEFFERMAN. 2013. Trophic analysis of two subtropical South American freshwater crabs using stable isotope ratios. **Hydrobiologia 702**: 5-13. doi: 10.1007/s10750-012-1290-y.

BYRON, C.J. & K.A. WILSON. 2001. Rusty crayfish (*Orconectes rusticus*) Movement within and between habitats in Trout Lake, Vilas County, Wisconsin. **Journal of the North American Benthological Society 20** (4): 606-614.

COHEN, F.P.A.; B.F. TAKANO; R.M. SHIMIZU & S.L.S. BUENO. 2011. Life cycle and population structure of *Aegla paulensis* (Decapoda: Anomura: Aeglidae). **Journal of Crustacean Biology 31** (3): 389-395. doi: 10.1651/10-3415.1.

COGO, G.B. & S. SANTOS. 2013. The role of aeglids in shredding organic matter in Neotropical streams. **Journal of Crustacean Biology 33** (4): 519-526. doi: 10.1163/1937240X-00002165.

COLPO, K.D.; L.R. OLIVEIRA & S. SANTOS. 2005. Population biology of the freshwater anomuran *Aegla longirostri* (Crustacea, Anomura, Aeglidae) from Ibicuí-Mirim River, Itaára, RS, Brazil. **Journal of Crustacean Biology 25** (3): 495-499. doi: 10.1651/C-2543.

CRAWLEY, M.J. 2012. **The R book.** Chichester, John Wiley, 2nd ed.

DALOSTO, M. & S. SANTOS. 2011. Differences in oxygen consumption and diel activity as adaptations related to microhabitat in Neotropical freshwater decapods (Crustacea). **Comparative Biochemistry and Physiology, Part A 160**: 461-466. doi: 10.1016/j.cbpa.2011.07.026.

FERO, K. & P.A. MOORE. 2008. Social spacing of crayfish in natural habitats: what role does dominance plays? **Behavioral Ecology Sociobiology 62**: 1119-1125. doi: 10.1007/s00265-007-0540-x.

FRANSOZO, A.; R.C. COSTA; A.L.D. REIGADA & J.M. NAKAGAKI. 2003. Population structure of *Aegla castro* Schmitt, 1942 (Crustacea: Anomura: Aeglidae) from Itatinga (SP), Brazil. **Acta Limnologica Brasiliensia 15** (2): 13-20.

GONÇALVES, R.S.; D.S. CASTIGLIONI & G. BOND-BUCKUP. 2006. Ecologia populacional de *Aegla franciscana* (Crustacea, Decapoda, Anomura) em São Francisco de Paula, RS, Brasil. **Iheringia, Série Zoologia, 96** (1): 109-114. doi: 10.1590/S0073-47212006000100019.

GRABOWSKI, R.C.; S. SANTOS & A.L. CASTILHO. 2013. Reproductive ecology and size of sexual maturity in the anomuran crab *Aegla parana* (Decapoda: Aeglidae). **Journal of Crustacean Biology 33** (3): 332-338. doi: 10.1163/1937240X-00002148.

HONAN, J.A. & B.D. MITCHELL. 1995. Reproduction of *Euastacus bispinosus* Clark (Decapoda: Parastacidae), and trends in the reproductive characteristics of freshwater crayûsh. **Marine and Freshwater Research 46**: 485-499. doi: 10.1071/MF9950485.

LÓPEZ-GRECO, L.; V. VIAU; M. LAVOLPE; G. BOND-BUCKUP & E.M. RODRIGUEZ. 2004. Juvenile hatching and maternal care in *Aegla uruguayana* (Anomura, Aeglidae). **Journal of Crustacean Biology 24** (2): 309-313. doi: 10.1651/C-2441.

MARCHIORI, A.B.; M.L. BARTHOLOMEI-SANTOS & S. SANTOS. 2014. Intraspecific variation in *Aegla longirostri* (Crustacea: Decapoda: Anomura) revealed by geometric morphometrics: evidence of an ongoing speciation process. **Biological Journal of the Linnean Society 112** (1): 31-39. doi: 10.1111/bij.12256.

MOORE, P.A. 2007. Agonistic behavior in freshwater crayûsh: the inûuence of intrinsic and extrinsic factors on aggressive encounters and dominance, p. 90-114. *In*: J.E. DUFFY & M. THIEL (Eds). **Evolutionary ecology of social and sexual systems – crustaceans as model organisms.** Oxford, Oxford University Press.

NORO, C.K. & L. BUCKUP. 2002. Biologia reprodutiva e ecologia de *Aegla leptodactyla* Buckup & Rossi (Crustacea, Anomura, Aeglidae). **Revista Brasilcira de Zoologia 19** (4): 1063-1074. doi: 10.1590/S0101-81752002000400011.

NYSTRÖM, P. 2002. Ecology, p. 192-235. *In*: D.M. HOLDICH (Ed.). **Biology of Freshwater Crayfish**. Oxford, Blackwell Science.

OLIVEIRA, D. & S. SANTOS. 2011. Maturidade sexual morfológica de *Aegla platensis* (Crustacea, Decapoda, Anomura) no Lajeado Bonito, norte do estado do Rio Grande do Sul, Brasil. **Iheringia, Série Zoologia, 101** (1-2): 127-130. doi: 10.1590/S0073-47212011000100018.

PALAORO, A.V.; L. AYRES-PERES & S. SANTOS. 2013. Modulation of male aggressiveness through different communication pathways. **Behavioral Ecology and Sociobiology 67** (2): 283-292. doi: 10.1007/s00265-012-1448-7.

PÉREZ-LOSADA, M.; G. BOND-BUCKUP; C.G. JARA & K.A. CRANDALL. 2009. Conservation assessment of southern South American freshwater ecoregions on the basis of the distribution and genetic diversity of crabs from the genus *Aegla*. **Conservation Biology 23** (3): 692-702. doi: 10.1111/j.1523-1739.2008.01161.x.

R CORE TEAM. 2013. **R: A language and environment for statistical computing**. Vienna, R Foundation for Statistical Computing.

ROCHA, S.S.; R.M. SHIMIZU & S.L.S. BUENO. 2010. Reproductive biology in females of *Aegla strinatii* (Decapoda: Anomura: Aeglidae). **Journal of Crustacean Biology 30** (4): 589-596. doi: 10.1651/10-3285.1.

SANTOS, S.; C.G.JARA; M.L. BARTHOLOMEI-SANTOS; M. PÉREZ-LOSADA & K.A. CRANDALL. 2013. New species and records of the genus *Aegla* Leach, 1820 (Crustacea, Anomura, Aeglidae) from the West-Central region of Rio Grande do Sul, Brazil. **Nauplius 21** (2): 211-223.

SCHWARZBOLD, A. 2010. **Ciência & Ambiente n. 41 – Os Rios da América**. Santa Maria, Editora Universidade Federal de Santa Maria.

SIQUEIRA, A.F.; A.V. PALAORO & S. SANTOS. 2013. Mate preference in the neotropical freshwater crab *Aegla longirostri* (Decapoda: Anomura): does the size matter? **Marine and Freshwater Behaviour and Physiology 46** (4): 219-227. doi: 10.1080/10236244.2013.808832.

SWIECH-AYOUB, B.P. & S. MASUNARI. 2001. Flutuações temporal e espacial de abundância e composição de tamanho de *Aegla castro* Schmitt (Crustacea, Anomura, Aeglidae) no Buraco do Padre, Ponta Grossa, Paraná, Brasil. **Revista Brasileira de Zoologia 18** (3): 1003-1017. doi: 10.1590/S0101-81752001000300032.

TEODÓSIO, E.A.O. & S. MASUNARI. 2009. Estrutura populacional de *Aegla schmitti* (Crustacea: Anomura: Aeglidae) nos reservatórios dos Mananciais da Serra, Piraquara, Paraná, Brasil. **Zoologia 26** (1): 19-24. doi: 10.1590/S1984-46702009000100004.

TREVISAN, A. & S. SANTOS. 2012. Morphological sexual maturity, sexual dimorphism and heterochely in *Aegla manuinflata* (Anomura). **Journal of Crustacean Biology 32** (4): 519-527. doi:10.1163/193724012X635944.

TREVISAN, A.; M.Z. MAROCHI; M. COSTA; S. SANTOS & S. MASUNARI. 2012. Sexual dimorphism in *Aegla marginata* (Decapoda: Anomura). **Nauplius 20**: 75-86.

ZAR, J. 2010. **Biostatistical analysis**. New Jersey, 5th ed., Prentice Hall.

5

Dimorphism and allometry of *Systaltocerus platyrhinus* and *Hypselotropis prasinata* (Coleoptera: Anthribidae)

Ingrid Mattos[1], José Ricardo M. Mermudes[1,3] & Mauricio O. Moura[2]

[1] *Laboratório de Entomologia, Departamento de Zoologia, Universidade Federal do Rio de Janeiro. Caixa Postal 68044, 21941-971 Rio de Janeiro, RJ, Brazil.*
[2] *Departamento de Zoologia, Universidade Federal do Paraná. Caixa Postal 19020, 81531-980 Curitiba, PR, Brazil.*
[3] *Corresponding author. E-mail: jrmermudes@gmail.com*

ABSTRACT. Males of sexually dimorphic anthribid species display structural modifications that suggest sexual selection. Polyphenism, which is expressed through morphological and behavioral novelties, is an important component of the evolutionary process of these beetles. In this study, we endeavored to ascertain the presence of variations in selected monomorphic traits, polyphenism in males, and variation in structures associated with sexual dimorphism and allometric patterns in two species: *Systaltocerus platyrhynus* Labram & Imhoff, 1840 and *Hypselotropis prasinata* (Fahraeus, 1839). To that end, we used Principal Components Analysis (PCA) and Canonical Variate analysis (CVA) to statistically analyze 26 measurements of 91 specimens. The PCA discriminated three groups (females, major, and minor males) for *S. platyrhinus*, but only two groups (males and females) for *H. prasinata*. The same groups discriminated by the PCA for *Systaltocerus* were confirmed by the CVA analysis, indicating a highly significant variation separating the three groups. We also analyzed positive allometry with respect to prothorax length – independent variable by Reduced Major Axis (RMA). The allometric pattern indicated by most of the linear measurements was strong and corroborates a possible relationship between male polyphenism and the reproductive behavior of major and minor males. We believe that these patterns, in species that show both sexual dimorphism and male polyphenism, are associated with the behavior of defending the female during oviposition, performed by major males.

KEY WORDS. Anthribinae; morphometry; polyphenism; sexual dimorphism.

Male sexual dimorphism and polyphenism are ubiquitous in several species of Coleoptera (EMLEN *et al.* 2005, KAWANO 2006). These phenotypic differences are thought to be linked to fitness, since they influence reproductive success (EBERHARD & GUTIEREZ 1991, EMLEN & NOJHOUT 2000, EMLEN 1994, 1996, 2008, EMLEN *et al.* 2005, 2007, KAWANO 2006). In insects, body size is an important phenotypic trait which often corresponds to adaptations (POSSADAS *et al.* 2007). Some species of Coleoptera, for instance beetles with horns (e.g., Scarabaeidae, Dynastinae) and those with oversized mandibles (Cerambycidae, Prioninae, and Lucanidae) are model systems for studies on the evolution of sexual dimorphism and polyphenism (EBERHARD & GUTIEREZ 1991, KAWANO 2000, SHIOKAWA & IWAHASHI 2000). Moreover, Anthribidae species show both sexual dimorphism and polyphenism (MERMUDES 2002, YOSHITAKE & KAWASHIMA 2004).

Fungus weevils (Anthribidae: Curculionoidea) comprise about 370 genera and at least 3,900 species (SLIPINSK *et al.* 2011). Most species of Anthribinae have remarkable sexual dimorphism, particularly with respect to the size of the rostrum and antennae (HOLLOWAY 1982, MERMUDES 2002, 2005, MERMUDES & NAPP 2006). Anthribidae females have toothed sclerotized plates at the apex of the ovipositor, which bear conchoidal projections that are used to excavate plant tissues for oviposition. This behavior is unique and distinct among Curculionoidea, which use only the rostrum to dig plant tissues (HOWDEN 1995).

Although sexual dimorphism in size and polyphenism in male size are widespread in Anthribinae (MERMUDES 2002, 2005, MERMUDES & NAPP 2006, MERMUDES & MATTOS 2010), detailed information about it is only available for a few species (HOLLOWAY 1982). YOSHITAKE & KAWASHIMA (2004) and MATSUO (2005) demonstrated that in large, intermediate, and small males of the Japanese fungus weevil *Exechesops leucopis* Jordan, 1928 the length of the eyestalks, which are associated with the agonistic behavior males use to protect females against other males on fruits of *Styrax japonica* Siebold & Zuccarini (Styracaceae) differs. Large males that have more developed cephalic eyestalks win the disputes, indicating that sexual dimorphism and polyphenism in males are under sexual selection. However, smaller males (without developed eyestalks) can copulate in the absence of competition when females are not accompanied by larger males, which may partly explain the sneaky behavior of small males described by YOSHITAKE & KAWASHIMA (2004).

Agonist behavior in Anthribidae was also observed by
THOMPSON (1963) and HOWDEN (1992). Thompson reported that
guarding males of *Deuterocrates longicornis* (Fabricius, 1781), a
species from West Africa, defend females and engage in fights
with other males using their mandibles. HOWDEN (1992) recorded that males of *Ptychoderes rugicollis* Jordan, 1895, a Neotropical species, use their antennae and rostrum to protect
females while they lay eggs on dead trees.

Considering the past detection of polyphenism in size
in two species of Neotropical Anthribinae, *Systaltocerus
platyrhinus* Labram & Imhoff, 1840 (variations in the length
and shape of the rostrum; MERMUDES 2002) and *Hypselotropis
prasinata* (Fahraeus, 1839) (different length of rostrum and
antennae; MERMUDES 2005, MERMUDES & RODRIGUES 2010), we
endeavored to determine whether there is variation in monomorphic characters (such as eyes, prothorax, and elytra),
polyphenism in males, variation in sexually dimorphic structures (rostrum, antennae, and ventrites) and allometric patterns. This study contributes to the understanding of patterns
of dimorphism and polyphenism in Anthribidae and evaluates structures that are likely to interfere with body size and/or
with the relative size of other structures in the two species.
However, whether agonistic interactions occur between males
in those species remains unknown.

MATERIAL AND METHODS

In this study, we used a sample of 34 specimens (25 males
and 9 females) of *S. platyrhinus* and 57 specimens (32 males
and 25 females) of *H. prasinata* loaned from three collections
(curators between parenthesis): MNRJ, Museu Nacional,
Universidade Federal do Rio de Janeiro, Rio de Janeiro (M.
Monné); AMCT, American Coleoptera Museum, San Antonio,
Texas (J. Wappes); and DZUP, Coleção Padre Jesus S. Moure,
Departamento de Zoologia, Universidade Federal do Paraná,
Curitiba (L. Marinoni).

All individuals were measured using the standard image-
analysis software Moticam 1000, or in the case of elytral length, a
digital caliper. Before each trait was measured, the specimen was
oriented so that the trait of interest was as closely parallel to the
plane of the objective lens as possible. The anatomical landmarks
measured follow MERMUDES & NAPP (2006) with some modifications.
These modifications, defined in Table I, are based on characters
that display variation among males and the sexes, independently
of geographical locality. The 26 traits (measurements in millimeters) used were log-transformed (Table I and Figs 1-7).

Linear models and cluster analysis were performed in
PAST version 2.0 (HAMMER *et al.* 2001). Multivariate analyses
(PCA and CVA) were run in vegan (OKSANEN *et al.* 2013) and
Morph (SCHLAGER 2013). Both packages were implemented in R
(R CORE TEAM 2013).

Variations in phenotypic traits between and within sexes
were accessed through the coefficient of variation (CV).

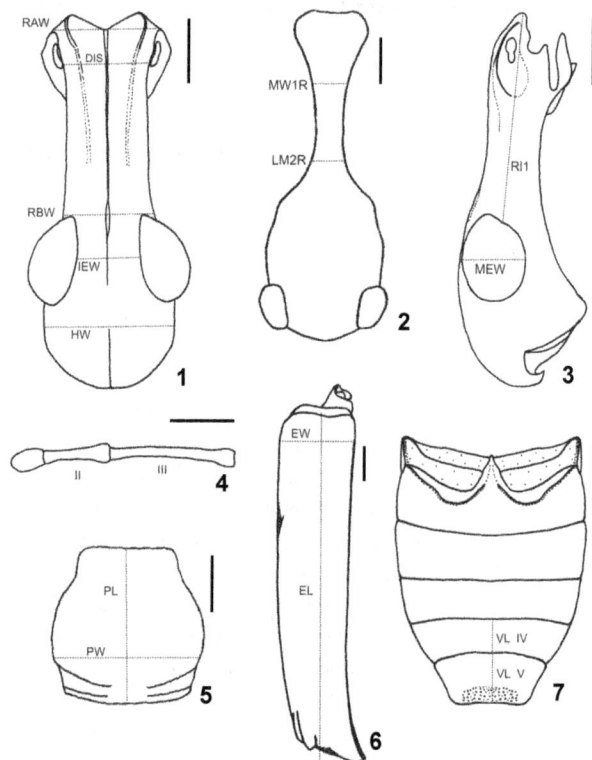

Figures 1-7. Diagram of the morphological traits measured: (1)
Hypselotropis prasinata, head, dorsal; (2) *Systaltocerus platyrhinus*,
head, frontal; (3-7) *H. prasinata*: (4) antennal segments I-III; (5)
prothorax, dorsal; (6) elytron, dorsal; (7) abdomen, ventral. For
abbreviations see Material and methods. Scale bars: 1 mm.

A cluster analysis with Ward's methods (based on Eu-
clidean distance) was carried out with 1,000 Bootstrap replicates (VALENTIN 2000). In this analysis, missing data were
replaced by the column average. Additionally, a Principal Components Analysis (PCA) of the covariance-variance matrix of
all variables was performed to reduce the dimension of the
data matrix and to visualize possible differences among groups
and characters that contributed the most to these differences.
The first two component axes were then used as variables in a
Canonical Variate Analysis (CVA) to test morphometric differences among groups.

The analyses were designed to test the relationship between body size (prothorax length = PL) and all other variables.
For this reason we used the allometric function $y = ax^b$ (HUXLEY
1932, 1950). However, the data was log-transformed and expressed by: $\log y = \log a + b (\log x)$, to fit a straight line (GOULD
1966).

Body size (prothorax length = PL) was used as a predictor variable and all other measurements were considered as
response variables. However, in allometric studies, no variable

Table I. Measurements obtained from each part of the body.

Measures and abbreviation	Description
Rostral length 1 (RL 1)	laterally between the anterior margin of the eye and the apex of the rostrum
Apical width of rostrum (RAW)	dorsally at the apical margin of the rostrum
Basal width of rostrum (RBW)	measured dorsally at the base of the rostrum
Medial width 1 of rostrum (MW1R)	dorsally in the rostrum, only in Systaltocerus platyrhinus (modified from Mermudes 2002)
Medial width 2 of rostrum (MW2R)	dorsally in the rostrum, only in Systaltocerus platyrhinus (modified from Mermudes 2002)
Head width (HW)	dorsally between the lateral margins of the head
Antennal segments, length = seven variables (II, III, IV, V, VI, VII, VIII)	along the midline of each segment
Antennal segments of club, length = three variables (IX, X, XI)	along the midline
Inter-eye width (IEW)	maximum distance measured between the inner eye margins
Maximum eye width (MEW)	laterally between the outer eye margins
Inter-scrobal distance (DIS)	maximum width between the inner margins
Prothorax length (PL)	dorsally along the midline between the anterior and posterior margins
Prothorax width (PW)	dorsally near the antebasal carina (Fig. 5)
Elytra length (EL)	dorsally between the anterior margin and the apical margin
Elytra width (EW)	dorsally across the humeri
Total body length (TL)	sum of PL, EL, and RL 1
Ventrite length IV (VL IV)	along the midline
Ventrite length V (VL V)	along the midline

can be considered independent (Gould 1966). Therefore, we decided to fit a model II regression, or Reduced Major Axis regression (RMA). This allows the combined variation of the two variables to be better described because there are associated errors in both.

The slope (b) of the model II regression is the allometric constant that expresses the relationship between two variables and it has been used as an indication of the allometric pattern (Emden 2008). Therefore, when b equals 0 there is no allometric relationship. However, when b = 1 the relationship is isometric, b < 1 determines a negative allometry, and b > 1 describes positive allometry. The level of statistical significance was set at 0.05 in all analyzes.

RESULTS

The mean and standard deviation of all measurements were given in the Appendixes 1 and 2. The amplitude of total body length (TL) and the coefficient of variation (CV) for *S. platyrhinus* and *H. prasinata* were summarized in Table II.

Sexual dimorphism. Males of *H. prasinata* (Fig. 8) are relatively larger than females (Fig. 9). Major males of *S. platyrhinus* (Fig. 10) are similar to females in size, whereas minor males (Fig. 11) of this species are smaller than their female counterparts (Fig. 12). Males and females of *H. prasinata* and *S. platyrhinus* did not differ in the following variables that correspond to monomorphic characters in both species: apical width of rostrum (RAW), basal width of rostrum (RBW), head width (HW), prothorax length (PL), prothorax width (PW), elytra

Table II. Amplitude of the total length (mm) for males and females of *S. platyrhinus* and *H. prasinata* (n = 34, males = 25 and females = 9).

Species	Groups	CV	TL
S. platyrhinus	Males	0.17	6.37-12.81
	Females	0.12	8.22-11.36
S. platyrhinus	Major males		10.32-12.95
	Minor males		6.37-9.51
H. prasinata	Males	0.15	10.86-19.60
	Females	0.16	8.51-17.96

length (EL), elytra width (EW), inter-scrobal distance (DIS), and inter-eye width (IEW), as detailed in Appendix 1.

The independent t test for sexual dimorphism of all variables is shown for the two species analyzed (Appendix 1). Males and females of the two species did not differ only in the maximum eye width (MEW). Based on the RMA results for *S. platyrhinus* (Table III), the elytral length and width did not show allometry. These results showed that all other structures are indicative of sexual dimorphism, as previously suggested by Holloway (1982) and Mermudes (2002).

Polyphenism in males. In *S. platyrhynus*, major and minor males differ significantly in almost all variables, with the exception of antennomeres VII, VIII, and IX and ventrite V. The result of the independent t test for the polyphenism in males of the two species analyzed is shown in Table III. The presence of two groups of males in *S. platyrhinus*, relatively discrete in size, indicates size polyphenism (Table III and Ap-

Figures 8-12. Dorsal habitus. (8-9) *Hypselotropis prasinata*: (8) male; (9) female. (10-12) *Systaltocerus platyrhynus:* (10) major male; (11) minor male; (12) female. Scale bar = 2 mm.

pendix 2). In *H. prasinata*, although there is no evidence of major and minor males, we found intermediate males, suggesting a continuous variation in size (Table IV). Therefore, there were no discrete groups, rejecting the hypothesis of size polyphenism for *H. prasinata* males.

Multivariate analysis. Cluster Analyses with the Ward's Method, considered very efficient (VALENTIN 2000), identified different groups for each species analyzed. Bootstrap support values for these groups are shown within parentheses: three groups found for *S. platyrhinus* (Fig. 13): major males (76), minor males (75), and females (99); and two for *H. prasinata* (Fig. 14): males (respectively 29, 23) and females (74).

The Principal Components Analysis (PCA) indicated that size has a greater influence on the identification of groups (major males, minor males, and females) of *S. platyrhynus* (Table V and Fig. 15). The separation of the groups was evident by the analysis of the axes of components 1 and 2, which explain more than 80% of total variance. In the first axis (PC1), two groups were identified: males and females. The second axis (PC2) shows the separation between major and minor males. For *H. prasinata*, the principal components analysis indicated that size contributes to the differentiation of groups (Table V). However, there is no evidence of polyphenism in males (Fig. 16). The first and second components explained 87% of total evidence.

Canonical Variate Analysis (CVA), together with MANOVA, confirmed that there are three different morphotypes in *S. platyrhynus* (MANOVA CVA: Wilks' Lambda = 0.000194; df1 = 50; df2 = 14; F = 19.82; p < 0.0001) with correct allocation of

specimens exceeding 90%. The separation of groups in *S. platyrhynus* (Fig. 17) was evident through the first two axes, of which the first CV provided information for the separation of males and females and the second CV distinguished major and minor males. This separation is obtained essentially by a size contrast among head width (HW), prothorax width (PW), elytra width (EW), and length of antennal segment VII. CVA was not undertaken for *H. prasinata* because it is only recommended when there are more than two groups (HAMMER 2002).

Allometry and sexual dimorphism. Results of analysis by the RMA in *S. platyrhynus* males, without separating major and minor groups (Table IV), showed positive allometry between the independent variable PL (prothorax length) and each of the six variables connected with the rostrum (rostral length 1, apical width of rostrum, medial width 1 and 2 of rostrum, basal width of rostrum, and inter-scrobal distance). Even within the analysis of males, only one variable of the head, inter-eye width (IEW), and three antennal segments (the proximal III-V), did not fit an allometric pattern, differing from females of *S. platyrhynus* in this respect (Table IV).

Differing from the results above, evidence of sexual dimorphism with allometric patterns was confirmed only for females of *H. prasinata* in the following characters: width of head, prothorax, and elytra. Males of *H. prasinata* (Table IV) showed positive allometry for only one trait in the antennae (segment III). Males and females of this species, however, showed positive allometry in thirteen measurements, whereas females showed exclusive positive allometry in five traits.

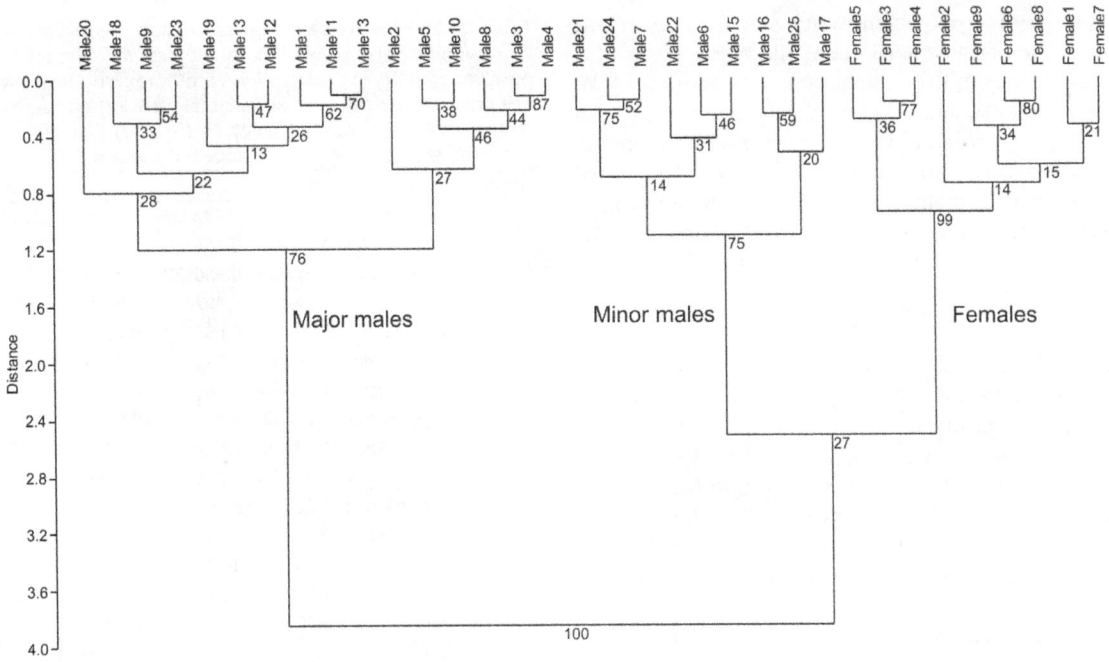

Figure 13. Dendogram obtained with Ward's Cluster Analysis methods for *S. platyrhinus*. 1,000 bootstrap replicates.

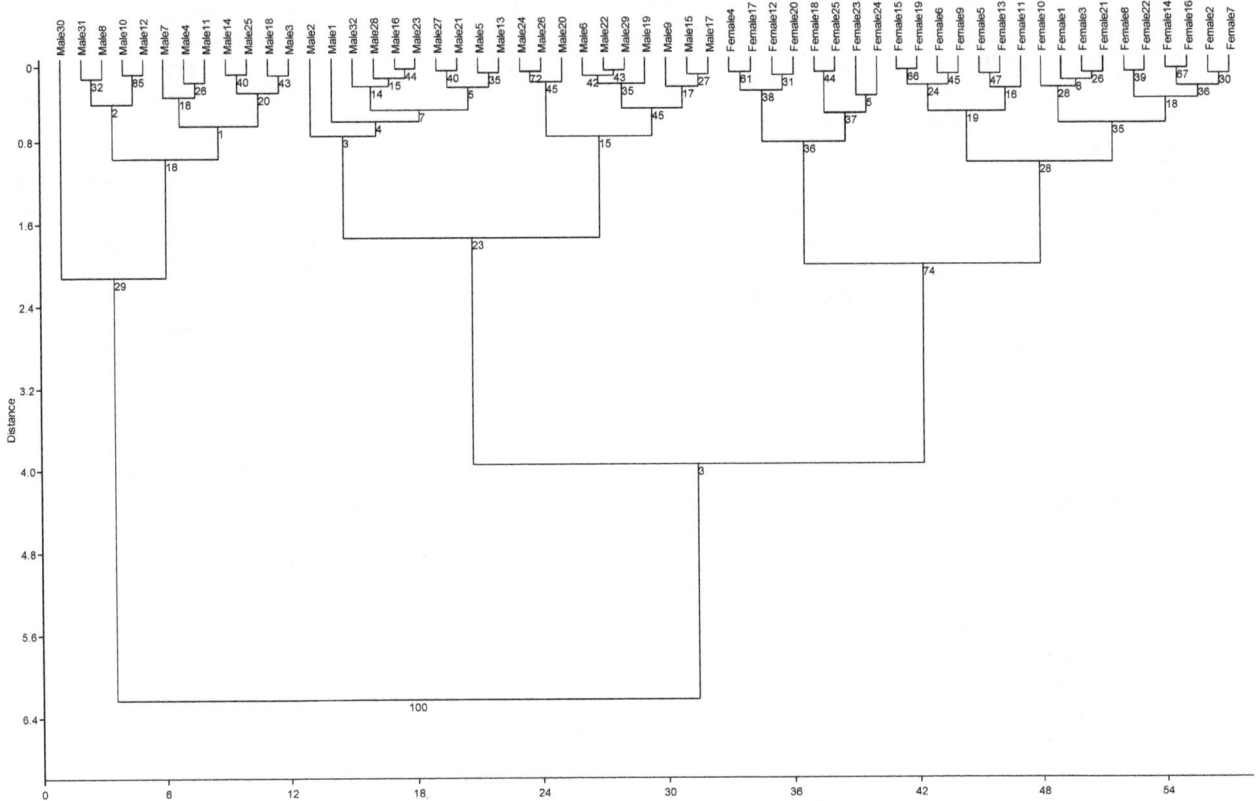

Figure 14. Dendogram obtained with Ward's Cluster Analysis methods for *H. prasinata*. 1,000 bootstrap replicates.

Allometry and polyphenism in males. In reviewing the evidence of allometry between major and minor males of *S. platyrhynus* (Table III), five variables from the rostrum (RAW, MW1R, MW2R, RBW, and DIS), two of antennomeres (III) and only one from the ventrite (VL IV) have positive allometry only in major males. On the other hand, minor males showed positive allometry in prothorax and elytral width, and length of rostrum and antennomeres II, III, and VI. This demonstrates a clear morphological plasticity between males.

DISCUSSION

According to Kawano (2006), body size is the most appropriate morphological trait for allometric analyses because it depends on the quality of the nutrition received by an individual during growth. However, other traits have been used as allometric predictors, for instance elytra length (Clark 1977, Goldsmith 1985), elytra width (Eberhard 1980, Cook 1987), and pronotum width (Emlem 1994, 1996, Eberhard *et al.* 1998, Emlen *et al.* 2005, Tomkins *et al.* 2005). Kawano (2006) stated that in sexually dimorphic beetles characters such as elytra length (EL), elytra width (EW), and prothorax width (PW) are not adequate allometric predictors because they do not represent a true measurement of body size (see Moczek *et al.* 2002 for a different view). Notably, we know that nutritional and environmental factors influence body size and also dimorphic structures, and that body size, as a dimorphic structure, is determined by endocrine mechanisms regulating development (Emlen *et al.* 2005).

We did not use the TL as an allometric predictor because it is a composite measure formed by the sum of rostral length (RL1), prothorax length (PL), and elytra length (EL). Moreover, all these traits exhibited positive allometry (see discussion below), which could lead, through high correlations, to an indirect correlation. In addition, the elytra width (EW) showed positive allometry for males of *S. platyrhinus* and *H. prasinata*, and females of *H. prasinata*. Thus, only the prothorax length (PL) was used as a proxy for the true size of the body in the subsequent allometric analysis.

The results on variation in monomorphic characters also corroborated the work of Holloway (1982), who postulated that the inter-scrobal distance (DIS) is a diagnostic character for genera in Anthribidae. Our results also showed that the DIS did not differ between sexes because it is a monomorphic character. Additionally, the results showed that, in a supposedly dimorphic structure such as the rostrum, there are measurements that do not vary between the sexes (monomorphic). For *H. prasinata* there were no differences between the sexes in rostral length (RL1). For *S. platyrhinus*, there were no differences in: (1) rostrum, the average width 1 (MW1B) and width 2 (MW2R); (2) antenna, length of antennomere II; and (3) abdomen, length of ventrite IV (VL IV).

Sexual dimorphism was discussed by Holloway (1982) and Mermudes & Napp (2006); it occurs in the relative length of the

Table III. The results based on regression. Reduced Major Axis regression (RMA) of pairs of morphological variables selected with positive allometry in *S. platyrhinus*. Prothorax length (PL) was used as a predictor variable, and all variables were log-transformed prior to the two male groups (major and minor). Values of positive allometry shown in bold, with slope >1 and p ≤ 0.05.

Measures	Males	Slope	r	p	95% IC
RL1	Major Males	-1.48550	-0.093925	0.7255	-2.2580; 1.7650
	Minor Males	**1.81560**	**0.758530**	**0.0181**	**1.0100; 2.8050**
RAW	Major Males	**1.22850**	**0.656570**	**0.0053**	**0.7506; 1.7230**
	Minor Males	0.96200	0.899870	0.0020	0.6779; 1.2760
MW1R	Major Males	**1.22850**	**0.656570**	**0.0053**	**0.7506; 1.7230**
	Minor Males	1.55180	0.218300	0.5583	-2.2640; 2.5990
MW2R	Major Males	**1.75670**	**0.762370**	**0.0005**	**1.0840; 2.4550**
	Minor Males	1.33680	0.802830	0.0143	0.8597; 1.9060
RBW	Major Males	**1.39400**	**0.652370**	**0.0054**	**0.6713; 1.9830**
	Minor Males	1.39230	0.815180	0.0089	0.8295; 1.9170
HW	Major Males	0.96648	0.565020	0.0229	0.6182; 1.3610
	Minor Males	0.85738	0.742430	0.0168	0.4798; 1.2800
PW	Major Males	0.93269	0.678800	0.0030	0.5844; 1.2920
	Minor Males	**1.08050**	**0.842200**	**0.0026**	**0.6262; 1.5120**
EL	Major Males	0.80780	0.619160	0.0099	0.5425; 1.0200
	Minor Males	0.77868	0.619390	0.0818	-0.6077; 1.1640
EW	Major Males	0.99183	0.657430	0.0059	0.6287; 1.3850
	Minor Males	**1.09810**	**0.819920**	**0.0034**	**0.6483; 1.7040**
MEW	Major Males	1.14760	0.363730	0.1763	-1.3830; 1.6280
	Minor Males	0.75328	0.862790	0.0029	0.5855; 0.9414
DIS	Major Males	**3.04740**	**0.530800**	**0.0355**	**1.4770; 4.4360**
	Minor Males	1.66940	0.474440	0.1888	-1.7580; 2.6060
IEW	Major Males	1.23250	0.534230	0.0283	0.5754; 1.9310
	Minor Males	1.15860	0.813750	0.0102	0.7183; 1.5460
II	Major Males	2.61660	0.277900	0.3112	-1.9190; 4.4160
	Minor Males	**1.40660**	**0.894920**	**0.0016**	**0.8673; 1.8060**
III	Major Males	**2.63410**	**0.577120**	**0.0139**	**1.1790; 4.4660**
	Minor Males	**1.23150**	**0.930700**	**0.0007**	**0.9571; 1.5440**
IV	Major Males	1.70980	0.494290	0.0512	0.9584; 2.6600
	Minor Males	1.90930	0.661310	0.0546	1.1760; 2.8730
V	Major Males	1.13960	0.434660	0.0868	-0.5834; 1.8540
	Minor Males	1.75840	0.644540	0.0652	0.9586; 2.5610
VI	Major Males	0.74156	0.419830	0.1016	-0.6722; 1.0240
	Minor Males	**1.88570**	**0.672400**	**0.0489**	**1.0130; 2.8120**
VII	Major Males	0.73473	0.286160	0.2817	-0.9168; 0.9983
	Minor Males	1.59160	0.596570	0.0893	-1.0050; 2.3940
VIII	Major Males	1.15560	0.397810	0.1253	-1.4790; 1.6420
	Minor Males	1.37890	0.456590	0.2185	-1.0770; 2.2130
IX	Major Males	1.60190	0.347030	0.1804	-1.9980; 2.3000
	Minor Males	1.19260	0.657950	0.0557	0.6023; 1.7510
X	Major Males	1.77750	0.281690	0.3055	-2.8070; 2.7750
	Minor Males	0.94794	0.432220	0.2407	-0.8429; 1.3880
XI	Major Males	1.37250	0.502660	0.0498	0.8856; 2.0770
	Minor Males	1.13450	0.572490	0.1132	-0.3999; 1.7300
VL IV	Major Males	**1.64070**	**0.701600**	**0.0017**	**0.9728; 2.4760**
	Minor Males	0.91300	0.015324	0.9692	-1.2600; 1.4550
VL V	Major Males	1.80020	0.219740	0.4421	-1.2380; 3.1420
	Minor Males	1.34300	0.466900	0.2049	-0.8419; 2.1480

Table IV. Reduced Major Axis regression (RMA) results between pairs of morphological variables selected with positive allometry in *S. platyrhinus* and *H. prasinata*. Prothorax length (PL) was used as a predictor variable. All variables were log-transformed prior to males and females. (***) for p ≤ 0.0001.

Measures	Sex	S. platyrhinus Slope b	r	p	I.C.	H. prasinata Slope b	r	p	I.C.
RL1	Males	1.90000	0.699050	***	1.4610; 2.3560	1.06040	0.931790	***	0.9302; 1.1910
	Females	0.59137	0.440290	0.2449	-0.4048; 1.800	1.10600	0.940410	***	0.9485; 1.3050
RAW	Males	1.16070	0.875300	***	1.0080; 1.3720	-4.99440	-0.058107	0.7606	-9.096; 4.3820
	Females	-0.41585	-0.078700	0.8496	-0.7777; 1.1800	0.98130	0.919110	***	0.8439; 1.2070
MW1R	Males	1.68740	0.549920	0.0064	1.0480; 2.5660	–	–	–	–
	Females	-0.32354	-0.012680	0.9765	-0.4807; 0.9039	–	–	–	–
MW2R	Males	1.60640	0.878990	***	1.3820; 1.9150	–	–	–	–
	Females	0.41981	0.038053	0.9181	-0.5954; 1.2100	–	–	–	–
RBW	Males	1.50130	0.863070	***	1.2590; 1.7940	1.17200	0.873160	***	0.8929; 1.4110
	Females	0.53790	0.031836	0.9298	-1.0410; 1.6160	1.02050	0.972850	***	0.9115; 1.1610
HW	Males	1.03900	0.823670	***	0.8540; 1.2510	0.91548	0.953520	***	0.8060; 1.0180
	Females	0.47593	0.043096	0.9097	-0.9619; 1.4070	1.02800	0.975950	***	0.9448; 1.1690
PW	Males	1.04600	0.875940	***	0.8642; 1.2310	0.99005	0.961220	***	0.8782; 1.0860
	Females	0.41758	0.088446	0.8306	-0.7940; 1.1870	1.10450	0.985070	***	1.0230; 1.2320
EL	Males	0.90022	0.807660	***	0.7675; 1.0580	0.98289	0.967430	***	0.8957; 1.0540
	Females	0.43514	0.348030	0.3544	-0.1564; 1.1820	1.05030	0.988480	***	0.9686; 1.1080
EW	Males	1.05590	0.860950	***	0.8797; 1.2260	0.98049	0.974080	***	0.8989; 1.0570
	Females	0.45187	0.155270	0.6926	-0.6374; 1.370	1.08220	0.983880	***	1.0170; 1.1800
MEW	Males	0.82335	0.695740	***	0.6704; 1.0870	0.78105	0.814940	***	0.6461; 0.9124
	Females	-0.34873	-0.10388	0.8103	-0.4492; 0.9072	0.87605	0.936000	***	0.7657; 1.0860
DIS	Males	2.07910	0.649830	0.0008	1.5690; 2.8290	1.12190	0.872020	***	0.8799; 1.3570
	Females	0.45967	0.465290	0.2183	0.1739; 1.2460	1.04790	0.970210	***	0.9672; 1.2050
IEW	Males	1.35670	0.835360	***	1.1270; 1.6580	1.15460	0.889160	***	0.9594; 1.3300
	Females	0.59589	0.116280	0.723	-0.8943; 1.8360	1.04290	0.957510	***	0.9386; 1.1890
II	Males	1.71650	0.620900	0.0005	0.9697; 2.6140	1.82900	0.917990	***	1.5820; 2.0320
	Females	-0.94552	-0.729110	0.0761	-1.1480; 1.3230	1.03330	0.908060	***	0.9029; 1.3000
III	Males	1.69210	0.740040	***	1.1920; 2.5400	1.92700	0.959180	***	1.7420; 2.1320
	Females	-1.07410	-0.089660	0.8043	-1.2830; 2.8000	1.01900	0.933010	***	0.9035; 1.2060
IV	Males	1.68500	0.749730	***	1.2730; 2.1020	2.24450	0.972940	***	2.0800; 2.4190
	Females	-0.44250	0.378060	0.3446	-1.2100; 0.8810	1.07030	0.939310	***	0.9352; 1.2770
V	Males	1.54270	0.761590	***	1.1060; 1.9080	6.15970	0.526750	***	2.5750; 10.8700
	Females	0.49559	0.074484	0.8421	-1.2560; 1.4010	1.07280	0.916670	***	0.9190; 1.2640
VI	Males	1.51620	0.764820	***	0.9724; 1.9400	2.99950	0.944720	***	2.8050; 3.2500
	Females	-0.45601	-0.077770	0.8245	-0.8710; 1.4290	1.02590	0.879440	***	0.8348; 1.2660
VII	Males	1.34630	0.727210	***	0.9090; 1.7510	2.83610	0.916910	***	2.5600; 3.1810
	Females	-0.44280	-0.023300	0.9504	-0.8681; 1.3760	0.95728	0.824990	***	0.8010; 1.2830
VIII	Males	1.27890	0.686200	0.0002	0.8871; 1.6360	2.56950	0.924410	***	2.248; 2.9630
	Females	0.56072	0.210630	0.5876	-0.9753; 1.6330	1.12210	0.613780	0.0013	0.8289; 1.7310
IX	Males	1.19600	0.641790	0.0005	0.8995; 1.5690	2.70990	0.836270	***	2.2630; 3.1940
	Females	-0.97553	-0.262960	0.406	-1.7080; 2.6140	1.02260	0.415650	0.0415	-0.9485; 1.4340
X	Males	1.07710	0.400380	0.0479	0.5625; 1.6440	1.66220	0.743030	***	1.3390; 2.0690
	Females	-0.96886	-0.548470	0.1422	-2.400; 1.5710	1.28970	0.746940	***	1.0500; 1.6130
XI	Males	1.06410	0.671810	0.0004	0.8033; 1.3920	2.27550	0.836560	***	1.8510; 2.7770
	Females	-0.86524	-0.41590	0.2525	-2.3730; 0.7534	1.52600	0.665790	0.0001	1.1680; 2.0830
VL IV	Males	1.24970	0.665430	***	0.9291; 1.7370	0.79050	0.705380	***	0.6296; 0.9581
	Females	-0.62952	-0.237740	0.546	-1.3870; 1.7320	0.95457	0.932760	***	0.8309; 1.0880
VL V	Males	1.26970	0.484400	0.0123	0.6120; 1.9140	0.88102	0.814830	***	0.6817; 1.0890
	Females	-1.07080	-0.285230	0.3561	-2.9930; 1.1460	0.85026	0.925360	***	0.7052; 0.9477

Table V. Loadings of the morphometric variables in the first two components of the Principal Components Analysis (PCA). Variables not measured marked with an asterisk.

Measures	S. platyrhinus		H. prasinata	
	PC1	PC2	PC1	PC2
RL1	-0.544	-0.114	0.135	-0.183
RAW	-0.150	-0.240	0.108	-0.182
MW1R	0.015	-0.318	*	*
MW2R	-0.353	-0.225	*	*
RBW	-0.375	-0.213	0.095	-0.214
HW	-0.156	-0.222	0.079	-0.200
PL	-0.098	-0.229	0.124	-0.178
PW	-0.153	-0.248	0.100	-0.206
EL	-0.127	-0.234	0.110	-0.191
EW	-0.156	-0.259	0.091	-0.207
MEW	-0.027	-0.126	0.073	-0.148
DIS	-0.295	-0.224	0.088	-0.210
IEW	-0.199	-0.204	0.115	-0.199
II	-0.076	-0.052	0.398	-0.096
III	-0.401	0.066	0.565	-0.009
IV	-0.551	0.158	0.613	-0.027
V	-0.600	0.174	1.018	0.087
VI	-0.583	0.187	0.845	0.048
VII	-0.561	0.211	0.864	0.104
VIII	-0.459	0.157	0.815	0.115
IX	-0.227	0.043	0.583	0.010
X	-0.102	0.018	0.308	-0.109
XI	-0.174	0.011	0.554	-0.045
VL IV	-0.030	0.117	-0.002	-0.199
VL V	0.037	-0.134	-0.056	-0.235

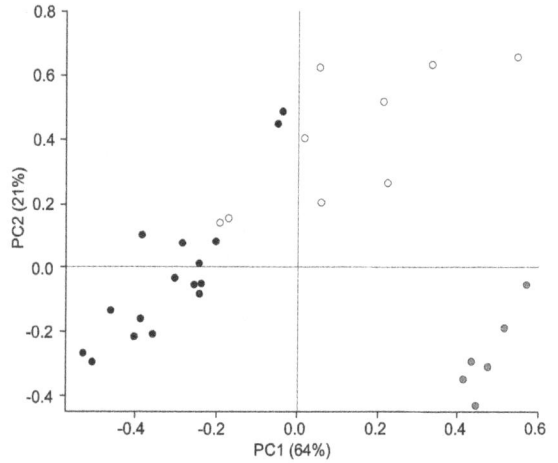

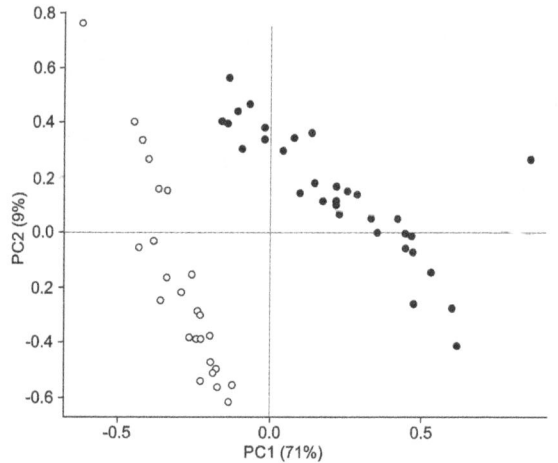

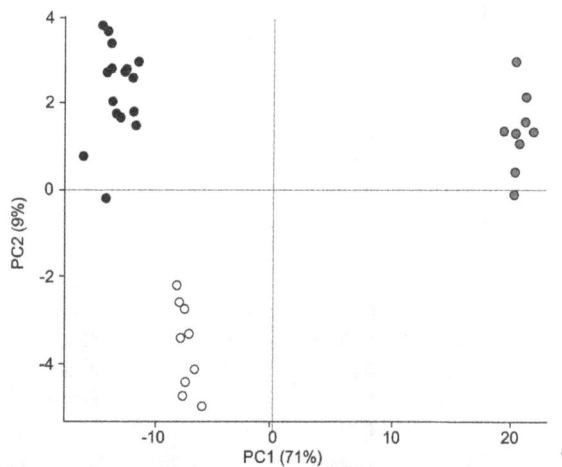

Figures 15-17. (15-16) Principal Components Analysis (PCA) for: (15) S. platyrhynus: females (●), major males (●), and minor males (light gray); (16) H. prasinata: females (○) and males (●); (17) Canonical Variate Analysis (CVA) of S. platyrhinus: females (●), major males (●), and minor males (○).

antenna of males of some Anthribinae. Also, MERMUDES (2005) used a relationship between the length of ventrites IV and V to distinguish between males and females of *Hypselotropis* Jekel, 1855. Our results here show that ventrite V is always longer than ventrite IV in both males and females of *H. prasinata*. This differs from the opinion of MERMUDES (2005) who believes that ventrite V in males is always slightly shorter than ventrite IV. However, our results confirmed this relationship for females. The large number of variables with values also tested by PCA and CVA (Figs 15-17) suggest that there is marked sexual dimorphism in some structural characters that had not been previously investigated.

The multivariate analysis (PCA) indicated that two relatively discrete groups of males of *S. platyrhinus* exist with respect to size (major males, minor males and females, Fig. 15), which was confirmed by CVA (Fig. 17), revealing the presence of polyphenism in males of this species. The analysis suggested that the allometric component contributes to the differentiation of groups, but there is no evidence of polyphenism in males of *H. prasinata* (Fig. 16), rejecting, at least in this analysis, the hypothesis of size polyphenism in this species.

Considering together all the results on allometry and sexual dimorphism, we conclude that, in the case of *S. platyrhinus*, the dimensions of the rostrum, antennal segments, and ventrites IV and V indicate that sexual dimorphism is in place, as previously suggested by Mermudes (2002). Traits that exhibited positive allometry are a strong indication of sexual dimorphism (Emlen 1996, Moczek *et al.* 2002, Matsuo 2005, Moczek 2006). It is worth noting that the variables with positive allometry in *S. platyrhinus* are at the anterior part of body (rostrum and frons). In this species, the rostrum and forehead are vertical (hypognathous), providing evidence that such structures are subject to sexual selection and are probably associated with male fighting, similar to the condition found in *Exechesops leucopis*.

Despite the fact that we have analyzed only two species, our results emphasize allometric patterns in structures with sexual dimorphism that can be highly variable within *S. platyrhynus* males. Almost all species of Anthribidae that show sexual dimorphism lack structures known as weapons (e.g., horns). Polyphenism in males was also found to be present in *S. platyrhinus*, making it possible to infer that many traits related to dimorphism could play a role in tactical alternatives that minor males developed when confronted with major males, as reported by Howden (1992), Yoshitake & Kawashima (2004), and Matsuo (2005).

Initially, the allometric variation could be derived from either behavioral differences between major or minor males or from a threshold size to developing weapons (horns or mandibles) with exaggerated sizes (Moczek & Emlen 2000, Moczek *et al.* 2002, Yoshitake & Kawashima 2004, Matsuo 2005). In males of some species of Anthribidae (which do not have horns = weapons), sexually dimorphic traits exhibit positive allometry with body size, whereas isometry or negative allometry is detected when sexually monomorphic traits are considered (or which are not associated with dimorphism) (Matsuo 2005).

The behavioral relationship involves male-male competition for females, but it does not eliminate the interactions between minor males when they meet, as well as alternative tactics developed by minor males to copulate (Emlen 1994, Moczek *et al.* 2002, Matsuo 2005, Tomkins & Moczek 2009).

Finally, it is possible that the morphological patterns of Anthribidae are linked to the protection of the female, which is secured by males during oviposition, and that a relationship between reproductive behavior and alternative morphologies exists (as noted by Howden 1992). This behavioral pattern can be elucidated in further studies on *S. platyrhinus*.

ACKNOWLEDGMENTS

We are grateful to the curators who made material available and to two anonymous reviewers for comments. This research was partially supported by grants from CNPq (processes 470980/2011-7, 475461/2007, and 312357/2006), FAPERJ (processes 101.476/2010 and 100.927/2011), and Programa de Pós-Graduação em Biociências da Universidade do Estado do Rio de Janeiro.

LITERATURE CITED

Clark, J.T. 1977. Aspects of variation in the stag beetle *Lucanus cervus* (L.) (Coleoptera: Lucanidae). **Systematic Entomology** 2: 9-16. doi: 10.1111/j.1365-3113.1977.tb00350.x.

Cook, D. 1987. Sexual selection in dung beetles. I. A multivariate study of the morphological variation in two species of *Onthophagus* (Scarabaeidae: Onthophagini). **Australian Journal of Zoology** 35: 123-132. doi: 10.1071/ZO9870123.

Eberhard, W.G. 1980. Horned beetles. **Scientific American** 242: 166-182.

Eberhard, W.G. & E.E. Gutiérez. 1991. Morphometric Variability in Continental and Atlantic Island Populations of Chaffinches *Fringilla coelebs*. **Evolution** 45 (1): 29-39.

Eberhard, W.G.; B.A. Huber; R.L. Rodriguez; R.D. Briceno; I. Salas & V. Rodriguez. 1998. One size fits all? Relationships between the size and degree of variation in genitalia and other body parts in twenty species of insects and spiders. **Evolution** 52: 415-431.

Emlen, D.J. 1994. Environmental control of horn length dimorphism in the beetle *Onthophagus acuminatus* (Coleoptera: Scarabaeidae). **Proceedings of the Royal Society B.** 256: 131-136. doi: 10.1098/rspb.1994.0060.

Emlen, D.J. 1996. Artificial selection on horn length-body size allometry in the horned beetle *Onthophagus acuminatus* (Coleoptera:Scarabaeidae). **Evolution** 50: 1219-1230.

Emlen, D.J. 2008. The evolution of animal weapons. **Annual Review of Ecology and Systematics** 39: 387-413. doi: 10.1146/annurev.ecolsys.39.110707.173502.

Emlen, D.J. & H.F. Nijhout. 2000. The Development and Evolution of Exaggerated Morphologies in Insects. **Annual Review of Entomology** 45: 661-708. doi: 10.1146/annurev.ento.45.1.661.

Emlen, D.J.; J. Hunt & L.W. Simmons. 2005. Evolution of sexual dimorphism and male dimorphism in the expression of beetle horns: phylogenetic evidence for modularity, evolutionary lability, and constraint. **The American Naturalist** 166 (Suppl.): S42-S68.

Emlen, D.J.; L.C. Lavine & B. Ewen-Campen. 2007. On the origin and evolutionary diversification of beetle horns. **Proceedings of the National Academy of Sciences** 104: 8661-8668.

Goldsmith, S.K. 1985. Male Dimorphism in *Dendrobias mandibularis* Audinet-Serville (Coleoptera: Cerambycidae). **Journal of the Kansas Entomological Society** 58: 534-538.

Gould, S.J. 1966. Allometry and size in ontogeny and phylogeny. **Biological Reviews** 41: 587-640. doi: 10.1111/j.1469-185X.1966.tb01624.x

Hammer, O. 2002. Morphometrics – brief notes50p. Available at http://folk.uio.no/ohammer/past/morphometry.pdf [Accessed: February 2014].

HAMMER, O.; D.A.T. HARPER & P.D. RYAN. 2001. Past: Palaeontological Statistics Software Package for Education and Data Analysis. **Palaeontological Electronica 4** (1): 9p. Available online at: http://palaeo-electronica.org/2001_1/past/past.pdf [Accessed: March 2011].

HOLLOWAY, B.A. 1982. **Anthribidae (Insecta: Coleoptera). Fauna of New Zealand 3.** Wellington, DSIR, 269p.

HOWDEN, A.T. 1992. Oviposition Behavior and Associated Morphology of the Neotropical Anthribid *Ptychoderes rugicollis* Jordan (Coleoptera: Anthribidae). **Coleopterists Bulletin 46:** 20-27.

HOWDEN, A.T. 1995. Structures related to oviposition in Curculionoidea. **Memoirs of the Entomological Society of Washington 14:** 53-100.

HUXLEY, J.S. 1932. **Problems of relative growth.** London, Methuen, 276p.

HUXLEY, J.S. 1950. Relative growth and form transformation. **Proceedings of the Royal Society London 137:** 465-469. doi:10.1098/rspb.1950.0055.

KAWANO, K. 2000. Genera and Allometry in the Stag Beetle Family Lucanidae, Coleoptera. **Annals of the Entomological Society of America 93:** 198-207. doi: http://dx.doi.org/10.1603/0013-8746(2000)093[0198:GAAITS]2.0.CO;2.

KAWANO, K. 2006. Sexual Dimorphism and the Making of Oversized Male Characters in Beetles (Coleoptera). **Annals of the Entomological Society of America 99:** 327-341. doi: doi: http://dx.doi.org/10.1603/0013-8746(2006)099[0327:SDATMO]2.0.CO;2.

MATSUO, Y. 2005. Extreme Eye Projection in the Male Weevil *Exechesops leucopis* (Coleoptera: Anthribidae): Its Effects on Intrasexual Behavioral Interferences. **Journal of Insect Behavior 18:** 465-477. doi: 10.1007/s10905-005-5605-y.

MERMUDES, J.R.M. 2002. *Systaltocerus platyrhinus* Labram & Imhoff, 1840: redescrições e considerações sobre a sinonímia com *Homalorhamphus vestitus* Haedo Rossi & Viana, 1957 (Coleoptera, Anthribidae, Anthribinae). **Revista Brasileira de Entomologia 46:** 579-590. doi: http://dx.doi.org/10.1590/S0085-56262002000400013.

MERMUDES, J.R.M. 2005. Revisão sistemática, análise cladística e biogeografia dos gêneros *Tribotropis e Hypselotropis* (Coleoptera, Anthribidae, Anthribinae, Ptychoderini). **Revista Brasileira de Entomologia 49:** 465-511. doi: http://dx.doi.org/10.1590/S0085-56262005000400009.

MERMUDES, J.R.M. & D.S. NAPP. 2006. Revision and cladistic analysis of the genus *Ptychoderes* Schoenherr, 1823 (Coleoptera, Anthribidae, Anthribinae, Ptychoderini) **Zootaxa 1182:** 1-130.

MERMUDES, J.R.M. & I. MATTOS. 2010. Description of Males of *Ptychoderes brevis* and *Ptychoderes jekeli*, with a cladistical reanalysis of *Ptychoderes* (Coleoptera: Anthribidae). **Annals of the Entomological Society of America 105:** 523-531. doi: http://dx.doi.org/10.1603/AN10016.

MERMUDES, J.R.M. & J.M.S. RODRIGUES. 2010. Description of two new species of *Hypselotropis* Jekel with a revised key and phylogenetic reanalysis of the genus (Coleoptera, Anthribidae, Anthribinae). **Zootaxa 2575:** 49-62.

MOCZEK, A.P. 2006. A matter of measurements: challenges and approaches in the comparative analysis of static allometries. **American Naturalist 167:** 606-611.

MOCZEK, A.P. & D.J. EMLEN. 2000. Male horn dimorphism in the scarab beetle, *Onthophagus taurus*: do alternative reproductive tactics favour alternative phenotypes? **Animal Behaviour 59:** 459-466. doi: 10.1006/anbe.1999.1342.

MOCZEK, A.P.; J. HUNT; D.J. EMLEN & L.W. SIMMONS. 2002. Threshold evolution in exotic populations of a polyphenic beetle. **Evolutionary Ecology Research 4:** 587-601.

OKSANEN, J.; F.G. BLANCHET; R.P. LEGENDRE; P.R. MINCHIN; R.B. O'HARA; G.L. SIMPSON; P. SOLYMOS; M.H.H. STEVENS & H. WAGNER. 2013. **Vegan: Community Ecology Package. R package version 2.0-7.** Available online at: http://cran.r-project.org/web/packages/vegan/index.html [Accessed: July 2013]

POSADAS, P.; E. ORTIZ-JAUREGUIZAR & M.E. PÉREZ. 2007. Dimorfismo sexual y variación morfométrica geográfica en *Hybreoleptops aureosignatus* (Insecta: Coleoptera: Curculionidae). **Anales de la Academia *Nacional* de Ciencias Exactas, Físicas y Naturales 59:** 141-150.

R CORE TEAM. 2013. **R: A language and environment for statistical computing.** Vienna, R Foundation for Statistical Computing. Available online at: http://www.R-project.org/ [Accessed: July 2013]

SCHLAGER, S. 2013. **Morpho: Calculations and visualizations related to Geometric Morphometrics.** R package version 0.25. Available at http://sourceforge.net/projects/morpho-rpackage [Accessed: July 2013]

SHIOKAWA, T. & O. IWAHASHI. 2000. Mandible dimorphism in males of a stag beetle, *Prosopocoilus dissimilis okinawanus* (Coleoptera: Lucanidae). ***Applied Entomology* and Zoology 35** (4): 487-494. doi: 10.1303/aez.2000.487.

SLIPINSKI, S.A.; R.A.B. LESCHEN & J.F. LAWRENCE. 2011. Order Coleoptera Linnaeus, 1758. *In*: Z.Q. ZHANG (Ed.). Animal biodiversity: An outline of higher-level classification and survey of taxonomic richness. **Zootaxa 3148:** 203-208.

THOMPSON, G.H. 1963. **Forest Coleoptera of Ghana. Biological notes and host trees.** Oxford, Forestry Memoirs 24, 78p.

TOMKINS, J.L.; J.S. KOTIAHO & N.R. LEBAS. 2005. Matters of scale: Positive allometry and the evolution of male dimorphisms. **American Naturalist 165:** 389-402.

TOMKINS J.L. & A.P. MOCZEK. 2009. Patterns of threshold evolution in polyphenic insects under different developmental models. **Evolution 62:** 459-468. doi: 10.1111/j.1558-5646.2008.00563.x.

VALENTIN, J.L. 2000. **Ecologia numérica: Uma introdução à análise multivariada de dados ecológicos.** Rio de Janeiro, Interciência, 117p.

YOSHITAKE, H. & I. KAWASHIMA. 2004. Sexual Dimorphism and Agonistic Behavior of *Exechesops leucopis* (Jordan) (Coleoptera: Anthribidae: Anthribinae). **The Coleopterists Bulletin 58:** 77-83.

Appendix 1. Mean, standard deviations, and variance measures of morphological characters of males and females of *S. platyrhinus* and *H. prasinata*. The result of the independent T test for sexual dimorphism is also shown. (***) for p ≤ 0.0001.

Measures	S. platyrhinus							H. prasinata						
	Males			Females			T test	Males			Females			T test
	Mean	SD	Variance	Mean	SD	Variance		Mean	SD	Variance	Mean	SD	Variance	
RL1	0.54	0.13	0.02	0.41	0.06	0.00	0.0091	0.45	0.01	0.07	0.44	0.01	0.09	0.4807
RAW	0.21	0.08	0.01	0.23	0.04	0.00	0.3405	0.26	0.12	0.34	0.21	0.01	0.08	0.6784
MW1R	0.03	0.12	0.01	0.16	0.03	0.00	0.0041	–	–	–	–	–	–	–
MW2R	0.30	0.11	0.01	0.23	0.04	0.00	0.1091	–	–	–	–	–	–	–
RBW	0.38	0.10	0.01	0.31	0.05	0.00	0.0688	0.09	0.01	0.08	0.12	0.01	0.08	0.2619
HW	0.30	0.07	0.01	0.32	0.05	0.00	0.4375	0.26	0.00	0.06	0.29	0.01	0.08	0.1792
PL	0.31	0.07	0.00	0.34	0.10	0.01	0.2870	0.55	0.00	0.07	0.55	0.01	0.08	0.7185
PW	0.37	0.07	0.01	0.40	0.04	0.00	0.3047	0.49	0.00	0.07	0.51	0.01	0.09	0.4486
EL	0.65	0.06	0.00	0.68	0.04	0.00	0.3094	0.91	0.00	0.07	0.91	0.01	0.08	0.8098
EW	0.42	0.07	0.01	0.45	0.05	0.00	0.2969	0.58	0.00	0.07	0.60	0.01	0.09	0.3034
MEW	-0.03	0.06	0.00	0.02	0.04	0.00	0.0231	0.08	0.00	0.05	0.09	0.00	0.07	0.6900
DIS	0.11	0.14	0.02	0.05	0.05	0.00	0.2983	0.09	0.01	0.08	0.11	0.01	0.08	0.2525
IEW	0.09	0.09	0.01	0.09	0.06	0.00	0.8413	0.07	0.01	0.08	0.07	0.01	0.08	0.8187
II	-0.56	0.12	0.01	-0.60	0.10	0.01	0.4182	-0.18	0.02	0.13	-0.36	0.01	0.08	***
III	-0.05	0.12	0.01	-0.35	0.11	0.01	***	0.13	0.02	0.13	-0.19	0.01	0.08	***
IV	0.02	0.12	0.01	-0.41	0.04	0.00	***	0.08	0.02	0.15	-0.25	0.01	0.09	***
V	0.06	0.11	0.01	-0.40	0.05	0.00	***	0.14	0.18	0.43	-0.33	0.01	0.09	***
VI	0.04	0.10	0.01	-0.42	0.05	0.00	***	0.09	0.04	0.21	-0.39	0.01	0.08	***
VII	0.01	0.09	0.01	-0.49	0.04	0.00	***	0.09	0.04	0.20	-0.42	0.01	0.08	***
VIII	-0.07	0.09	0.01	-0.49	0.06	0.00	***	0.07	0.03	0.18	-0.41	0.01	0.09	***
IX	-0.23	0.08	0.01	-0.44	0.10	0.01	***	-0.05	0.03	0.19	-0.35	0.01	0.08	***
X	-0.47	0.07	0.01	-0.61	0.10	0.01	0.0002	-0.48	0.01	0.11	-0.61	0.01	0.10	***
XI	-0.28	0.07	0.01	-0.42	0.09	0.01	***	-0.09	0.02	0.16	-0.39	0.02	0.12	***
VL_IV	-0.40	0.09	0.01	-0.34	0.06	0.00	0.0725	-0.22	0.00	0.05	-0.15	0.01	0.08	0.0002
VL_V	-0.36	0.09	0.01	-0.23	0.11	0.01	0.0010	-0.10	0.00	0.06	0.02	0.00	0.07	***

Appendix 2. Mean, standard deviations, and variance measures of the morphology of major males and minor males of *S. platyrhinus*. The result of T test for independent polyphenism in males is also shown. (***) for p ≤ 0.0001.

Measures	Major Males			Minor Males			T test
	Mean	Variance	SD	Mean	Variance	SD	
RL1	4.17	0.43	0.65	2.67	0.45	0.67	***
RAW	1.78	0.05	0.21	1.37	0.04	0.19	0.0002
MW1R	1.17	0.07	0.27	0.99	0.04	0.20	0.1450
MW2R	2.29	0.15	0.38	1.60	0.10	0.31	0.0002
RBW	2.74	0.13	0.36	1.94	0.16	0.39	***
HW	2.20	0.05	0.22	1.71	0.05	0.21	***
PL	2.21	0.05	0.23	1.79	0.07	0.26	0.0002
PW	2.59	0.06	0.24	2.06	0.10	0.32	***
EL	4.88	0.16	0.40	3.92	0.20	0.45	***
EW	2.86	0.08	0.28	2.29	0.14	0.37	***
MEW	0.99	0.01	0.11	0.87	0.01	0.10	0.0109
DIS	1.50	0.16	0.40	1.06	0.07	0.26	0.0108
IEW	1.41	0.03	0.17	1.02	0.03	0.18	***
II	0.31	0.01	0.11	0.24	0.00	0.05	0.0181
III	1.01	0.05	0.23	0.77	0.02	0.14	0.0195

Continues

Appendix 2. Continued.

Measures	Major Males			Minor Males			T test
	Mean	Variance	SD	Mean	Variance	SD	
IV	1.19	0.04	0.19	0.87	0.05	0.23	0.0013
V	1.30	0.02	0.14	0.94	0.05	0.23	***
VI	1.25	0.01	0.09	0.91	0.05	0.23	***
VII	1.15	0.01	0.09	0.86	0.04	0.19	***
VIII	0.96	0.01	0.11	0.73	0.02	0.14	***
IX	0.63	0.01	0.10	0.52	0.01	0.09	0.0129
X	0.36	0.00	0.06	0.32	0.00	0.04	0.2229
XI	0.57	0.01	0.08	0.48	0.01	0.08	0.0132
VL_IV	0.44	0.01	0.07	0.34	0.00	0.05	0.0011
VL_V	0.47	0.01	0.11	0.40	0.01	0.07	0.0460

6

Estimating cyclopoid copepod species richness and geographical distribution (Crustacea) across a large hydrographical basin: comparing between samples from water column (plankton) and macrophyte stands

Gilmar Perbiche-Neves[1], Carlos E.F. da Rocha[1] & Marcos G. Nogueira[2]

[1] Departamento de Zoologia, Instituto de Biociências, Universidade de São Paulo. Rua do Matão, travessa 14, 321, 05508-900 São Paulo, SP, Brazil. Email: gilmarpneves@yahoo.com.br
[2] Departamento de Zoologia, Instituto de Biociências, Universidade Estadual Paulista. Distrito de Rubião Júnior, 18618-970 Botucatu, SP, Brazil.

ABSTRACT. Species richness and geographical distribution of Cyclopoida freshwater copepods were analyzed along the "La Plata" River basin. Ninety-six samples were taken from 24 sampling sites, twelve sites for zooplankton in open waters and twelve sites for zooplankton within macrophyte stands, including reservoirs and lotic stretches. There were, on average, three species per sample in the plankton compared to five per sample in macrophytes. Six species were exclusive to the plankton, 10 to macrophyte stands, and 17 were common to both. Only one species was found in similar proportions in plankton and macrophytes, while five species were widely found in plankton, and thirteen in macrophytes. The distinction between species from open water zooplankton and macrophytes was supported by non-metric multidimensional analysis. There was no distinct pattern of endemicity within the basin, and double sampling contributes to this result. This lack of sub-regional faunal differentiation is in accordance with other studies that have shown that cyclopoids generally have wide geographical distribution in the Neotropics and that some species there are cosmopolitan. This contrasts with other freshwater copepods such as Calanoida and some Harpacticoida. We conclude that sampling plankton and macrophytes together provided a more accurate estimate of the richness and geographical distribution of these organisms than sampling in either one of those zones alone.

KEY WORDS. La Plata River basin; reservoirs; rivers, zooplankton.

Freshwater copepods are a link in the trophic web, connecting producers to consumers (PERBICHE-NEVES et al. 2007); they inhabit lakes, rivers, pools, caves, humid rocks, etc. (BOXSHALL & DEFAYE 2008), where it is easy to find free living copepods of the order Cyclopoida.

In large spatial scales, cyclopoids are less endemic than other copepods, for instance diaptomids. Many cyclopoid species in *Mesocyclops*, *Metacyclops*, *Eucyclops*, etc. are widely distributed in the Neotropical region and in the world (REID 1985, SILVA 2008). However, there are exceptions to this rule, for instance species of *Thermocyclops* (REID 1989), which occur in the South hemisphere, and differ between the Afrotropical and Neotropical biogeographical regions. In contrast to the patterns of distribution of cyclopoids in large bio geographical areas, there are no clear patterns in the spatial distribution of these organisms among river basins in South America, as observed for diaptomids (SUÁREZ-MORALES et al. 2005). The low endemism of cyclopoids can be in part explained by their efficient dispersion (e.g., by birds, fishes, humans) and their recent colonization of many parts of the world (BOXSHALL & JAUME 2000, SUÁREZ-MORÁLES et al. 2004). Comparing among the main rivers of the La Plata Basin, the composition of cyclopoid species is similar (PAGGI & JOSÉ DE PAGGI 1990, LANSAC-TÓHA et al. 2002).

Cyclopoid copepods of inland waters are more diverse than in the littoral, which can be colonized by aquatic macrophytes. For example, two studies sampling the two types of habitats, open water and macrophyte stands have documented this trend for lakes in Brazil (LANSAC-TÓHA et al. 2002, MAIA-BARBOSA et al. 2008). The habitat complexity provided by aquatic macrophytes (GENKAI-KATO 2007, LUCENA-MOYA & DUGGAN 2011) also allow several species to be more abundant in them (GERALDES & BOAVIDA 2004).

Most of zooplankton horizontal migration between limnetic zones and macrophyte stands in lentic environments can be attributed to predation pressure by planktivorous fish (GENKAI-KATO 2007, FANTIN-CRUZ et al. 2008). Lower richness of cyclopoid species tends to be found in limnetic waters, where generally few abundant species dominate. Clear tendencies in some ecological attributes can be observed for copepods and other crustaceans in reservoirs (SILVA & MATSUMURA-TUNDISI 2002, NOGUEIRA et al. 2008).

There are no comparative studies of copepod richness between habitats in large geographical scales, only in lotic stretches or in lakes (LANSAC-TÔHA *et al.* 2002, MAIA-BARBOSA *et al.* 2008). This study is the first to compare copepod species richness in a large hydrographic basin, the fourth largest in the world. We simultaneously sampled plankton and macrophyte stands in rivers and reservoirs. Based on references, we tested the alternative hypotheses that in limnetic zones there is greater richness of species in macrophytes than in zooplankton. Additionally, we tried to pinpoint particular species in each kind of habitat, and to ascertain how sampling effort can determine the species that are found.

MATERIAL AND METHODS

The "La Plata" River basin crosses Argentina, Brazil, Bolivia, Paraguay and Uruguay. Samples were taken in the summer (from January to March 2010) and in the winter (from June to July 2010), periods that have different mean temperatures and precipitation. It was established that a minimum of two sampling trips were necessary to estimate richness. Altogether 24 sites were sampled, which included 12 reservoirs in the high Paraná and Uruguay rivers (because they are dominant in this stretch), and 12 lotic stretches (in Paraguay, middle and low Paraná and Uruguay rivers (Fig. 1).

The main rivers of the La Plata" River are the Paraná, Paraguay and Uruguay rivers. In each river, we chose sampling sites that were deemed representative of three stretches of the river: high, middle and low. In the main river of the basin, the Paraná River, we sampled the first reservoir (Ilha Solteira Reservoir) after it had been built. We also sampled one more reservoir (Itaipu Reservoir) at the end of the high stretch (680 km apart from each other), and another reservoir at the end of the Paraná River in the beginning of the middle stretch (1,000 away from the first sampling site). After this last reservoir (Yacyreta Reservoir), at the middle stretch of the river, there was a long lotic stretch, where we sampled six sites, each being approximately 250 km apart from the other, until we reached the mouth of the "La Plata" River, between Buenos Aires and Uruguay.

Beyond the three main rivers of the "La Plata" basin (Paraná, Paraguay and Uruguay), we sampled five main tributary rivers of the Paraná River (Grande, Paranaiba, Tiete, Paranapanema, and Iguaçu rivers), because collectively they amount to a large area of the basin in the high stretch. All tributaries are totally dammed, with a long cascade formed by a series of reservoirs. In each tributary river, we sampled the first and the last large reservoir. In general, the first reservoir of these tributaries has a dendritic shape, a large area (more than 200 km²), is deep and has high volume and high water retention time, functioning as a regulator for the downstream reservoir series (AGOSTINHO *et al.* 2007, NOGUEIRA *et al.* 2012). In the Uruguay River, many large reservoirs have been constructed, especially in the high stretch, but there is an old reservoir in the low

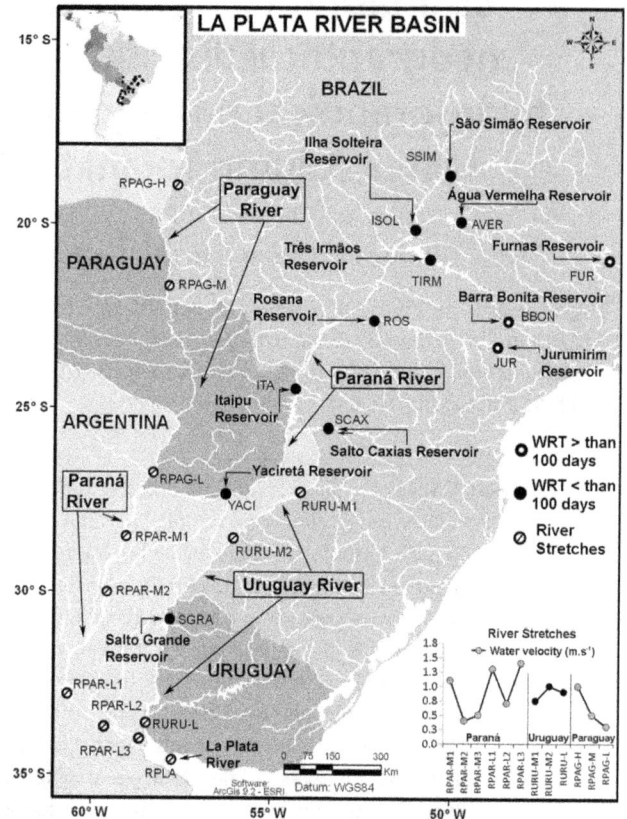

Figure 1. Sampling site map of zooplankton and macrophytes in La Plata river basin, divided in reservoirs (the first and the last in each regulated river), and free-damming lotic stretches. There are codes for abbreviation of the name of reservoirs, as also for high (H), middle (M) and low (L) stretches in lotic environments studied. Water retention time (WRT) and water flow data are given.

stretch. We sampled two reservoirs (the first and the last reservoirs in this river), and three sites in lotic stretches (two in the middle stretch, 260 km away from each other), and one in the low stretch, 5 km from the delta of Paraná River.

In the Paraguay River there are no reservoirs, only lotic stretches. Thus, the high, middle (250 km apart) and low stretches (650 km below the middle station) were sampled in the main channel. In each sampling site, we obtained zooplankton samples from open waters (limnetic region) and also zooplankton samples within macrophyte stands. In total, ninety-six samples were obtained.

As far as zooplankton are concerned, the sampling was obtained from the main river channels and from upstream zones of the reservoir, all of these with water retention time (WRT) longer than 15 days (see AGOSTINHO *et al.* 2007, ZALOCAR DE DOMITROVIC *et al.* 2007, BOLTOVSKOY *et al.* 2013), which is considered as the lower time threshold for the development of a copepod life cycle (RIETZLER *et al.* 2002). The approximate WRT of each reservoir is indicated in Fig. 1, to highlight differences between storage reservoirs and run-of-river or intermediate reservoirs. Water flow values in lotic sites are also shown in Fig. 1.

Zooplankton samples were taken by vertical hauls through water column (from close to the bottom to the surface) at each station. Values of water filtered varied between 706 L and 2,826 L, and the average value was 1,766 L. Conical plankton nets, 0.30 m mouth diameter per 0.90m side length, and 68 ìm mesh size, were used after being modified with a mouth reducing cone (Tranter & Smith 1979), a kind of anti-reflux bulkhead with another 0.50 m diameter circle. In deeper stations the vertical hauls were extended to a depth of 40m. In rivers, vertical hauls were taken from a drifting boat in order to ensure that the hauls were not excessively oblique. The volume of water filtered was estimated by the cylinder volume formula, using: pi * radius of the mouth net^2 * the length of the haul.

For samples obtained inside macrophyte stands, organisms were sampled with conical plankton nets of 68 ìm of mesh size, adapted with a 2 m drive cable of aluminum and with a steel screen to avoid excessive macrophyte intake. This net was passed between and alongside the macrophyte banks at standardized time of five minutes to obtain good amount of qualitative material.

Samples were fixed with 4% formalin solution. In the laboratory, they were analysed in their totality. Male and female copepods were identified to species. Copepods were examined under stereo- and compound microscopes, and identified using specialized taxonomic references (Reid 1985, Einsle 1996, Rocha 1998, Karaytug 1999, Alekseev 2002, Ueda & Reid 2003, Silva & Matsumura-Tundisi 2005). Species richness was considered according to the number of species, in the most robust way as possible. The samples obtained in this study are deposited in the "Collection of microcrustaceans of continental waters" of "Universidade Estadual Paulista" (in Botucatu, Brazil). Vouchers are also deposited in at the Museu de Zoologia da Universidade de São Paulo MZUSP) (e.g., *Macrocyclops albidus* (Jurine, 1820) (MZUSP30601), *Eucyclops neumani* (Pesta, 1927) (MZUSP30602), and *Microcyclops ceibaensis* (Marsh, 1919) (MZUSP30603).

The non-parametric Man Whitney U test (for non-parametric data) was used to compare species richness between zooplankton and macrophytes stand. We used R Cran Project (R Development Core Team 2012) for this test.

A non-metric multidimensional scaling (NMDS) analysis using Bray-Curtis dissimilarity was applied for spatial ordination of the data, aiming to verify differences among sampling sites, and taking the species into consideration. We used the Vegan and Mass packages for software R Cran Project (R Development Core Team 2012), according to Oksanen (2013). The iterative search was carried out using the "meta MDS" function, by several random starts, and selecting among similar solutions with the smallest stresses. For scaling we used centering, PC rotation and half change scaling.

RESULTS

We identified 32 species of cyclopoid copepods (Table I), 23 of which were found in zooplankton and 26 in macrophytes. Six species were exclusive to plankton, 10 to macrophyte stands, and 17 were common to both. Only *Mesocyclops meridianus* (Kiefer, 1926) was found in similar proportions in plankton (50% from total samples) and in macrophytes (54%), while five species (e.g., *Thermocyclops*) were widely found in plankton, and thirteen (e.g., *Microcyclops* and *Eucyclops*) in macrophytes.

Table I. List of cyclopoid species with respective abbreviations (Ab) for NMDS, and number of occurrences in sampling sites at plankton (P) of limnetic zones (from a total of 24 sites) and within macrophytes stands (M) samples (from a total of 24 sites). In bold occurrences only in plankton or macrophytes.

Cyclopoida	Ab.	P	M
Acanthocyclops robustus (Sars, 1863)	Arob	4	6
Ectocyclops herbsti Dussart, 1984	Eher	0	**11**
Ectocyclops rubescens Brady, 1904	Erub	1	2
Eucyclops elegans (Herrick, 1884)	Eele	**1**	0
Eucyclops ensifer Kiefer, 1936	Eens	**1**	0
Eucyclops serrulatus (Fischer, 1851)	Eser	0	**10**
Eucyclops solitarius Herbst, 1959	Esol	**1**	0
Eucyclops prionophorus Kiefer, 1931	Epri	0	**11**
Eucyclops leptacanthus Kiefer, 1956	Elep	0	**4**
Eucyclops neumani (Pesta, 1927)	Eneu	0	**6**
Homocyclops ater (Herrick, 1882)	Hat	0	**3**
Macrocyclops albidus (Jurine, 1820)	Malb	4	14
Megacyclops viridis (Jurine, 1820)	Mvir	0	**1**
Mesocyclops aspericornis (Daday, 1906)	Masp	1	2
Mesocyclops ellipticus Kiefer, 1936	Mell	**1**	0
Mesocyclops longisetus curvatus Dussart, 1987	Mloc	1	10
Mesocyclops longisetus longisetus (Thiébaud, 1912)	Mlol	**1**	0
Mesocyclops meridianus (Kiefer, 1926)	Mmer	12	13
Mesocyclops ogunnus Onabamiro, 1957	Mogu	6	3
Metacyclops laticornis (Lowndes, 1934)	Mlat	**3**	0
Metacyclops leptopus (Kiefer, 1927)	Mlep	0	**1**
Metacyclops mendocinus (Wierzejski, 1892)	Mmen	1	2
Microcyclops anceps anceps (Richard, 1897)	Manc	8	20
Microcyclops ceibaensis (Marsh, 1919)	Mcei	3	7
Microcyclops finitimus Dussart, 1984	Mfin	2	15
Microcyclops mediasetosus Dussart & Frutos, 1985	Mmed	8	10
Paracyclops chiltoni (Thomson, 1883)	Pchil	4	6
Thermocyclops decipiens (Kiefer, 1929)	Tdec	23	3
Thermocyclops inversus Kiefer, 1936	Tinv	**11**	0
Thermocyclops minutus (Lowndes, 1934)	Tmin	16	1
Tropocyclops prasinus meridionalis (Kiefer, 1931)	Tpram	2	1
Tropocyclops prasinus prasinus (Fischer, 1860)	Tprap	0	**1**

In general, higher median value of species richness was found in macrophyte than in plankton samples (Fig. 2), with a significant difference. For zooplankton in the summer we found a mean ± standard-deviation of 2.75 ± 1.03 species and 3.12 ± 1.74 species. For macrophytes, we found 5.08 ± 1.62 species in the summer and 5.41 ± 2.20 species per sample.

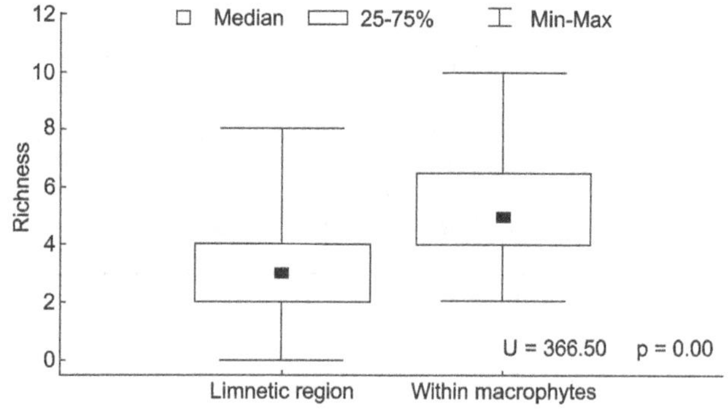

Figure 2. Median ± minimum maximum values of species richness between zooplankton and macrophytes samples.

After 20 attempts we found two convergent solutions for NMDS. This analysis (Fig. 3) classified the kind of habitat sample in basically three groups: zooplankton, macrophytes, and a median line where zooplankton and macrophyte samples were located close to each other, supporting the result shown in Table I. Geographical distribution was strongly affected by the habitat sampled. There was no clear spatial pattern of endemicity for cyclopoid species within the La Plata basin.

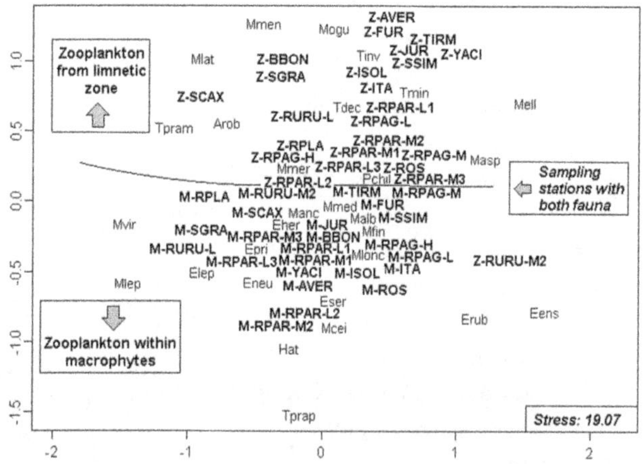

Figure 3. NMDS analysis for ordination of sampling stations and cyclopoid species in our study. See three groups of sampling sites and species: zooplankton, macrophytes and a median group where species occurred were found in zooplankton and macrophytes. For codes of species see Table I.

DISCUSSION

Our results were similar to those obtained by Maia-Barbosa et al. (2008), which found major richness of zooplanktonic species in places with aquatic macrophytes, pointing to the heterogeneity of habitats as the responsible factor. Thus, the hypothesis tested in our study was corroborated. The model of Genkai-Kato (2007) supports this statement, pointing out macrophyte areas as a refuge for zooplankton species to avoid predation of planktivorous fish.

The greatest number of exclusive species was found in macrophytes. Our results show that some genera, as *Eucyclops* and *Ectocyclops* are more frequent in macrophytes than in zooplankton, where *Thermocyclops* is common. This result agrees with Lansac-Toha et al. (2002) and Maia-Barbosa et al. (2008). Lansac-Toha et al. (2002) found 12 species of cyclopoids in the flood plain of the upper Paraná River in aquatic macrophytes. Of these species, a few were in plankton, being found more frequently and sometimes exclusively in macrophytes, similar to the findings of Maia-Barbosa et al. (2008). These authors found13 species of cyclopoid copepods in litoranean regions with aquatic macrophytes in Lake Dom Helvécio (State of Minas Gerais, Brazil), but only 7 of these were found in areas without macrophytes.

We did not find any clear pattern of geographical distribution for the cyclopoid species of our study. Many species are well distributed and some of them are cosmopolitan (Ueda & Reid 2003, Boxshall & Defaye 2008). According to Selden et al. (2010) the recent evolution of the order Cyclopoida in the Neogene (± 15 M.A.) is another possible factor for its wide distribution. The results for Cyclopoida contrasts with the results of other freshwater copepods such as diaptomids and some harpacticoids, according to the literature.

In our results, the most widely distributed species were *Ectocyclops herbsti* Dussart, 1984, *Eucyclops serrulatus* (Fischer, 1851), *Macrocyclops albidus* (Jurine, 1820), *Microcyclops anceps* (Richard, 1897), *Microcyclops finitimus* Dussart, 1984, *Microcyclops ceibaensis* (Marsh, 1919), *Thermocyclops decipiens* (Kiefer, 1929), *Thermocyclops minutus* (Lowndes, 1934), and *Thermocyclops inversus* Kiefer, 1936. It is important to highlight that some species, for instance *Mesocyclops aspericornis* (Daday, 1906) and *Mesocyclops ogunnus* Onabamiro, 1957, are invasive from Africa (Ueda & Reid 2003).The species cited above are widely distributed in the Neotropical Region (Reid 1985). The most common species, in addition to those that occur in both environments sampled, are adapted to different ecosystems. This contrasts with species that are from litoranean zones or macrophyte stands. Examples are *Homocyclops ater* (Herrick, 1882) and *E. herbsti*.

Compared to other studies in the La Plata basin (e.g., Paggi & José de Paggi 1990, Lansac-Toha et al. 2002, Silva & Matsumura-Tundisi 2002, Nogueira et al. 2008), the number of species found in this study is considerably high, and this was attributed to the combination of sampling in a large area and in two types of habitat (plankton and macrophytes), allowing the capture of exclusive species from specific habitats. Our richness of cyclopoids is also high if compared with other basins in the world (TASH 1971, KOBAYASHI et al. 1998).

The geographical distribution of cyclopoid copepods was highly influenced by the type of habitat sampled, as shown by the NMDS. This result confirms that some species are almost restricted to open waters or to macrophytes stands. Other species are common in several kinds of habitats, or can be found accidently in them, when removed by the water flow, for example. For a more accurate estimate of the diversity of these copepods it is necessary to sample at least two kinds of habitats, open waters and macrophyte stands.

ACKNOWLEDGEMENTS

We thank FAPESP (2008/02015-7; 2009/00014-6; 2011/18358-3) for financial support.

LITERATURE CITED

AGOSTINHO, A.A.; L.C. GOMES & F.M. PELICICE. 2007. **Ecologia e manejo de recursos pesqueiros em reservatórios do Brasil.** Maringá, Eduem, 501p.

ALEKSEEV, V.R. 2002. Copepoda, p. 123-188. *In*: C.H. FERNANDO (Ed.). **A guide to tropical freshwater zooplankton: Identification, ecology and impact on fisheries.** London, Backhuys Publishers.

BOLTOVSKOY, D.; N. CORREA; F. BORDET; V. LEITES & D. CATALDO. 2013. Toxic *Microcystis* (cyanobacteria) inhibit recruitment of the bloom-enhancing invasive bivalve *Limnoperna fortunei*. **Freshwater Biology 58** (9): 1968-1981. doi:10.1111/fwb.12184

BOXSHALL, G.A. & D. DEFAYE. 2008. Global diversity of Copepods (Crustacea: Copepoda) in freshwater. **Hydrobiologia 595:** 195-207.

BOXSHALL, G.A. & D. JAUME. 2000. Making Waves: The Repeated Colonization of Fresh Water by Copepod Crustaceans. **Advances in Ecological Research 31:** 61-79.

EINSLE, U. 1996. Copepoda Cyclopoida: Genera *Cyclops, Megacyclops, Acanthocyclops*, p. 1-82. *In*: H.J.F. DUMONT (Ed.). **Guides to the identification of the microinvertebrates of the continental waters of the world.** The Netherlands, Backhuys Publishers.

FANTIN-CRUZ, I.; K.K. TONDATO; J.M.F. PENHA; L.A.F. MATEUS; P. GIRARD & R. FANTIN-CRUZ. 2008. Influence of fish abundance and macrophyte cover on microcrustacean density in temporary lagoons of the Northern Pantanal-Brazil. **Acta Limnologica Brasiliensia 20** (4): 339-344.

GENKAI-KATO, M. 2007. Macrophyte refuges, prey behaviour and trophic interactions: consequences for lake water clarity. **Ecology Letters 10:** 105-114.

GERALDES, A.M. & M.J. BOAVIDA. 2004. Do Littoral Macrophytes Influence Crustacean Zooplankton distribution? **Limnetica 23** (1-2): 57-64.

KARAYTUG, S. 1999. Genera *Paracyclops, Ochridacyclops* and key to the Eucyclopinae. *In*: H.J.F. DUMONT (Ed.). **Guides to the identification of the microinvertebrates of the continental waters of the world.** The Netherlands, Backhuys Publishers.

KOBAYASHI, T.; R.J. SHIEL; P. GIBBS & P.I. DIXON. 1998. Freshwater zooplankton in the Hawkesbury-Nepean River: comparison of community structure with other rivers. **Hydrobiologia 377:** 133-145.

LANSAC-TÔHA, F.A.; L.F.M. VELHO; J. HIGUTI & E.M. TAKAHASHI. 2002. Cyclopidae (Crustacea, Copepoda) from the upper Paraná River floodplain, Brazil. **Brazilian Journal of Biology 62** (1): 125-133.

LUCENA-MOYA, P. & I.C. DUGGAN. 2011. Macrophyte architecture affects the abundance and diversity of littoral microfauna. **Aquatic Ecology 45:** 279-287.

MAIA-BARBOSA, P. M.; R.S. PEIXOTO & A.S. GUIMARÃES. 2008. Zooplankton in littoral Waters of a tropical lake: a revisited biodiversity. **Brazilian Journal of Biology 68** (4): 1061-1067.

NOGUEIRA, M.G.; P.C. REIS-OLIVEIRA & Y.T. BRITTO. 2008. Zooplankton assemblages (Copepoda and Cladocera) in a cascade of reservoirs of a large tropical river (SE Brazil). **Limnetica 27** (1): 151-170.

NOGUEIRA, M.G.; G. PERBICHE-NEVES & D.A.O. NALIATO. 2012. Limnology of two contrasting hydroelectric reservoirs (storage and run-of-river) in southeast Brazil, p. 167-184. *In*: H. SAMADI-BOROUJENI (Ed.). **Hydropower.** Croatia, Intech.

OKSANEN, J. 2013. **Multivariate Analysis of Ecological Communities in R: vegan tutorial.** Available online at: http://cc.oulu.fi/~jarioksa/opetus/metodi/vegantutor.pdf [Accessed: 14/VII/2013].

PAGGI, J.C. & S. JOSÉ DE PAGGI. 1990. Zooplankton de ambientes lóticos e lênticas do rio Paraná médio. **Acta Limnologica Brasiliensia 3:** 685-719.

PERBICHE-NEVES, G.; M. SERAFIM-JÚNIOR; A.R. GHIDINI & L. BRITO. 2007. Spatial and temporal distribution of Copepoda (Cyclopoida and Calanoida) of an eutrophic reservoir in the basin of upper Iguaçu River, Paraná, Brazil. **Acta Limnologica Brasiliensia 19** (4): 393-406.

R DEVELOPMENT CORE TEAM. 2012. **A language and environment for statistical computing.** Vienna, Austria, R Foundation for Statistical Computing ISBN 3-900051-07-0, URL. Available online at: http://www.R-project.org [Accessed 10.XII.2012].

REID, J.W. 1985. Chave de identificação e lista de referências bibliográficas para as species continentais sulamericanas de vida livre da Ordem Cyclopoida (Crustacea, Copepoda). **Boletim de Zoologia, Universidade de São Paulo 9:** 17-143.

REID, J.W. 1989. The distribution of species of the genus *Thermocyclops* (Copepoda, Cyclopoida) in the western hemisphere, with description of *T. parvus*, new species. **Hydrobiologia 175:** 149-174.

RIETZLER, A.C.; T. MATSUMURA-TUNDISI & J.G. TUNDISI. 2002. Life cycle, feeding and adaptive strategy implications on the co-occurence of *Argyrodiaptomus furcatus* and *Notodiaptomus iheringi* in Lobo-Broa Reservoir (SP, Brazil). **Brazilian Journal of Biology 62**: 93-105.

ROCHA, C.E.F. 1998. New morphological characters useful for the taxonomy of genus *Microcyclops* (Copepoda, Cyclopoida). **Journal of Marine Systems 15**: 425-431.

SELDEN, P.A.; R. HUYS; M.H. STEPHENSON; A.P. HEWARD & P.N. TAYLOR. 2010. Crustaceans from bitumen clast in Carboniferous glacial diamictite extend fossil record of copepods. **Nature Communications 1** (50): 1-6.

SILVA, W.M. & T. MATSUMURA-TUNDISI. 2002. Distribution and abundance of Cyclopoida populations in a cascade of reservoir of the Tietê River (São Paulo State, Brazil).**Verhandlungen Internationale Vereinigung für Theoretische und Angewandte Limnologie 28**: 667-670.

SILVA, W.M. & T. MATSUMURA-TUNDISI. 2005. Taxonomy, ecology, and geographical distribution of the species of the genus *Thermocyclops* Kiefer, 1927 (Copepoda, Cyclopoida) in São Paulo State, Brazil, with description of a new species. **Brazilian Journal of Biology 65** (3): 521-31.

SILVA, W.M. 2008. Diversity and distribution of the free-living freshwater Cyclopoida (Copepoda:Crustacea) in the Neotropics. **Brazilian Journal of Biology 68** (4): 1099-1106.

SUÁREZ-MORALES, E.; J.W. REID; F. FIERS & T.M. ILIFFE. 2004. Historical biogeography and distribution of the freshwater cyclopine copepods (Copepoda, Cyclopoida, Cyclopinae) of the Yucatan Peninsula, Mexico. **Journal of Biogeography 31** (7): 1051-1063.

SUÁREZ-MORALES, E.; J.W. REID & M. ELÍAS-GUTIÉRREZ. 2005. Diversity and Distributional Patterns of Neotropical Freshwater Copepods (Calanoida: Diaptomidae). **International Review of Hydrobiology 90** (1): 71-83.

TASH, J.C. 1971. Some Cladocera and Copepoda from the Upper Klamath River Basin. **Northwest Science 45** (4): 239-243.

TRANTER, D.J. & P.E. SMITH. 1979. Filtration performance, p. 27-56. *In*: D.J. TRANTER (Ed.). **Zooplankton sampling.** Paris, Imprimerie Rolland, 3rd ed.

UEDA, H. & J.W. REID. 2003. Copepoda: Cyclopoida – Genera *Mesocyclops* and *Thermocyclops*, p. 1-316. *In*: H.J.F. DUMONT (Ed.). **Guides to the identification of the microinvertebrates of the continental waters of the world.** The Netherlands, Backhuys Publishers.

ZALOCAR DE DOMITROVIC, Y.; A.S.G. POI DE NEIFF & S.L. CASCO. 2007. Abundance and diversity of phytoplankton in the Paraná River (Argentina) 220 km downstream of the Yacyretá Reservoir. **Brazilian Journal of Biology 67** (1): 53-63.

7

Ecomorphological relationships among four Characiformes fish species in a tropical reservoir in South-eastern Brazil

Débora de S. Silva-Camacho[1,2], Joaquim N. de S. Santos[1],
Rafaela de S. Gomes[1] & Francisco G. Araújo[1]

[1] Laboratório de Ecologia de Peixes, Universidade Federal Rural do Rio de Janeiro. Antiga Rodovia Rio-SP km 47,
23851-970 Seropédica, RJ, Brazil.
[2] Corresponding author. E-mail: debora_desouza@yahoo.com.br

ABSTRACT. The aim of this study was to assess the ecomorphological patterns and diet of four Characiformes fish species in a poorly physically structured tropical reservoir. We tested the hypothesis that body shape and diet are associated, because environmental pressure acts on the phenotype, selecting traits according to the available resources. Ten ecomorphological attributes of 45 individuals of each species – *Astyanax* cf. *bimaculatus* (Linnaeus, 1758), *Astyanax parahybae* Eigenmann, 1908, *Oligosarcus hepsetus* (Cuvier, 1829), and *Metynnis maculatus* (Kner, 1858) –, collected between February and November 2003, were analyzed, and the patterns were assessed using Principal Components Analysis (PCA). Diet similarity among fish species was assessed using cluster analysis on feeding index. The first two axes from PCA explained 61.73% of the total variance, with the first axis being positively correlated with the compression index and relative height, whereas the second axis was positively correlated with the pectoral fin aspect. Two well-defined trophic groups, one herbivorous/specialist (*M. maculatus*) and the other formed by two omnivorous/generalist (*A.* cf. *bimaculatus*, *A. parahybae*) and one insectivorous-piscivorous (*O. hepsetus*) were revealed by the cluster analysis. *Astyanax.* cf. *bimaculatus* and *A. parahybae* differed. The first has comparatively greater relative height, relative length of the caudal peduncle and lower caudal peduncle compression index. However, we did not detect a close correspondence between diet and body shape in the reservoir, and inferred that the ecomorphological hypothesis of a close relationship between body shape and diet in altered systems could be not effective.

KEY WORDS. Body shape; diet; freshwater fishes; morphological diversity; niche overlap.

Ecomorphological studies of fishes aim to understand the patterns of association between the morphology of these organisms and their resource use. A major focus of some of these studies is the relationship between morphological variables and feeding behavior, or habitat use (TEIXEIRA & BENNEMANN 2007, OLIVEIRA *et al*. 2010, FAYE *et al*. 2012). One of the first studies on the ecomorphological patterns of fishes was published by KEAST & WEEB (1966), who observed that specializations of the oral apparatus determined habitat preferences and contributed to lessening interspecific competition among the fish species of the Opinicon Lake, Canada. Coexistence among several species in fish communities are facilitated because morphological segregation enables spatial and feeding partitioning (WIKRAMANAYAKE 1990).

Morphological divergences among species can be assessed by certain indices that can be interpreted as indicators of life strategies for habitat colonization (FREITAS *et al*. 2005) and use of food resources (WAINGHWRITH & RICHARD 1995). Correlations between the diversity of morphological patterns and resource partitioning have been used to test ecomorphological hypo-

theses in fish communities (WIKRAMANAYAKE 1990, HUGENY & POUILLY 1999). However, in the current literature there is no consensus about the direct relationship between ecology and morphology. Some studies maintain that there is a strong relationship between the morphology of organisms and their use of resources (MOYLE & SENANAYAKE 1984, WIKRAMANAYAKE 1990), whereas others have not found support for this relationship (GROSSMAN 1986, MOTTA *et al*. 1995), or have found only a weak correlation (CLIFTON & MOTTA 1998).

The use of ecomorphological approaches in the Neotropical region may be particularly relevant to address questions on niches and shared resources, since the region is characterized by a high diversity of fish (WINEMILLER 1991). This approach has been taken to study fish species inhabiting reservoirs, and which possess morphological and reproductive plasticity to adapt to modified environmental conditions (DUARTE *et al*. 2011). Moreover, impoundments can cause changes in the feeding strategies of species (HAHN & FUGI 2007), through selection: traits that are more suited for the colonization and use of available resources in the altered environment will be selected (CUNICO &

AGOSTINHO 2006). This study describes the ecomorphological patterns of four Characiformes fish species in the Lajes Reservoir, and evaluates the relationship between their diet and ecomorphological variables associated with their feeding behavior, locomotion, and their use of the water column. We tested the hypothesis that the shape of the body and diet of fishes are closely associated, since the environmental pressure acts on the phenotype, selecting traits according to the available resources.

MATERIAL AND METHODS

Fishes were collected using gillnets between February and November 2003, from the Lajes Reservoir (22°42'-22°50'S, 43°53'-44°05'W) and were fixed in 10% formalin during 48 hours and transferred to 70% ethanol. Four abundant Characiformes species were analyzed and forty-five individuals of each species were selected: *Astyanax* cf. *bimaculatus* (Linnaeus, 1758), Standard Length average (SL) = 96.65 ± 9.69 mm standard deviation, *Astyanax parahybae* Eigenmann, 1908, SL = 104.37 ± 7.27 mm, *Oligosarcus hepsetus* (Cuvier, 1829), SL = 144.28 ± 23.5 mm, and *Metynnis maculatus* (Kner, 1858) SL = 110.11 ± 6,81 mm. Voucher specimens were deposited in the fish collection of the Laboratory of Fish Ecology, Universidade Federal Rural do Rio de Janeiro, Brazil under numbers 1636 to 1641.

The Lajes Reservoir is a major impoundment in the State of Rio de Janeiro and was built between 1905 and 1908 by damming small streams. Initially built to generate hydroelectric power, the reservoir today is a strategic water supply for the municipality of Rio de Janeiro because it provides good quality water. This reservoir, situated at 415 m a.s.l., has a poorly structured physical habitat, lacking routes for fish migration because the tributaries that it receives in the slopes of the Serra do Mar are small (ARAÚJO & SANTOS 2001). Water levels range between 5-9 m during the year (ARAÚJO & SANTOS 2001, SANTOS *et al.* 2004).

Thirteen morphometric measurements were taken for each individual: standard length, body height, body width, head height, head length, pectoral fin length, pectoral fin width, caudal peduncle length, caudal peduncle height, caudal peduncle width, eye height, mouth height and mouth width. Absolute measurements were taken on the left side of each specimen with a caliper accurate to 0.01 mm. Only adults above L_{50} (average length of first gonadal maturation, according to data from VAZZOLER 1996) were used in order to avoid eventual allometric effects.Ecomorphological attributes of compression index (CI), relative height (RH), relative length of the caudal peduncle (RLP), caudal peduncle compression index (CPC), pectoral fin aspect (PFA), eye position (EP), relative length of the head (RLH), relative width of the mouth (RWM), relative height of the mouth (RHM) and mouth aspect (MA) were calculated based on morphometric measurements interpreted as indicators of lifestyle or adjustments to the different habitats and diets, as described in several studies (GATZ 1979a, b, MAHON 1984, WATSON & BALON 1984, BALON *et al.* 1986, BARRELLA *et al.* 1994, FREIRE & AGOSTINHO 2001).

Diet analysis was based on stomach contents examined under a stereoscopic microscope and identified to the lowest taxonomic level. Food items were identified according to BRUSCA & BRUSCA (2007) and MUGNAI *et al.* (2010). Empty stomachs were excluded from the analyses. The feeding index (IAi according to KAWAKAMI & VAZZOLER 1980) was calculated to obtain the relative importance of each food item –, using the wet weight of each item.

Principal Components Analysis (PCA) was performed using a correlation matrix of the 10 ecomorphological attributes of the four species to characterize them according to their morphological characteristics. The broken-stick criterion (JACKSON 1993) was used to select the most significant axes in the PCA. This criterion selects axes with eigenvalues higher than those expected by chance (KING & JACKSON 1999). We performed a Discriminant Function Analysis (DFA) to identify the ecomorphological attributes that best discriminated morphological characteristics among *A.* cf. *bimaculatus* and *A. parahybae*. This procedure was performed using Statistica 7.0 software.

Fish diets were analyzed by cluster analysis on the feeding index based on wet weight of food items, using the UPGMA linkage method and the Euclidian Distance matrix. To perform this analysis we used the software Primer 6.0. The amplitude of the trophic niche of each species was estimated using the Shannon's index of niche breadth (H' $\log_2$). This index varies from 0 (only one kind food item is present in the species' diet) to 1 (similar quantities of many food items are present in the species' diet). Feeding overlap among species was assessed using PIANKA's (1973) index, with the help of the software EcoSim700. This index varies from zero (signifying no overlap) to one (complete overlap).

The relationships between diet and ecomorphological attributes among the fishes were explored using Canonical Correspondence Analysis (CCA), a direct gradient analysis technique that ordinates a set of observations (in this case species) by directly relating them to two series of associated variables (ecomorphological attributes and diet) (TER BRAAK 1986). CCA is a modification of Correspondence Analysis that adds a multiple regression step and simultaneously relates the primary set of variables (in this case diet) with the secondary variables (ecomorphological attributes), constraining the ordination in such a manner that scores represent the maximum correlation between diet and morphology. A permutation test was used to assess the statistical significance of the relationship. The analysis was performed on log10-transformed ecomorphological attributes and dietary data with the CANOCO software, version 4.5.

RESULTS

Principal Component Analysis of the ecomorphological attributes produced two axes with eigenvalues higher than those expected by chance, explaining 61.73% of the total variance (Table I). The first component (PC1) explained 42.37% of the variability and was positively correlated with the compression index and relative height, and negatively correlated with the relative length of the caudal peduncle, relative height of the mouth and aspect of the mouth. *Metynnis maculatus* was positively correlated with axis 1 and was characterized by a laterally compressed and high body, whereas *O. hepsetus* was characterized by large mouth opening and elongated caudal peduncle, being negatively correlated with axis 1. The species *A.* cf. *bimaculatus* and *A. parahybae* positioned close to the origin of the ordination diagram and were characterized by a fusiform body. Despite this similarity, *A.* cf. *bimaculatus* had a comparatively wider mouth and more compressed caudal peduncle, whereas *A. parahybae* had a longer caudal peduncle, higher head and higher and narrow mouth. The second component (PC2) explained 19.36% of the variability and was positively correlated with the aspect of the pectoral fin. This axis was negatively correlated with *A.* cf. *bimaculatus*, which had a shorter and wider pectoral fin and was positively correlated with *M. maculatus*, *O. hepsetus* and *A.* cf. *bimaculatus*, which had a longer and narrower pectoral fin (Table I, Fig. 1).

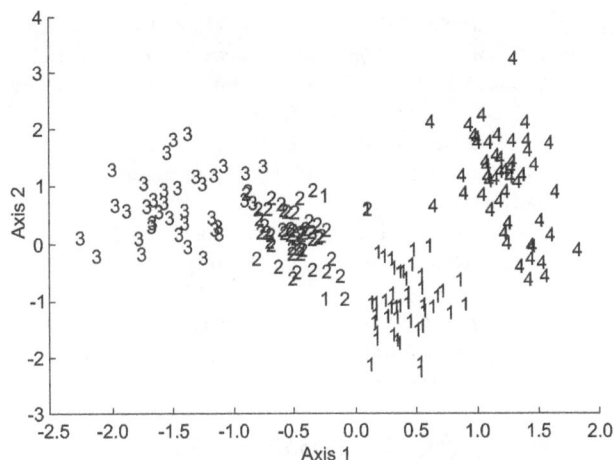

Figure 1. Ordination diagram from two principal components on ecomorphological attributes of four studied species: (1) *A.* cf. *bimaculatus*; (2) *A. parahybae*; (3) *O. hepsetus*; (4) *M. maculatus*.

Table I. Factor loads from principal components analysis on ecomorphological attributes of four examined Characiformes fish species in Lajes Reservoir, Rio de Janeiro, Brasil.

Ecomorphological attributes	Axis 1	Axis 2
Compression Index (CI)	**0.80**	0.43
Relative Height (RH)	**0.88**	0.33
Relative Length of the Caudal Peduncle (RLP)	**-0.77**	0.36
Caudal Peduncle Compression Index (CPC)	0.60	-0.53
Pectoral Fin Aspect (PFA)	0.00	**0.73**
Eye Position (EP)	-0.40	-0.40
Relative Length of the Head (RLH)	-0.14	0.51
Relative Width of the Mouth (RWM)	-0.01	-0.45
Relative Height of the Mouth (RHM)	**-0.93**	0.00
Mouth Aspect (MA)	**-0.92**	0.13
Percentage of explained variance (%)	42.37	19.36
Eigenvalues	4.24	1.98
Broken-stick eigenvalues	2.93	1.93

According to the Discriminant Function Analyses, there are significant morphological differences between *A. parahybae* and *A.* cf. *bimaculatus* (Wilks' $\lambda = 0.147$, $p < 0.0001$). The most important attributes for discriminating among the studied species were relative height, relative length of the caudal peduncle, and caudal peduncle compression index.

Astyanax parahybae (H' = 0.25), *A.* cf. *bimaculatus* (H' = 0.57) and *O. hepsetus* (H' = 0.31) clustered in a single group in the results of the cluster analysis of the diet (Index of Feeding Importance), whereas *M. maculatus* (H' = 0.03) formed a separated branch (Fig. 2, Table II). The diets of the former three species consist mainly of insects and the items in the diets of

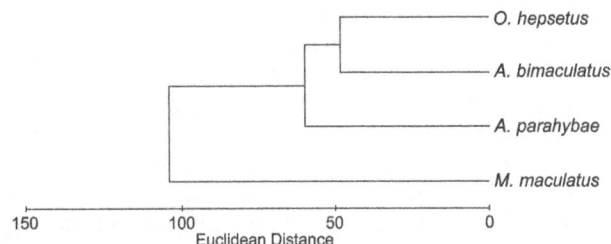

Figure 2. Dendrogram from cluster analysis on Index of Feeding Importance for the four examined Characiformes fish species in Lajes Reservoir.

Table II. Index of Feeding Importance for feeding item of the four examined Characiformes fish species in Lajes Reservoir. Number of examined individuals for each species = 45. Main items are show in bold.

Items	*A.* cf. *bimaculatus*	*A. parahybae*	*O. hepsetus*	*M. maculatus*
Hymenoptera	2.97	0.00	4.57	0.00
Coleoptera	0.83	0.00	0.00	0.00
Diptera	0.25	0.79	0.00	0.00
Odonata	0.00	0.79	5.14	0.22
Insects remains	**75.39**	**41.23**	**41.13**	0.20
Bivalvia	0.02	0.00	0.00	0.00
Nematoda	0.00	0.00	0.99	0.00
Fish	2.12	0.00	**31.07**	0.00
Algae	8.42	7.29	3.99	**89.54**
Seeds	2.21	**49.87**	0.00	**10.02**
Larval fish	0.38	0.00	0.00	0.00
Fish eggs	0.00	0.00	**12.62**	0.00
Vegetal remains	0.66	0.00	0.00	0.00
Detritus and Sediment	6.65	0.00	0.45	0.00

A. cf. *bimaculatus* and *O. hepsetus*, which have a wider trophic niche, overlap by 78% (Table III). Moreover, seeds are an important food item in the diet of *A. parahybae*, which have a comparatively narrower trophic niche than *A.* cf. *bimaculatus*. On the other hand, the diet of *M. maculatus* consists mainly on vegetal fragments (algae), and the trophic niche of this species is narrower.

The Canonical Correspondence Analysis did not show significant correlation between diet and morphology (p = 0.567, test with 1000 permutations). The sum of the eigenvalues for all axes defined by CCA was only 0.011, suggesting that differences in dietary composition were poorly explained by variations in morphological variables.

DISCUSSION

The four studied species in the Lajes Reservoir presented clear differentiated ecomorphological traits, suggesting that these species potentially use different resources. We found a lack of correspondence between diet and morphology, although the patterns of habitat use (locomotion and position in the water column) were consistent with the ecomorphological hypothesis. This finding suggests that the fish species were more influenced by spatial than by trophic structure.

Metynnis maculatus presented higher CI and RH than the other species examined. These attributes are associated with the use of structured habitats that have a variety of shelters and different types of substrate, where fish with high and laterally compressed bodies can perform maneuvers such as pitch or yaw (Alexander 1967). Structured habitats are limited in the Lajes Reservoir, especially during the dry season, when most of the shoreline is exposed. Then, structured areas, which are favorable to the herbivorous feeding habits of *M. maculatus*, are restricted to some shallow bays, which house shrubs and grasses. Conversely, during the wet season, the submerged vegetation in the shoreline is well structured and should allow for the maneuvers of *M. maculatus*. The herbivorous habit of *M. maculatus* inhabiting reservoirs of the middle and lower Tiete River was described by Smith *et al.* (2003). It is made possible by the small terminal mouth of this species, which enables the capture of plant fragments in the water column. Moreover, herbivorous feeding habits suit slow swimmers that have the ability to perform vertical movements in habitats where dynamism is low (Balon *et al.* 1986), such as reservoirs. Their results were corroborated by our cluster analysis, in which *M. maculatus* formed an isolated branch in the dendrogram based on its herbivorous feeding habits (mostly algae and seeds), thus characterizing this species as the most specialist among the species studied.

In contrast to *M. maculatus*, *O. hepsetus* had lower CI and RH, and higher RHM, MA and RLP, indicating a narrow mouth with large gap and a longer head and caudal peduncle. This set of attributes are typical of fishes that feed on large prey items that have good swimming capacity, for instance piscivorous

predators that inhabit benthic habits, dwelling in more dynamic environments (Gatz 1979a, b, Watson & Balon 1984, Balon *et al.* 1986, Freire & Agostinho 2001). Although fishes were the main prey of this species, insects and fish eggs were also consumed, suggesting that the trophic plasticity of *O. hepsetus* is high. The insectivorous-piscivorous habits of *O. hepsetus* at the Lajes Reservoir were described by Araújo *et al.* (2005).

The morphological attributes related to body height and apparatuses to capture food of *Astyanax* cf. *bimaculatus* and *A. parahybae* were intermediate between *M. maculatus* and *O. hepsetus*. The scores obtained for individuals of *Astyanax*, close to the origin in the ordination diagram, indicate a more fusiform body, a characteristic that is associated with generalist feeding habits (Oliveira *et al.* 2010). According to Uieda (1984) and Casatti *et al.* (2001), species of *Astyanax* have a continuous swimming behavior across different parts of the water column, which is favored by their laterally positioned eyes and small and relatively elongate pectoral fins. Such characteristics suggest a lack of specialization to explore a particular feeding resource. These species, which are similar in phenotypic traits, present similar adaptations to use resources and thus have a strong potential to compete with one another (Wootton 1990). However, in the tropics, competition seems to be reduced, owing to the feeding plasticity of most species, resource partitioning (Araújo-Lima *et al.* 1995) and resource availability (Winemiller & Jepsen 1998), as well as the phenotypic characteristics of each species (Wainwright & Richard 1995, Labropoulou & Eleftheriou 1997, Bellwood & Wainwright 2001). Contrarily to the classical results of the traditional niche theory, species that use similar resources may coexist if they are sufficiently similar in their skills to compete for the limiting resources (Fagerström 1988). This pattern was confirmed for the Lajes Reservoir because of the high trophic niche overlap of *Astyanax* species, although their ecomorphological differences allow the exploration of differentiated resources that probably decrease competition.

In spite of exhibiting a more generalist body shape, the two *Astyanax* species have morphological differences that enable them to use resources differently. This pattern was also found by Santos *et al.* (2011) for the same species in the Funil Reservoir. *Astyanax parahybae* fish have phenotypic characteristics that enable them to capture larger prey items, or prey items that have faster movements, than the items eaten by *A.* cf. *bimaculatus*. This is possible because *A. parahybae* has narrower mouth, greater gap and longer caudal peduncle, whereas *A.* cf. *bimaculatus* has a higher RH, CI and CPC, and a wider mouth. *Astyanax* cf. *bimaculatus* is more frequent and abundant than *A. parahybae* in the Lajes Reservoir (Araújo & Santos 2001), presenting wide niche breadth. We speculate that the generalist feeding habits and the morphological characteristics of *A.* cf. *bimaculatus* are more suitable for standing waters and enable them to be more successful exploring a wide range of trophic resources compared with *A. parahybae*.

According to the cluster analysis, A. cf. *bimaculatus* presented a diet more similar to *O. hepsetus* than to *A. parahybae* because they have a comparatively wider niche breadth and prey mainly on insects, whereas *A. parahybae* eats a great amount of seeds. Although the two species of *Astyanax* had high niche overlap, they seem to concentrate on different food items. While A. cf. *bimaculatus* use a variety of food items, *A. parahybae* concentrates on a few food items such as insects and seeds. These findings show the trophic plasticity of *A. parahybae*, since they use other kinds of food items in other systems (Hirt et al. 2011). Moreover, A. cf. *bimaculatus* seems to be better adapted to lentic environments (e.g. height body, compressed peduncle), which help this species to better explore different food items, while *A. parahybae* seems to be less efficient in obtaining access to a wide range of food resources, thus restricting its diet to a fewer number of items.

No significant association was found between the ecomorphological attributes and diet of these four Characiformes species in the reservoir. The lack of correspondence between morphology and diet can be related to the trophic plasticity of species such as *O. hepsetus*, A. cf. *bimaculatus* and *A.parahybae*. As these species present great flexibility in their feeding behavior, it is unlikely that we will observe a strong link between diet and morphology, because they belong to the same broad trophic category, as indicated by cluster analysis. Moreover, Hugueny & Pouilly (1999) suggested that food availability may weaken the relationship between diet and morphology because stomach contents probably reflect food availability more than morphological adaptation.

Trophic plasticity can also be associated with morphological divergence of fish species in reservoirs. Santos et al. (2011) suggested that there is morphological divergence between A. cf. *bimaculaus* and *A. parahybae* in the Funil Reservoir, Southeastern Brazil, as consequence of impoundment. Similarly, Franssen (2011), studying fish species of Central Plains of the USA, found that, although the components of body shape are plastic, anthropogenic habitat modifications may drive trait divergence in native populations in reservoir-altered habitats. According to Langerhans (2008), phenotypic variation can be adaptive in conditions of low flow and in habitats with high densities of predators, and these two factors could be driving the observed morphological shifts. Morphological divergence between phylogenetically related species has been reported to occur when morphological variations are found as responses to selective pressures of the environment (Casatti & Castro 2006, Oliveira et al. 2010). Thus, the mechanism involved in competition in altered environments may suggest an intensification of generalist habits, rather than the specialization predicted by the classical competition theory (Gabler & Amundsen 2009). Therefore, altered systems like the Lajes Reservoir, which have a simple habitat structure, may favor trophic generalist fish species and contribute to the lack of correlation between morphology and diet.

We conclude that the ecomorphological hypothesis of a close relationship between fish body shape and diet was not corroborated in the present study for the altered system studied. However, the patterns of habitat use of the four species studied corresponded to the expectation of the ecomorphological approach. In altered systems, the patterns of resource use can be broad enough to allow fish species to change their choices and to respond to local biotic and/or abiotic conditions, contradicting the traditional ecological theories. Further studies are necessary to give a more detailed picture on the relationship between body shape and diet, especially in dammed rivers, because the species that persist in this new system had to adapt to new environmental constraints.

ACKNOWLEDGMENTS

Specimens of fish used in this study were obtained from the Project "Fish ecology in Lajes Reservoir", which is financially support by a Research & Development Program by ANNEEL/LIGHT/UFRRJ consortium. FAPERJ – Rio de Janeiro Carlos Chagas Filho Research Support Agency also supplied financial support for this project.

LITERATURE CITED

Alexander, R.M. N. 1967. **Functional design in fishes.** London, Hutchinson University Library, 160p.

Araújo, F.G. & L.N. Santos. 2001. Distribution and composition of fish assemblages in Lajes Reservoir, Rio de Janeiro, Brazil. **Revista Brasileira de Biologia 61**: 563-576.

Araújo, F.G.; C.C. Andrade; R.N. Santos; A.F.G.N. Santos & L.N. Santos. 2005. Spatial and seasonal changes in the diet of *Oligosarcus hepsetus* (Characiformes: Characidae) in a Brazilian reservoir. **Revista Brasileira de Biologia 65**: 1-8.

Araújo-Lima, C.A.R.M.; A.A. Agostinho & N.N. Fabré. 1995. Trophic aspects of fish communities in Brazilian rivers and reservoirs, p. 105-136. *In*: J.G. Tundisi; C.E.M. Bicudo & T. Matsumura-Tundisi (Eds). **Limnology in Brazil.** Rio de Janeiro, ABC/SBL, 376p.

Balon, E.K.; S.S. Crawford & A. Lelek. 1986. Fish communities of the upper Danube River (Germany, Austria) prior to the new Rhein-Main-Donau connection. **Environmental Biology of Fishes 15**: 243-271.

Barrella, W.; A.C. Beaumord & M. Petrere-Jr. 1994. Comparison between the fish communities of Manso river (MT) and Jacare Pepira river (SP), Brazil. **Acta BiologicaVenezuelica 15**: 11-20.

Bellwood, D.R. & P.C. Wainwright. 2001. Locomotion in labrid fishes: implications for habitat use and cross-shelf biogeography on the Great Barrier Reef. **Coral Reefs 20** (2): 139-150.

Brusca, R.C. & G.J. Brusca. 2007. **Invertebrados.** 2nd. ed. Rio de Janeiro, Guanabara Koogan, XXII+968p.

Casatti, L. & R.M.C. Castro. 2006. Testing the ecomorphological hypothesis in a headwater riffles fish assemblage of the rio São Francisco, southeastern Brazil. **Neotropical Ichthyology 4** (2): 203-2014.

Casatti, L.; F. Langeani & R.M.C. Castro. 2001. Peixes de riacho do parque estadual morro do Diabo, bacia do alto rio Paraná, SP. **Biota Neotropica 1** (1): 1-15.

Clifton K.B. & P.J. Motta. 1998. Feeding morphology, diet and ecomorphological relationships among five caribbean labrids (Teleostei, Labridae). **Copeia** (4): 953-966.

Cunico, A.M. & A.A. Agostinho. 2006. Morphological patterns of fish and their relationships with reservoirs hydrodynamics. **Brazilian Archives of Biology and Technology 49** (1): 125-134.

Duarte, S.; F.G. Araújo; N. Bazzoli. 2011. Reproductive plasticity of *Hypostomus affinis* (Siluriformes: Loricariidae) as a mechanism to adapt to a reservoir with poor habitat complexity. **Zoologia 28** (5): 577-586.

Fagerström, T. 1988. Lotteries in communities of sessile organisms. **Trends in Ecology and Evolution 3**: 303-306.

Faye, D.; F. Le Loc'h; O.T. Thiaw & L.T. Morais. 2012. Mechanisms of food partitioning and ecomorphological correlates in ten fish species from a tropical estuarine marine protected area (Bamboung, Senegal, West Africa). **African Journal of Agricultural Research 7** (3): 443-455.

Franssen, N.R. 2011. Anthropogenic habitat alteration induces rapid morphological divergence in a native stream fish. **Evolutionary Applications 4** (6): 791-804.

Freire, A.G. & A.A. Agostinho. 2001. Ecomorfologia de oito espécies dominantes da ictiofauna do reservatório de Itaipu (Paraná/Brasil). **Acta Limnologica Brasiliensia 13** (1): 1-9.

Freitas, C.E.C.; E.L. Costa & M.G.M. Soares. 2005. Ecomorphological correlates of thirteen dominant fish species of Amazonian floodplain lakes. **Acta Limnologica Brasiliensia 17** (3): 339-347.

Gatz Jr, A.J. 1979a. Ecological morphology of freshwater stream fishes. **Tulane Studies of Zoology and Botany 21**: 91-124.

Gatz Jr, A.J. 1979b. Community organization in fishes as indicated by morphological features. **Ecology 60**: 711-718.

Gabler, H.M. & P.A. Amundsen. 2009. Feeding strategies, resource utilisation and potential mechanisms for competitive coexistence of Atlantic salmon and alpine bullhead in a sub-Arctic river. **Aquatic ecology 44** (2): 325-336.

Grossman, G.D. 1986. Food resource partitioning in a rocky intertidal fish assemblage. **Journal of Zoology 1**: 317-355.

Hahn, N.S. & R. Fugi. 2007. Alimentação de peixes em reservatórios brasileiros: alterações e conseqüências nos estágios iniciais do represamento. **Oecologia Brasiliensis 11** (4): 469-480.

Hirt, L.M.; P.R. Arayala & S.A. Flores. 2011. Population structure, reproductive biology and feeding of *Astyanax fasciatus* (Cuvier, 1819) in an Upper Paraná River Tributary, Misiones, Argentina. **Acta Limnologica Brasiliensia 23** (1): 1-12.

Hugueny, B. & M. Pouilly. 1999. Morphological correlates of diet in an assemblage of West African freshwater fishes. **Journal of Fish Biology 54** (6):1310-1325.

Jackson, D.A. 1993. Stopping rules in principal components analysis: a comparison of heuristical and statistical approaches. **Ecology 74**: 2201-2214.

Kawakami, E. & G. Vazzoler. 1980. Método gráfico e estimativa de índice alimentar aplicado no estudo de alimentação de peixes. **Boletim do Instituto Oceanográfico 29** (2): 205-207.

Keast, A. & D. Webb. 1966. Mouth and body form relative to feeding ecology in the fish fauna of a small lake, Lake Opinicon, Ontario. **Journal of the Fisheries Research Board of Canada 23**: 1846-1874.

King, J.R. & D.A. Jackson. 1999. Variable selection in large environmental data sets using principal components analysis. **Environmetrics 10**: 67-77.

Labropoulou, M. & A. Eleftheriou. 1997. The foraging ecology of two pairs of congeneric demersal fish species: importance of morphological characteristics in prey selection. **Journal of Fish Biology 50**: 324-340.

Langerhans, R. B. 2008. Predictability of phenotypic differentiation across flow regimes in fishes. **Integrative and Comparative Biology 48**: 750-768.

Mahon, R. 1984. Divergent structure in fish taxocenes of north temperate stream. **Canadian Journal of Fisheries and Aquatic Sciences 41**: 330-350.

Motta, J.P.; K.B. Clifton; P. Hernandez & B.T. Eggold. 1995. Ecomorphological correlates in ten species of subtropical seagrass fishes: diet and microhabitat utilization. **Environmental Biology of Fishes 44**: 37-60.

Moyle, P.B. & F.R. Senanayake. 1984. Resource partitioning among the fishes of rainforest streams in Sri Lanka. **Journal of Zoology 202**: 195-223.

Mugnai, R.; J.L. Nessimian & D.F. Baptista. 2010. **Manual de identificação de macroinvertebrados aquáticos do Estado do Rio de Janeiro**. Rio de Janeiro, Technical Books, 174p.

Oliveira, E.F.; E. Goulart; L. Breda; C.V. Minte-vera, L.R.S. Paiva & M.R. Vismara. 2010. Ecomorphological patterns of the fish assemblage in a tropical floodplain: effects of trophic, spatial and phylogenetic structures. **Neotropical Ichthyology 8** (3): 569-586.

Pianka, E.R. 1973. The structure of lizard communities. **Annual Review of Ecology and Systematics 4**: 53-74.

Santos, A.B.I; F.L. Camilo; R.J. Albieri & F.G. Araújo. 2011. Morphological patterns of five fish species (four characiforms, one perciform) in relation to feeding habits in a tropical reservoir in south-eastern Brazil. **Journal of Applied Ichthyology 27**: 1360-1364.

Santos, A.F.G.N.; L.N. Santos & F.G. Araújo. 2004. Water level influences on body condition of *Geophagus brasiliensis* (Perciformes, Cichlidae) in a Brazilian oligotrophic reservoir. **Neotropical Ichthyology 2**: 151-156.

SMITH, W.S.; C.C.G. F. PEREIRA; E.L.G. ESPÍNDOLA & O. ROCHA. 2003. A importância da zona litoral para a disponibilidade de recursos alimentares à comunidade de peixes em reservatórios, p. 233-248. *In*: R. HENRY (Ed.). **Nas interfaces dos ecossistemas aquáticos.** São Carlos, Rima, 350p.

TEIXEIRA, I. & S.T. BENNEMANN. 2007. Ecomorfologia refletindo a dieta dos peixes em um reservatório no sul do Brasil. **Biota Neotropica 7** (2): 67-76.

TER BRAAK, C.J.F. 1986. Canonical correspondence analysis: a new eigenvector technique for multivariate gradient analysis. **Ecology 67**: 1167-1179.

UIEDA, V.S. 1984. Ocorrência e distribuição dos peixes em um riacho de água doce. **Revista Brasileira de Biologia 44**: 203-213.

VAZZOLER, A.M.A. DE M. 1996. **Biologia da reprodução de peixes teleósteos: teoria e prática.** Maringá, EDUEM-SBI, 169p.

WAINWRIGHT, P.C. & B.A. RICHARD. 1995. Predicting patterns of prey use from morphology of fishes. **Environmental Biology of Fishes 44**: 97-113.

WATSON, D.J. & E.K. BALON. 1984. Ecomorphological analysis of taxocenes in rainforest streams of northern Borneo. **Journal of Fish Biology 25**: 371-384.

WIKRAMANAYAKE, E.D. 1990. Ecomorphology and biogeography of a tropical stream fish assemblage: evolution of assemblage structure. **Ecology 71**: 1756-1764.

WINEMILLER, K.O. 1991. Ecomorphological diversification in lowland freshwater fish assemblages from five biotic regions. **Ecological Monographs 61**: 343-365.

WINEMILLER, K. & D.B. JEPSEN. 1998. Effects of seasonality and fish movement on tropical river food webs. **Journal of Fish Biology 53** (Suppl. A): 267-296.

WOOTTON, R.J. 1990. **Ecology of teleost fishes.** London, Chapman and Hall, 404p.

8

Size-selective predation of the catfish *Pimelodus pintado* (Siluriformes: Pimelodidae) on the golden mussel *Limnoperna fortunei* (Bivalvia: Mytilidae)

João P. Vieira & Michelle N. Lopes

Laboratório de Ictiologia, Universidade Federal do Rio Grande. Avenida Itália km 8, 96201-900 Rio Grande, RS, Brazil.
E-mail: vieira@mikrus.com.br

ABSTRACT. This paper describes the size-selective predation on *Limnoperna fortunei* (Dunker, 1857) by *Pimelodus pintado* (Azpelicueta, Lundberg & Loureiro, 2008) from the time it arrived at the Mirim Lagoon basin (2005). Sampling was carried out using bottom trawl in depths of 3-6 m, from January to November 2005, and from October to November 2008. *Pimelodus pintado* began to prey upon *L. fortunei* soon after its arrival (austral spring of 2005). On the spring of 2008, *L. fortunei* was found to be the most important food item of *P. pintado*. The variation in length of the mussels (0.7-3.2 cm, with a mode of 1.3 cm) indicates that the species is now fully established in the system. Our data indicates that large individuals of *P. pintado* incorporate more mussels in their diets than small individuals. However, regardless of their size, *P. pintado* individuals predate only on small (<1.4 cm) representatives of *L. fortunei*. This prey size corresponds to a phase when the mussel is more mobile and readily available for fish. Larger, more aggregated prey groups that are attached to hard substrates are avoided by fish predators.

KEY WORDS. Diet; freshwater invasion; opportunistic predator; Mirim Lagoon.

The Asian freshwater golden mussel, *Limnoperna fortunei* (Dunker, 1857), was first recorded in the Americas in 1991, from the coast of the Rio de la Plata, by PASTORINO *et al.* (1993). *Limnoperna fortunei* has the capability of colonizing a wide range of habitats and has few natural predators, widely colonizing freshwater and estuarine environments in the Neotropical region. Since its introduction, the species has expanded its range to Argentina, Brazil, Paraguay and Uruguay (DARRIGRAN *et al.* 1998, DARRIGRAN & EZCURRA DE DRAGO 2000, DARRIGRAN 2002, OLIVEIRA *et al.* 2006).

In 1998, the Asian freshwater golden mussel was recorded at the northern reaches of the Patos Lagoon drainage basin (MANSUR *et al.* 1999, 2003) and, in the next years, it was found in the southern portion of the basin in densities of up to 140,000 ind/m². After that, population densities stabilized at averages 60,000 ind/m² (MANSUR *et al.* 2003). The golden mussel invaded the Mirim Lagoon in 2005 through the São Gonçalo Channel, which connects the Mirim and Patos Lagoons (LANGONE 2005, BURNS *et al.* 2006b, COLLING *et al.* 2012, LOPES & VIEIRA 2012).

There is little question now that catastrophic biological events like these can profoundly affect entire ecosystems to the point that the invader species monopolizes a large proportion of the resources available (SYLVESTER *et al.* 2005, DARRIGRAN & DAMBORENEA 2011). The cascading effects of such trophic web disruptions can be extremely important (POWER 1992, RUETZ *et al.* 2002, THORP & CASPER 2003). The high densities of golden mussel and the fact that individuals become fixed to the sub-strate by their byssal threads create a new microenvironment. This microenvironment, in turn provides a new habitat for some epifaunal species and, at the same time, can lead to the displacement of other benthic organisms (SANTOS *et al.* 2012). Since its invasion of South America, *L. fortunei* has threatened the survival and modified the natural occurrence and abundance of several native macroinvertebrate species (DARRIGRAN *et al.* 1998, DARRIGRAN 2002, SANTOS *et al.* 2012), including the Anomura crab *Aegla platensis* Schmitt, 1942 in the São Gonçalo Channel (LOPES *et al.* 2009).

The predation strength of fish upon the golden mussel ranges from negligible to efficient, when the total control of the mussel's population growth is achieved (STEWART *et al.* 1998, BARTSCH *et al.* 2005). In some cases, mussel predators showed increased productivity and growth as a result of the new food supply (PODDUBNYI 1966, BOLTOVSKOY *et al.* 2006, KARATAYEV *et al.* 2007). Ancillary data on Argentine freshwater fish yields support to the above conclusions, suggesting that *L. fortunei* may have had a positive effect on fish biomass in Neotropical systems (BOLTOVSKOY *et al.* 2006).

Catfish of the genus *Pimelodus* have omnivorous feeding habits (GARCIA *et al.* 2006, 2007 which are characterized by a generalist feeding behavior and the opportunistical exploitation of eventual peaks in prey abundance (BONETTO *et al.* 1963, MONTALTO *et al.* 1999, BRAGA 2000, GARCÌA & PROTOGINO 2005). The use of *L. fortunei* as a food source by *Pimelodus* spp. (Paraná River and Patos Lagoon basin) had been previously reported

(Montalto *et al.* 1999, Cataldo *et al.* 2002, Darrigran & Damborenea 2011). For this reason, we hypothesized that the catfish *Pimelodus pintado* (Azpelicueta, Lundberg & Loureiro, 2008) at the Mirim Lagoon would incorporate the Asian freshwater golden mussel in its diet.

The objective of this study was to describe the chronological incorporation of *L. fortunei* into the diet of *P. pintado*, from the time the prey arrived in the São Gonçalo Channel, at the Mirim Lagoon basin. Additionally, we show that *P. pintado* selects mussel prey within a certain size range, from the wide range of sizes available in the environment.

MATERIAL AND METHODS

The Mirim Lagoon basin is located between 31 and 34°S and 52 and 54°W in the eastern part of the South American central plains (Fig. 1). The basin area covers 62,250 km², with 29,250 km² in southern Brazil and the remaining 33,000 km² in eastern Uruguay. The main geographical feature of this basin is the Mirim lagoon itself, with an average area of 3.749 km² (Bracco *et al.* 2005). During periods of intense rainfall, water from the Mirim Lagoon and its adjacent wetland system drain through the natural São Gonçalo Channel (75 km long, 200 to 500 m wide, 6 m maximum depth) into the Patos Lagoon, which ultimately connects with the Atlantic Ocean through the Rio Grande channel. The São Gonçalo Channel Dam divides the São Gonçalo Channel into a freshwater environment to the south and an estuarine environment to the North (Burns *et al.* 2006a, Moura *et al.* 2012).

Bottom fauna samples were obtained from the limnetic region of the São Gonçalo Channel between 31°48'S, 52°23'W and 32°7'S, 52°35'W (Fig. 1). Bottom trawl was carried 3 to 6 m deep, from January to November 2005, and from October to November 2008. Every season, four samplings at three different sampling areas (Dam, Tigre and Piratini), totalizing twelve samplings per season, were carried out using a fishermen's wooden boat (10.9 m long with a 60-Hp engine). Five-minute tows (approximately 400 m) were performed using a 10.5 m (head rope) shrimp trawl (1.3 cm bar mesh wings and body with a 0.5 cm bar mesh cod end liner) and a pair of weighted outer doors (Lopes *et al.* 2009, Lopes & Vieira 2012).

Representatives of *P. pintado* collected in each tow were stored in separate plastic bags and fixed in 10% formaldehyde. Voucher specimens were deposited at "Coleção Ictiológica da FURG" number 6,056 (25) 6061 (6). At the laboratory, the total length (TL) of each individual was measured to the nearest millimeter. Stomachs were extracted by cutting out the esophagus and pylorus and fixed in 70% alcohol. Prey items were identified to the lowest taxonomic level, counted and grouped into categories. We determined Frequency of Occurrence (FO%), Percent Area (PA%) and Index of Relative Importance (IRI) of the *P. pintado* diet for samples collected during the austral summer (January and February), autumn (April), winter (July and

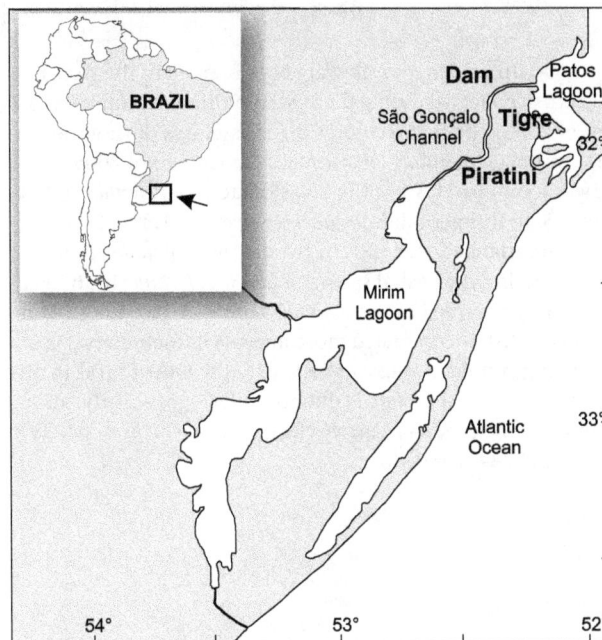

Figure.1 Mirim Lagoon (drainage basin, 62.250 km²) and the São Gonçalo channel that connects it with the Patos Lagoon. The limnetic region is localized southern the dam. The sampling areas are show (Dam, Tigre and Piratini).

August) and the austral spring (September, October and November) of 2005 and the austral spring (October and November) of 2008. IRI was determined as FO% x (PN%+PA%), where PN% are number percent (Hyslop 1980) of each food item. We compared de relative importance of the golden museel in the stomachs of catfish sampled in the spring of 2005 with those sampled in the spring of 2008 (three years after *L. fortunei* occurrence was recorded in the system).

Using an electronic caliper with accuracy of 0.01 mm, the maximum shell length of *L. fortunei* prey was measured using the individuals present in the stomachs of *P. pintado*, as well as using the specimens collected in the environment (bottom trawl, during 2008). The maximum length was considered as the distance from the anterior end, situated just above and ahead of the umbones, until the rear end of the shell (Mansur *et al.* 1987).

After logarithmic transformation ($\log_{10}+1$) the differences between the sizes of golden mussels found in the digestive tract of different sizes classes of *P. pintado* were tested using one-way analysis of variance (size class 10-15 cm TL, N = 6; size class 15-20 cm TL, N = 52; size class 20-25 cm TL, N = 51; size class 25-30 cm TL, N = 38) (Zar 1999). The differences between the sizes of golden mussels found in the digestive tract of all *P. pintado* and those collected in the environment were tested using the nonparametric Kolmogorov-Smirnov test (Sokal & Rohlf 1995).

RESULTS

Asian golden mussels were not observed in the stomachs of catfish collected during the austral summer, autumn or winter of 2005. The first record of the mussel was during the austral spring (September, October and November) of 2005, when it was found in 21.8% of the 180 stomachs of *P. pintado* (10-30 cm TL). In this period, *L. fortunei* represented 1.8% of the Relative Importance Index (IRI) in the diet to *P. pintado* (Table I).

During the austral spring of 2005, *L. fortunei* individuals were only observed in the stomachs of fish larger than 14 cm, and large fish incorporated more mussels in their diets (Fig. 2). The length golden mussel shells (0.3-1.4 mm) found in the digestive tracts of *P. pintado* did not differ significantly among the different predator length class analyzed (Fig. 3, ANOVA, $F_{3,143} = 0.9744$, p = 0.407).

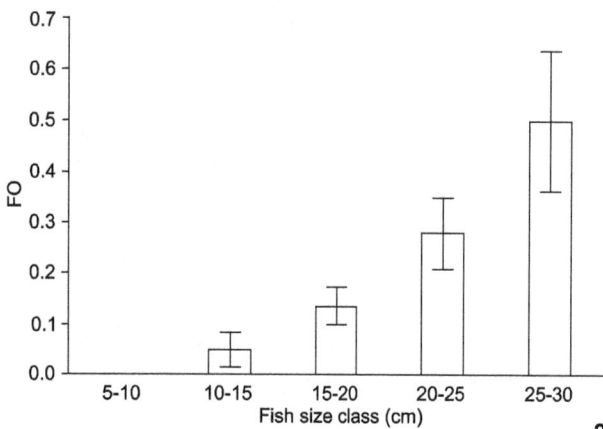

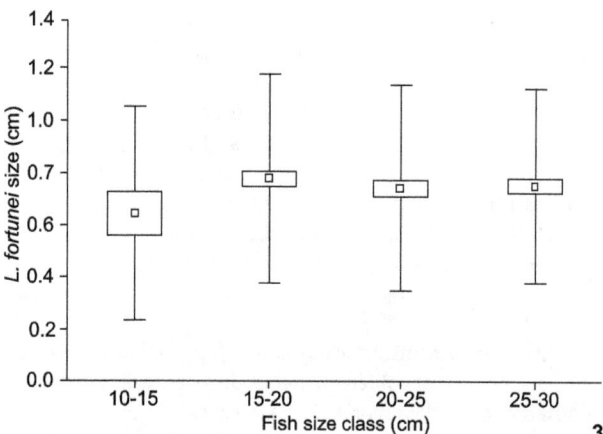

In the austral spring of 2008 (Table I), *L. fortunei* was found in 60.8% of the 30 stomachs of *P. pintado* analyzed and was the most important food item in the diet of the cat fish (IRI = 31.4%). Specimens of *L. fortunei* (N = 7789) collected by the shrimp trawl in 2008 were measured and compared with the 51specimens found in the stomachs of *P. pintado* (20.7-30.5 cm TL) sampled in the same area. The length range of fortune of the mussel in the environment (n = 7789) was 0.7 to 3.2 cm, with a mode of 1.3 cm TL, and the shell length of mussels found in the in the guts (n = 147) of *P. pintado* ranged from 0.3 to 1.4 cm (Fig. 4) and the nonparametric Kolmogorov-Smirnov test revealed a significant difference (Fig. 5, D = 0.82491, p < 0.001) between them.

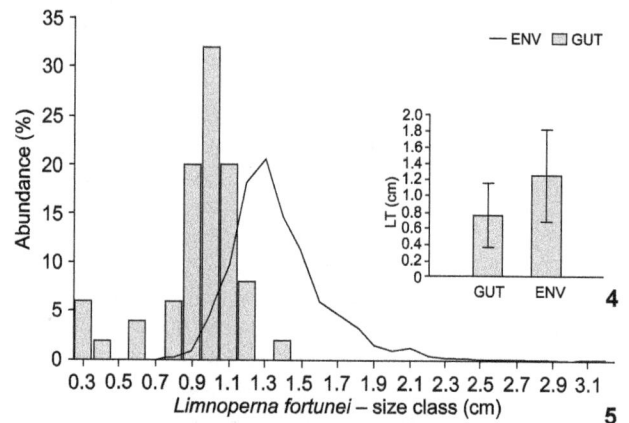

Figures 4-5. (4) Shell size distribution of *Limnoperna fortunei* collected in the environment (ENV), and in the guts (GUT) of *Pimelodus pintado* at the spring of 2008. (5) Mean size and 95% confidence interval of *Limnoperna fortunei*.

DISCUSSION

Boltovskoy & Cataldo (1999) showed that during the first year of the introduction of *L. fortunei*, mussels reached up to 2.0 cm in length, and by the end of their second year, some reached 3.0 cm. The maximum length of mussels in this study was 3.2 cm. Additionally, Burns *et al.* (2006b) recorded the species in the São Gonçalo Channel for the first time in 2005. Both facts suggest that the species was already fully established in São Gonçalo Channel in 2008 (Lopes & Vieira 2012).

The mean shell length of *L. fortunei* found in the digestive tracts of *P. pintado* were similar for the different predator length classes analyzed and reached a maximum size of 1.4 cm (Fig. 3). Montalto *et al.* (1999) and Lopes & Vieira (2012) showed that Asian golden mussels are predated in different development stages by several fish species in the Neotropical region, but small mussels (< 1.5 cm) were more frequent and abundant in fish guts. Cantanhêde *et al.* (2008) analyzed the diet of a

Figures 2-3. (2) Frequency of occurrence of *Limnoperna fortunei* in the digestive tracts of *P. pintado* at different fish size class analyzed in the spring of 2005. The line represent the standard deviation and the bars represent mean. (3) Box plot of *Limnoperna fortunei* shell size distribution by *Pimelodus pintado* size class in the spring of 2005.

Table I. Seasonal variation of the Frequency of Occurrence (FO%), Percentage by Number (PN%), Percentage by Area (PA%), and Relative Importance Index (IRI%) of items found in the diet of *Pimelodus pintado* during 2005 and 2008.

| Items | 2006 | | | | | | | 2008 | | | |
| | Summer | Autumn | Winter | Spring | | | | Spring | | | |
	FO%	FO%	FO%	FO%	PN%	PA%	IR%	FO%	PN%	PA%	IR%
Small crustaceans	13.51	47.89	65.48	74.15	63.77	40.07	67.76	17.39	7.04	3.64	1.97
Crabs	20.27	2.82	1.19	3.40	0.21	2.34	0.08	21.74	2.18	19.75	5.06
Insects	35.14	60.56	64.29	57.14	15.33	19.27	17.40	65.22	16.26	8.11	16.85
L. fortunei	–	–	–	**21.77**	5.16	4.49	1.85	**60.87**	18.83	34.77	31.37
Corbicula spp.	–	4.23	11.90	2.72	0.25	0.12	0.01	13.04	1.94	1.02	0.41
Heliobia spp.	27.03	12.68	7.14	16.33	3.16	0.87	0.58	47.83	51.70	9.20	30.88
Fish	70.27	49.30	16.67	21.29	4.28	7.42	2.17	4.35	0.24	2.68	0.13
Plant	2.70	35.21	15.48	39.46	3.23	22.98	9.10	52.17	3.88	17.52	11.84
Other	9.48	4.23	2.38	19.73	1.93	2.07	0.69	26.09	1.94	2.98	1.36
Mineral	–	9.86	4.76	13.61	2.70	0.37	0.37	8.70	0.97	0.32	0.12

predator fish larger than *P. pintado* (*Pterodoras granulosus*; 17-55 cm TL) and found similar results, whereas the size range of *L. fortunei* individuals ranged from 0.8 to 1.7 cm in mean, although the mussel is well known to reach more than 3 cm (BOLTOVSKOY & CATALDO 1999, MAROÑAS *et al.* 2003, MANSUR *et al.* 2008).

Young and adult *L. fortunei* individuals have a considerable ability to move to new locations with some taxis, suggesting that this represents an ability to avoid predation by hiding (LOPES & VIEIRA 2012). The distance that mussels are able to move decreases with increasing shell length, and they tend to aggregate after reaching 1.5 cm TL (URYU *et al.* 1996). The observations of the present study suggest that more mobile individuals of *L. fortunei* (smaller than <1.4 cm) are able to crawl over the bottom and are probably more readily available in the São Gonçalo Channel to fish such as *P. pintado* than larger and more aggregated mussels that get attached to the hard substrate. Fish with a generalist feeding behavior are good samplers of prey diversity and can easily detect new sources of food, and are flexible enough to exploit eventual peaks in prey abundance (GLOVA & SAGAR 1989, MENDOZA-CARRANZA & VIEIRA 2007). Then, independently from the size of individuals, the catfish under study consumes only golden mussels that are smaller than 1.4 cm TL, which implies that the predator prefers smaller prey because they are more readily available.

It is important to note that *L. fortunei* was first reported in the limnetic zone of the São Gonçalo Channel in January 2005 (BURNS *et al.* 2006b). Since the diet of *P. pintado* is influenced by the availability of food in the environment, the data presented here suggest that the trend observed in this study, i.e., the absence of *L. fortunei* in the stomachs of *P. pintado* from the austral summer to the winter and the presence of it in the spring 2005, is a good indicator that *L. fortunei* started to be abundant in the limnetic zone of the São Gonçalo Channel

during the austral spring of 2005. At the present time, *L. fortunei* is fully established in the São Gonçalo Channel (COLLING *et al.* 2012, LOPES & VIEIRA 2012).

Little empirical information is available to explain the success of invaders (MANSUR *et al.* 2012). The arrival and spread of *L. fortunei* at the Mirim Lagoon will probably bring rapid changes in the benthic community as well as the displacement of other mollusk species, as described by DARRIGAN (2002) and DARRIGRAN & DAMBORENEA (2005) for other South American fresh water habitats. Currently, unpublished data suggest that *L. fortunei* is much more abundant in the dense vegetated limnetic part of the São Gonçalo Channel (61 km long and 17 m wide) than in the wide open Mirim Lagoon itself (3,749 km² area) which has few hard subtracts for mussel fixation (LOPES *et al.* 2009).

With the invasion of *L. fortunei* in the Neotropical region, dietary changes have been noted in omnivorous fishes, which have switched from a low quality, predominantly plant-based diet, to an energetically rich diet dominated by invasive mollusks (MONTALTO *et al.* 1999, FERRIZ *et al.* 2000, BOLTOVSKOY *et al.* 2006, LOPES & VIEIRA 2012). With the invasion of *L. fortunei*, part of the organic matter in the turbid São Gonçalo Channel will be filtered and modified into a form available to organisms that cannot feed on small particles, like fishes.

LOPES & VIEIRA (2012) shows that 12 of 19 predators in the Mirin/São Gonçalo Channel feed on *L. fortunei*. Regardless of the size and foraging behavior of the predator, individuals smaller than 14 mm on average are preyed upon. The incidence of individuals of *L. fortunei* smaller than 14 mm in the diet of detritivorous fishes like *Rineloricaria microlepdogaster* and *R. strigilata*, which do are not adapted to predate on mollusks, confirms the hypothesis that individuals of golden mussel up to 14 mm TL are more vagile than larger individuals (URYU *et al.* 1996), and frequently move on the bottom of the São

Gonçalo Channel, being more available to fish predation than individuals larger than 14 mm TL, which tend to be clustered or hindered in crevices of the substrate (Lopes & Vieira 2012).

ACKNOWLEDGEMENTS

This work was supported by CNPq (Conselho Nacional de Desenvolvimento Científico e Tecnológico) and PELD Program (Pesquisas Ecológicas de Longa Duração). M.N.L was a former postgraduate student of Biologia de Ambientes Aquaticos Continentais, FURG.

LITERATURE CITED

Bartsch, M.R.; L.A. Bartsch & S. Gutreuter. 2005. Strong effects of predation by fishes on an invasive macroinvertebrate in a large floodplain river. **Journal North American Benthological Society** 24: 168-77.

Boltovskoy, D. & D.H. Cataldo. 1999. Population dynamics of *Limnoperna fortunei*, an invasive fouling mollusc, in the lower Parana River (Argentina). **Biofouling** 14: 255-263.

Boltovskoy, D.; N. Correa; D. Cataldo & F. Sylvester. 2006. Dispersion and ecological impact of the invasive freshwater bivalve *Limnoperna fortunei* in the Río de la Plata watershed and beyond. **Biological Invasions** 8: 947-963.

Bonetto, A.A.; C. Pignalberi & E. Cordiviola. 1963. Ecologia alimentaria del amarillo y moncholo, *Pimelodus clarias* (Bloch) y *Pimelodus albicans* (Valenciennes) (Pisces, Pimelodidae). **Physis** 24: 87-94.

Bracco, R.; L. Puerto; H. Inda & C. Castiñeira. 2005. Mid-late Holocene cultural and environmental dynamics in Eastern Uruguay. **Quaternary International** 132: 37-45.

Braga, F.M. de S. 2000. Biologia e Pesca de *Pimelodus maculatus* (Siluriformes, Pimelodidae) no reservatório de Volta Grande, Rio Grande (MG-SP). **Acta Limnologica Brasiliensia** 12: 1-14.

Burns, M.D.; A.M. Garcia; J.P. Vieira; M.A. Bemvenuti; D.M.L. Motta Marques & V. Condini. 2006a. Evidence of fragmentation affecting fish movement between Patos and Mirim coastal lagoons in southern Brazil. **Neotropical Icthiology** 4: 69-72.

Burns, M.D.; R.M. Geraldi; A.M. Garcia; C.E. Bemvenuti; R.R. Capitoli & J.P. Vieira. 2006b. Primeiro registro de ocorrência do mexilhão dourado *Limnoperna fortunei* na Bacia de drenagem da Lagoa Mirim, RS, Brasil. **Biociências** 14: 83-83.

Cantanhêde, G.; N.S. Hahn; E.A. Gubiani & R. Fugi. 2008. Invasive molluscs in the diet of *Pterodoras granulosus* (Valenciennes, 1821) (Pisces, Doradidae) in the Upper Paraná River floodplain, Brazil. **Ecology of Freshwater Fish** 17: 47-53.

Cataldo, D.; D.D. Boltovskoy; V. Marini & N. Correa. 2002. Limitantes de *Limnoperna fortunei* en la cuenca del Plata: la predación por peces. *In*: **Tercera jornada sobre conservación de la fauna íctica en el río Uruguay.** Paysandu, Uruguay, La Comisión Administradora de Río Uruguay.

Colling, L.A.; R.M. Pinotti & C.E. Bemvenuti. 2012. *Limnoperna fortunei* na Bacia da Lagoa dos Patos e Lagoa Mirim, p. 187-191. *In*: C.P. Santos; D. Pereira; I.C.P. Paz; L.M. Zurita; M.C.D. Mansur; M.T. Raya Rodriguez; M.V. Nerhke & P.A. Bergonci (Eds). **Moluscos límnicos invasores no Brasil: biologia, prevenção e controle.** Porto Alegre, Redes Editora, 412p.

Darrigran, G. 2002. Potencial impact of filter-feeding invaders on temperate inland freshwater environments. **Biological Invasions** 4: 145-156.

Darrigran, G. & E. de Drago. 2000. Invasion of *Limnoperna fortunei* (Dunker, 1857) (Bivalvia: Mytilidae) in South America. **Revista Nautilus** 114: 69-73.

Darrigran, G. & C. Damboronea. 2005. A bioinvasion history in South America. *Limnoperna fortunei* (Dunker, 1857), the golden mussel. **American Malacological Bulletin** 20: 105-112.

Darrigran, G. & C. Damboronea. 2011. Ecosystem engineering impacts of *Limnoperna fortunei* in South America. **Zoological Science** 28: 1-7.

Darrigran, G.; S.M. Martin; B. Gullo & L. Armendariz. 1998. Macroinvertebrate associated to *Limnoperna fortunei* (Dunker, 1857) (Bivalvia, Mytilidae). Río de La Plata, Argentina. **Hydrobiologia** 367: 223-230.

Ferriz, R.A.; C.A. Villar; D. Colautti & C. Bonetto. 2000. Alimentación de *Pterodoras granulosus* (Valenciennes) (Pisces, Doradidae) en la baja cuenca del Plata. **Revista del Museo Argentino de Ciencias Naturales** 2: 151-156.

Garcia, A.M.; D.J. Hoeinghaus; J.P. Vieira; K.O. Winemiller; D.M.L.M. Marques & M.A. Bemvenuti. 2006. Preliminary examination of food web structure of Nicola Lake (Taim Hydrological System, south Brazil) using dual C and N stable isotope analyses. **Neotropical Ichthyology** 4: 279-284.

Garcia, A.M.; D.J. Hoeinghaus; J.P. Vieira & K.O. Winemiller. 2007. Isotopic variation of fishes in freshwater and estuarine zones of a large subtropical coastal lagoon. **Estuarine Coastal and Shelf Science** 73: 399-408.

García, M.L.A. & L.C. Protogino. 2005. Invasive freshwater molluscs are consumed by native fishes in South America. **Journal of Applied Ichthyology** 21: 34-38.

Glova, G.J. & P.M. Sagar. 1989. Prey selection by *Galaxias vulgaris* in the Hawkins River, New Zealand. **New Zealand Journal of Marine and Freshwater Research** 23: 153-161.

Hyslop, E.J. 1980. Stomach contens analysis- a review of methods and their applications. **Journal Fish Biology** 17: 411-429.

Karatayev, A.Y.; D.K. Padilla; D. Minchim; D. Boltovskoy & L.E. Burlakova. 2007. Changes in global economies and trade: the potential spread of exotic freshwater bivalves. **Biological Invasions** 9: 161-180.

Langone, J.A. 2005. Notas sobre el mejillón dorado *Limnoperna fortunei* (Dunker, 1857) (bivalvia, mytilidae) en Uruguay. **Publicación extra del Museo Nacional de Historia Natural y Antropologia** 1: 17.

Lopes, M. & J.P. Vieira. 2012. Predadores potenciais para o controle do mexilhão-dourado, p. 357-363. *In*: C.P. Santos; D.

PEREIRA; I.C.P. PAZ; L.M. ZURITA; M.C.D. MANSUR; M.T. RAYA RODRIGUEZ; M.V NERHKE & P.A. BERGONCI (Eds). **Moluscos límnicos invasores no Brasil: biologia, prevenção e controle.** Porto Alegre, Redes Editora, 412p.

LOPES, M.N.; J.P. VIEIRA & M.D.M. BURNS. 2009. Biofouling of the golden mussel *Limnoperna fortunei* (Dunker, 1857) over the Anomura crab *Aegla platensis* Schmitt, 1942. **Pan-American Journal of Aquatic Sciences 4:** 222-225.

MANSUR, M.C.D.; C. SCHULZ & L.M.M.P. GARCES. 1987. Moluscos bivalves de água doce: Identificação dos gêneros do sul e leste do Brasil. **Acta Biológica Leopoldencia 9:** 181-202.

MANSUR, M.C.D.; L.M.Z. RICHINITTI & C.P. SANTOS. 1999. *Limnoperna fortunei* (Dunker, 1857) molusco bivalve invasor na bacia do Guaíba, Rio Grande do Sul, Brasil. **Biociências 7:** 147-149.

MANSUR, M.C.D.; C.P. SANTOS; G. DARRIGRAN; I. HEYDRICHT; C.T. CALLIL & F.R. CARDOSO. 2003. Primeiros dados quail-quantitativos do mexilhão dourado, *Limnoperna fortunei* (Dunker), no Delta do Jacuí, no lago Guaíba e na Laguna dos Patos, Rio Grande do Sul, Brasil e alguns aspectos de sua invasão no novo ambiente. **Revista Brasileira de Zoologia 20:** 75-84.

MANSUR, M.C.D.; H. FIGUEIRÓ; C.P. SANTOS; L. GLOCK; P.E.A. BERGONCI & D. PEREIRA. 2008. Variação espacial do comprimento e do peso úmido total de *Limnoperna fortunei* (Dunker, 1857) no delta do rio Jacuí e lago Guaíba (RS, Brasil). **Biotemas 21:** 49-54.

MANSUR, M.C.D.; C.P. SANTOS; D. PEREIRA; I.C.P. PAZ; M.L.L. ZURITA; M.T.R. RODRIGUEZ; M.V. NEHRKE & P.E.A. BERGONCI. 2012. **Moluscos Límnicos Invasores no Brasil: biologia, prevenção, controle.** Porto Alegre, Redes Editora, 412p.

MAROÑAS, M.E.; G.A. DARRIGRAN; E.D. SENDRA & G. BRECKON. 2003. Shell growth of the golden mussel, *Limnoperna fortunei* (Dunker, 1857) (Mytilidae), in the Río de la Plata, Argentina. **Hydrobiologia 495:** 41-45.

MENDOZA-CARRANZA, M. & J.P. VIEIRA. 2007. Whitemouth croaker *Micropogonias furnieri* (Desmarest, 1823) feeding strategies across four southern Brazilian estuaries. **Aquatic Ecology 42** (1): 83-93.

MONTALTO, L.; O.B. OLIVEROS; I. E. DE DRAGO & L.D. DEMONTE. 1999. Peces del rio Parana Medio predadores de una especie invasora: *Limnoperna fortunei* (Bivalvia, Mytilidae). **Revista de la Facultad de Bioquimica y Ciencias Biológicas de la Universidad Nacional del Litoral 3:** 85-101.

MOURA, P.M.; J.P. VIEIRA; A.M. GARCIA. 2012. Fish abundance and species richness across an estuarine freshwater ecosystem in the Neotropics. **Hydrobiologia 696:** 107-122.

OLIVEIRA, M.D.; A.M. TAKEDA; L.F. BARROS; S.D. BARBOSA; E.K. REZENDE. 2006. Invasion by *Limnoperna fortunei* (Dunker, 1857) (Bivalvia, Mytilidae) of the Pantanal wetland, Brazil. **Biological Invasions 8** (1): 97-104.

PASTORINO, G.; G. DARRIGRAN; S. MARTIN & L. LUNASCHI. 1993. *Limnoperna fortunei* (Dunker, 1857) (Mytilidae), nuevo bivalvo invasor em águas Del Rio de la Plata. **Neotropica 39:** 101-102.

PODDUBNY, A.G. 1966. Adaptive response of *Rutilus rutilus* to variable environmental conditions. **Trudy Instiyuta Biologii Vnutrennykh Vod Akademii Nauk 10:** 131-138.

POWER, M.E. 1992. Habitat heterogeneity and the functional significance of fish in River Food Webs. **Ecology 73:** 1675-1688.

RUETZ, C.R.; R.M. NEWMAN & B. VONDRACEK. 2002. Top-down control in a detritus-based food web: fish, shredders and leaf breakdown. **Oecologia 132:** 307-315.

SANTOS, S.B.; S.C. THIENGO; M.A. FERNANDEZ; I.C. MIYAHIRA; I.C.B. GONÇALVES; R.F. XIMENES; M.C.D. MANSUR & D. PEREIRA. 2012. Espécies de moluscos límnicos invasores no Brasil, p. 25-49. *In:* C.P. SANTOS; D. PEREIRA; I. C. P. PAZ; L.M. ZURITA; M.C.D. MANSUR; M.T. RAYA RODRIGUEZ; M.V NERHKE & P.A. BERGONCI (Ed.). **Moluscos límnicos invasores no Brasil: biologia, prevenção e controle.** Porto Alegre, Redes Editora, 412p.

SOKAL, R.R. & F.J. ROHLF. 1995. Biometry: **The Principles and Practice of Statistics in Biological Research.** New York, W.H. Freeman, 937p.

STEWART, T.W.; J.G. MINER & R.L. LOWE. 1998. A experimental analysis of crayfish (*Orconectes rusticus*) effects on a Dreissena dominated benthic macroinvertebrate community in western Lake Erie. **Canadian Journal of Fisheries and Aquatic Sciences 55:** 1043-1050.

SYLVESTER, F.; J. DORADO; D. BOLTOVSKOY; A. JUÀREZ & D. CATALDO. 2005. Filtration rates of the invasive pest bivalve *Limnoperna fortunei* as a function of size and temperature. **Hydrobiologia 534:** 71-80.

THORP, J.H. & A.F. CASPER. 2003. Importance of biotic interactions in large rivers: an experiment with planktivorous fish, dreissenid mussels and zooplankton in the St. Lawrence River. **River Research and Applications 19:** 265-279.

URYU, Y.; K. IWASAKY & M. HINQUE. 1996. Laboratory experiments on behaviour and movement of a freshwater mussel, *Limnoperna fortunei* (Dunker). **Journal of a Molluscan Studies 62:** 327-341.

ZAR, J.H. 1999. **Biostatistical Analysis.** New Jersey, Prentice Hall, 662p.

9

Morphology of the shell of *Happiella* cf. *insularis* (Gastropoda: Heterobranchia: Systrophiidae) from three forest areas on Ilha Grande, Southeast Brazil

Amilcar Brum Barbosa[1] & Sonia Barbosa dos Santos[1,2]

[1] *Laboratório de Malacologia Límnica e Terrestre, Departamento de Zoologia, Instituto de Biologia Roberto Alcantara Gomes, Universidade do Estado do Rio de Janeiro. Rua São Francisco Xavier 524, PHLC sala 525-2, 20550-900 Rio de Janeiro, RJ, Brazil.*
[2] *Corresponding author. E-mail: milkabrum@yahoo.com.br*

ABSTRACT. We conducted a study on shell morphology variation among three populations of *Happiella* cf. *insularis* (Boëttger, 1889) inhabiting different areas (Jararaca, Caxadaço, and Parnaioca trails) at Vila Dois Rios, Ilha Grande, Angra dos Reis, state of Rio de Janeiro, Brazil. Linear and angular measurements, shell indices representing shell shape, and whorl counts were obtained from images drawn using a stereomicroscope coupled with a camera lucida. The statistical analysis based on ANOVA (followed by Bonferroni's test), Pearson's correlation matrix, and discriminant analysis enabled discrimination among the populations studied. The variable that most contributed to discriminate among groups was shell height. Mean shell height was greatest for specimens collected from Jararaca, probably reflecting the better conservation status of that area. Good conservation is associated with enhanced shell growth. Mean measurements were smallest for specimens from Parnaioca, the most disturbed area surveyed. Mean aperture height was smallest for specimens from Parnaioca, which may represent a strategy to prevent excessive water loss. Discriminant analysis revealed that the snails from Jararaca differ the most from snails collected in the two other areas, reflecting the different conservation status of these areas: shells reach larger sizes in the localities where the humidity is higher. The similarities in shell morphology were greater between areas that are more similar environmentally (Caxadaço and Parnaioca), suggesting that conchological differences may correspond to adaptations to the environment.

KEY WORDS. Conchology; discriminant analysis; ecology; morphometry; threatened biome.

Land snails are exceptionally diverse in morphology, for instance they display great polymorphism in shell color and variations in shell dimensions. For this reason, they are a good subject for evolutionary biology studies (CLARKE *et al.* 1978). Differences in size, morphology and growth rates are associated with ecological conditions, natural selection, and phylogenetic history (VERMEIJ 1971, CLARKE *et al.* 1978, EMBERTON 1994, 1995b, COOK 1997, PARMAKELIS *et al.* 2003, TESHIMA *et al.* 2003). According to GOULD (1984), the low mobility of land snails influences character variability. The literature shows that habitat alterations, which result in fragmentation, are an important factor affecting shell morphological differentiation (COOK 1997, GOODFRIEND 1986, EMBERTON 1982, 1994), which can be accelerated in degraded environments (CHIBA 2004, CHIBA & DAVISON 2007).

Ilha Grande, a continental island in the southern portion of the state of Rio de Janeiro, harbors large, continuous and conserved fragments of Atlantic Forest (ROCHA *et al.* 2006), which is among the most threatened biomes in the world (MYERS *et al.* 2000). Over 50% of Ilha Grande is covered by ombrophilous dense forest, now at different levels of regeneration (ALHO *et al.* 2002, OLIVEIRA 2002, ALVES *et al.* 2005, CALLADO *et al.* 2009) from disturbances caused by a range of human activities over the past five decades, being now a natural laboratory to study shell morphological differentiation induced by in environment conditions.

The focus of this study was to investigate variations in the morphology of the shell of *Happiella* cf. *insularis* in three different environments (Table I). This species was described by BOËTTGER (1889) based on a single shell collected from the type locality, Ilha das Flores, São Gonçalo city, Rio de Janeiro, where additional specimens have not been found (SANTOS *et al.* 2010). BOËTTGER's (1889) description, which was not accompanied by illustrations, highlighted the following diagnostic features: maximum diameter with 5.25 mm, shell height 2 mm, large umbilicus, one-fourth the size of the shell base; shell pebble-shaped, thin, white, polished, spire apex slightly prominent, with ½ whorls, slightly convex; borders distinct, mildly striated, last border over the third, approximately as wide as shell, less arched at top than bottom, angled below central region;

suture deeply impressed. Aperture elliptical-lunular, with small slit, aperture height with 2 mm, aperture width with 2.25 mm, simple peristomatic edge, with curved, spherical, sub-angular syphunculus [sic] protruding to right side of base.

THIELE (1927), in addition to the type locality of *H.* cf. *insularis*, also listed it in Piracicaba (state of São Paulo), Blumenau (state of Santa Catarina) and Porto Alegre (state of Rio Grande do Sul); MORRETES (1949) also listed it only in Ilha das Flores and SIMONE (2007) to Xanxerê and São Carlos (state of Santa Catarina).

In the present study, we analyzed the shell morphology of three populations of *H.* cf. *insularis* subjected to different environmental conditions, with the goal to assess variability in shell morphology, as detailed morphology and range of variation can prove useful for refining species diagnoses.

MATERIAL AND METHODS

The specimens used in this study were collected from three areas, known as the Jararaca, Caxadaço, and Parnaioca trails, located in Vila Dois Rios, on the ocean side of Ilha Grande, Municipality of Angra dos Reis, southern region of the state of Rio de Janeiro (23°04'25" to 23°13'10"S, 44°05'35" to 44°22'50"W). In each collecting site (Fig. 1), a distinct level of forest regeneration (VERA-Y-CONDE & ROCHA 2006) can be found, making them suitable for investigations on the influence of environmental factors on shell morphology.

Table I contains a summary of the environmental parameters measured at the three areas studied.

We selected intact shells from 102 adults, grown to approximately three whorls and proportionally similar to each other. Thirty-three shells from the Jararaca Trail were selected, in addition to 34 and 35 shells from the Caxadaço and Parnaioca trails, respectively.

Material examined. *Happiella* cf. *insularis*. BRAZIL, *Rio de Janeiro*: Angra dos Reis, Ilha Grande, Vila Dois Rios, Trilha da Jararaca, 14.VI.1998, S.B. Santos *leg.* (Col. Mol. UERJ 942-1); 27.IX.1996, S.B. Santos *leg.* (Col. Mol. UERJ 977); 11.I.1996, V.C. Queiroz *leg.* (Col. Mol. UERJ 980-4 and 5); 12.I.1996, S.B. Santos *leg.* (Col. Mol. UERJ 990-1 and 2); 21.III.1997, S.B. Santos *leg.* (Col. Mol. UERJ 1132); 23.III.1997, S.B. Santos *leg.* (Col. Mol. UERJ 1133-1 and 2); 20.IX.1997, S.B. Santos *leg.* (Col. Mol.

UERJ 1155); 30.XI.1997, D.P. Monteiro *leg.* (Col. Mol. UERJ 1168-1 and 2); *ditto*, 26.VI.1999, S.B. Santos *leg.* (Col. Mol. UERJ 1241-1 and 2); 21.III.1997, S.B. Santos *leg.* (Col. Mol. UERJ 1252-2); 14.I.1998, D.P. Monteiro *leg.* (Col. Mol. UERJ 1617-2); 17.II.1998, A.S. Alencar *leg.* (Col. Mol. UERJ 1618-2); 17.II.1998, D.P. Monteiro *leg.* (Col. Mol. UERJ 1646); 15.I.1998, S.B. Santos *leg.* (Col. Mol. UERJ 1647-2); 17.II.1998, M.A. Fernandez *leg.* (Col. Mol. UERJ 1650); 14.I.1998, A.S. Alencar *leg.* (Col. Mol. UERJ 1651); 14.I.1998, D.P. Monteiro *leg.* (Col. Mol. UERJ 1653-2); 17.I.1998, S.B. Santos *leg.* (Col. Mol. UERJ 1656-2 and 3); 17.II.1998, D.P. Monteiro *leg.* (Col. Mol. UERJ 1658-2, 3, 4 and 6); 17.II.1998, S.B. Santos *leg.* (Col. Mol. UERJ 1659-1, 2, and 3). Vila Dois Rios, Trilha do Caxadaço, 19.X.1995, V.C. Queiroz *leg.* (Col. Mol. UERJ 999-2, 3, 4, 5, and 6); 30.V.1997, S.B. Santos *leg.* (Col. Mol. UERJ 1064-7 and 3); 28.XI.1997, D.P. Monteiro *leg.* (Col. Mol. UERJ 1110-1, 2, 3, 4, and 5); 15.VIII.1996, S.B. Santos *leg.* (Col. Mol. UERJ 1114); 19.X.1995, V.C. Queiroz *leg.* (Col. Mol. UERJ 1144-1, 2, 3, and 4); 08.VIII.1999, M. Sttorti *leg.* (Col. Mol. UERJ 1310); 21.X.2000, D.P. Monteiro *leg.* (Col. Mol. UERJ 2061-1, 2, and 3); 15.III.2001, S.B. Santos *leg.* (Col. Mol. UERJ 2156-2, 3, 4, 5, and 6); 28.X.2001, C.C. Siqueira *leg.* (Col. Mol. UERJ 2225-2, 3, 4, 5, and 6); 2.VIII.2005, A.B Barbosa, Lacerda, L.E.M., T.A. Viana *leg.* (Col. Mol. UERJ 7445-1, 2, and 3). Vila Dois Rios, Trilha da Parnaioca), 28.V.1997, N. Salgado

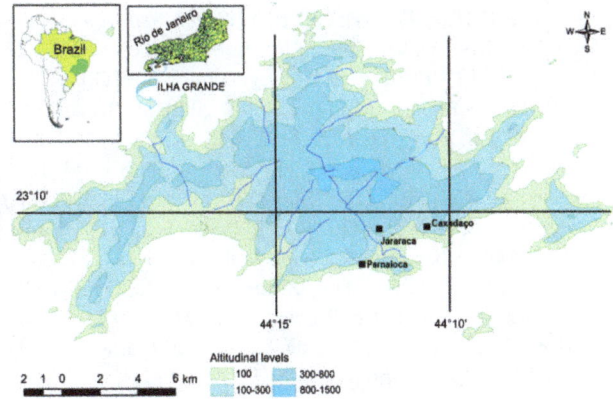

Figure 1. Location of Ilha Grande, in Angra dos Reis Municipality, state of Rio de Janeiro, Brazil, showing the Jararaca, Parnaioca, and Caxadaço trails. Map: Luiz E.M. Lacerda.

Table I. Summary of local environmental parameters from three areas (Jararaca, Caxadaço and Parnaioca trails) at Ilha Grande.

Area	Mean ambient temperature (a) (°C)	Mean ground temperature (a) (°C)	Mean leaf litter depth layer (a) (cm)	Mean relative air humidity (a) (%)	Canopy height (m)	Canopy closure	Elevation	Degree of impact	Time of regeneration
Jararaca Trail	22.46 ± 3.42	20.81 ± 3.42	7.17 ± 2.68	84.81 ± 9.27	33 (c)	Greater	250 m asl	Advanced stage of ecological succession	At least 90 years-old (e)
Caxadaço Trail	29.95 ± 2.09	21.01 ± 2.15	4.11 ± 1.64	83.82 ± 8.98	15-20 (b)	Intermediate	180 m asl	Early stage of ecological succession	At least 50 years-old (c)
Parnaioca Trail	25.29 ± 3.12	22.04 ± 2.16	5.87 ± 2.66	80.93 ± 10.51	15 (c)	Lower	At sea level	Early stage of ecological succession	5 to 25 years-old (c)

a) D.P. Monteiro (unpubl. data), b) ALHO *et al.* (2002), c) VERA-Y-CONDE & ROCHA (2006), d) CALLADO *et al.* (2009), e) SLUYS *et al.* (2012).

leg. (Col. Mol. UERJ 1129-1); 13.VIII.1996, S.B. Santos *leg.* (Col. Mol. UERJ 1139); 13.VIII.1996, S.B. Santos *leg.* (Col. Mol. UERJ 1177-1); 13.VIII.1996, S.B. Santos *leg.* (Col. Mol. UERJ 1175); 13.VIII.1996, S.B. Santos *leg.* (Col. Mol. UERJ 1178-1 and 2); 16.VI.2002, S.B. Santos *leg.* (Col. Mol. UERJ 1827-2); 02.II.2002, D.P. Monteiro *leg.* (Col. Mol. UERJ 2989-1); 01.II.2000, D.P. Monteiro *leg.* (Col. Mol. UERJ 3005-1, 2, and 3); 31.I.2000, D.P. Monteiro *leg.* (Col. Mol. UERJ 3006-1, 2, 3, 4, 5, 6, 7, 8, 10, and 11); 31.I.2000, S.B. Santos *leg.* (Col. Mol. UERJ 3007-1, 2, 3, 4, 5, 6, 7, and 8); 31.I.2000, D.P. Monteiro *leg.* (Col. Mol. UERJ 3288-2); 28.I.2000, P. Coelho *leg.* (Col. Mol. UERJ 3289); 3.VIII.2005, A.B. Barbosa, Lacerda, L.E.M., T.A. Viana *leg.* (Col. Mol. UERJ 7444-1, 2, and 3).

Drawings of the shells in apical, umbilical, and lateral views were made with the aid of a camera lucida under an Olympus SZH10 stereomicroscope. The drawings were used to obtain angular and linear measurements, establish the number of whorls, and calculate the ratios between measurements, according to the criteria proposed by DIVER (1931), PARODIZ (1951), SOLEM & CLIMO (1985) and FONSECA & THOMÉ (1994). The following angular measurements were considered: maximum angle (MA), columellar angle (CA), sutural angle (SA), lower sutural angle (SS'), and spire angle (SPA) (Figs 2 and 3). The linear measurements taken were: shell height (h), aperture height (ah), aperture width (aw), spire height (sh) (Fig. 2), first whorl diameter (1wd), maximum diameter (D), smaller diameter (d), first whorl width (1ww), and second whorl width (2ww) (Fig. 4), diameter umbilical (ud) (Fig. 5). The following ratios were calculated: shell height/maximum diameter (h/D), maximum diameter/umbilical diameter (D/ud), umbilical diameter/shell height (ud/h), aperture height/aperture width (ah/aw), aperture height/smaller diameter (ah/d), first whorl diameter/maximum diameter (1wd/D), maximum diameter/total number of whorls (D/NW), and first whorl width/second whorl width (1ww/2ww) (SOLEM & CLIMO 1985, FONSECA & THOMÉ 1994, EMBERTON 1995a). The method proposed by DIVER (1931) was applied to obtain the number of protoconch whorls (pW), total number of whorls (NW), and total number of teleoconch whorls (TW) (Fig. 4).

The analysis of shell morphological variation followed VALORVITA & VÄISÄNEN (1986), with some modifications. Descriptive statistics were performed for each variable in each group, and normality was tested using the skewness test. In cases when a given variable had asymmetrical distribution, the following transformation procedures were applied to normalize it as appropriate: e-base logarithm of X (Neperian logarithm) [1wd/D, NW], square root of X [D/ud], sin X [MA, ah], cos X [SA, 1ww/2ww], tan X [ah/aw, ah/d], and reciprocal of X (i.e., 1/X) [1ww, 2ww, h/D, D/NW] (KREBS 1998, ZAR 1999), where X is the variable considered.

After normalization, each variable was standardized by reduction (SPIEGEL 1993) and compared using analysis of variance (ANOVA) followed by the Bonferroni's test. Differences

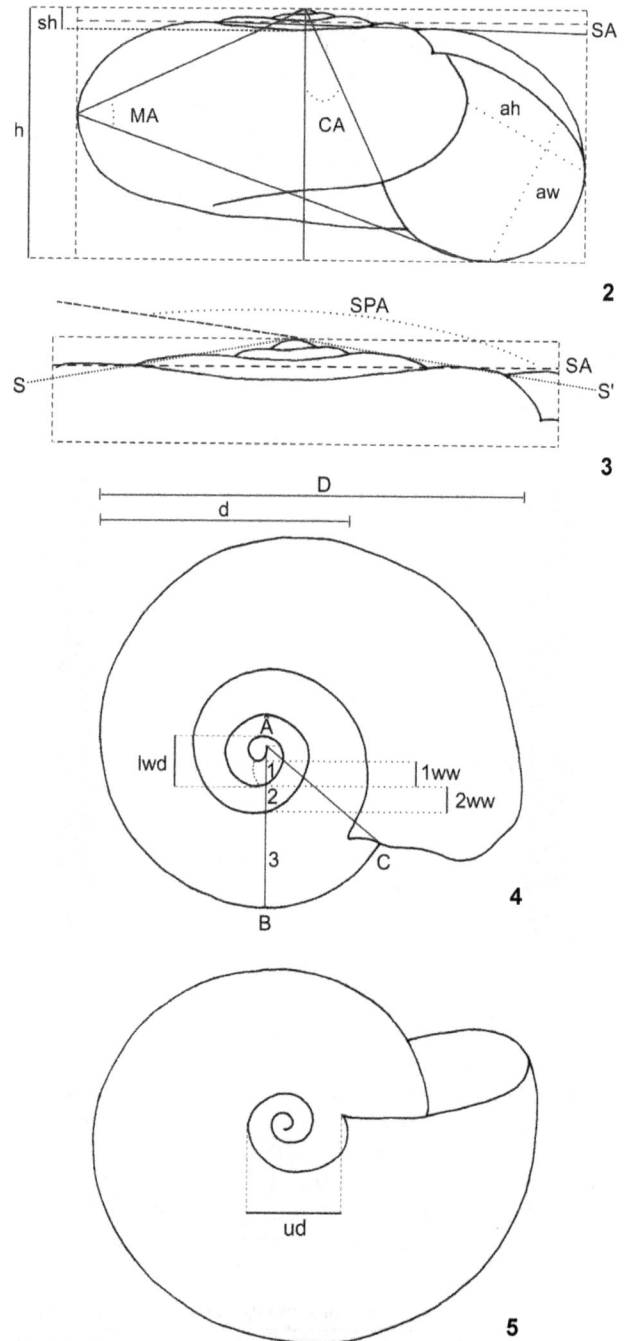

Figures 2-5. *Happiella* cf. *insularis*. Schematic depiction of shell, with angular, linear measurements, and number of whole whorls: (2) CA, columellar angle; MA, maximum angle; ah, aperture height; SA, sutural angle; sh, spire height; aw, aperture width. (3) SPA, spire angle; SA, sutural angle; SS', lower sutural angle. (4) D = maximum diameter; d, smaller diameter; 1ww, first whorl width; 2ww, second whorl width; 1wd, first whorl diameter; 1, 2, 3, number of whole whorls. (5) umbilical view: ud, umbilical diameter.

at p < 0.05 were considered statistically significant. The third decimal place was dropped and differences at p ≤ 0.05 were considered statistically significant.

After the exclusion of the highly correlated variables (KLECK 1982, ENGELMAN 1997), discriminant analysis was performed on 15 variables: MA, CA, SA, SS, SPA, 1wd,h, ah,aw,sh, 1ww, 2ww, pW, NW, and TW. Preliminary Pearson's correlation matrix revealed a high correlation (r ≥ 0.90) between the variables D, d, ud, and ah. These were removed from the analysis, and the variable h, representing all correlated variables excluded, was kept. Upon analysis, the variables MA, SA, sh, and 1ww were also removed, owing to their low contribution to group discrimination, as shown by the discriminant function coefficients. Statistical analyses were performed with the aid of the SYSTAT 7.0 statistical package (ENGELMAN 1997).

RESULTS

Shell morphology

Happiella cf. *insularis* (Fig. 6) has thin, translucent, depressed, shiny shells; periostracum color varies from yellow amber (when alive) to whitish yellow (when fixed in alcohol or in cases when only the shell is found *in situ*); spire slightly elevated (BOËTTGER 1889) (Fig. 9); under the optical microscope, the texture is smooth, with mildly marked growth lines (BOËTTGER 1889) (Fig. 7); under scanning microscopy, protoconch, more granular, and teleoconch, of rougher aspect (Fig. 10), umbilicus opened (Fig. 8) (BOËTTGER 1889); aperture wide, crescent-shaped, slightly oblique; peristome simple (BOËTTGER 1889), thin sharp edges, without teeth (Fig. 9) (THIELE 1931, ZILCH 1959, MONTEIRO & SANTOS 2001); suture not impressed (Fig. 7) (R.L. RAMÍREZ unpubl.

Figures 6-10. *Happiella* cf. *insularis*. (6) habitus; (7-10) specimen Col. Mol. UERJ 1653-2: (7) apical view; (8) umbilical view; (9) apertural view; (10) scanning electron microscopy of view's apical shell. Scale bar: 6-9 = 1 mm, 10 = 100 µm. Photos: 6 = Antônio C. de Freitas, 7-9 = Amilcar B. Barbosa, 10 = Alan C.N. de Moraes, LABMEL/UERJ.

data); body whorl rounded (Fig. 10) (R.L. Ramírez unpubl. data); rapid increment's shell growth (Fig. 7) (Emberton 1995a).

Shell morphometry

Table II shows the morphometric and meristic data of the 102 shells examined. The mean values of these features were lowest in specimens from the Parnaioca Trail.

The results of the ANOVA, distinguished among the three samples collected from Jararaca, Caxadaço, and Parnaioca, revealed significant differences in all linear and angular measure-

ments, except for the mean maximum angle. Specimens from the Jararaca and Parnaioca trails differed significantly in mean columellar and mean spire angles, but the differences in these measurements between samples from Jararaca and Caxadaço and from Parnaioca and Caxadaço were not statistically significant. The shells from Jararaca differed from those from the Caxadaço and Parnaioca trails in the mean sutural and lower sutural angles; shells from Caxadaço and Parnaioca, however, were statistically similar with regards to these two variables. The Bonferroni's test revealed differences between samples from Parnaioca and

Table II. Descriptive statistics of morphometric and meristic variables and ratios for *Happiella* cf. *insularis* collected from three areas on Ilha Grande. Linear measurements (cm): (D) maximum diameter, (d) smaller diameter, (ud) umbilical diameter, (1wd) first whorl diameter, (ah) aperture height, (sh) spire height, (h) shell height, (aw) aperture width, (1ww) first whorl width, (2ww) second whorl width. Angular measurements (degrees): (CA) columellar angle, (MA) maximum angle, (SPA) spire angle, (SA) sutural angle, (SS') lower sutural angle. Ratios: (D/ud) maximum diameter/umbilical diameter, (D/NW) maximum diameter/total number of whorls, (ud/h) maximum diameter/umbilical diameter, (1wd/D) first whorl diameter/maximum diameter, (h/D) shell height/maximum diameter, (aw/d) aperture height/smaller diameter, (ah/aw) aperture height/aperture width. Number of whorls: (pW) number of protoconch whorls, (TW) number of teleoconch whorls, (NW) total number of whorls. (N) sample size, (SD) standard deviation, (VAR) variance.

	Jararaca (N = 33)					Caxadaço (N = 34)					Parnaioca (N = 35)				
	Min.	Mean	Max.	SD	VAR	Min.	Mean	Max.	SD	VAR	Min.	Mean	Max.	SD	VAR
Linear measurements															
D	0.500	0.739	0.875	0.090	0.008	0.371	0.591	0.806	0.169	0.028	0.319	0.476	0.833	0.125	0.015
d	0.325	0.494	0.606	0.071	0.005	0.227	0.405	0.575	0.110	0.012	0.221	0.331	0.558	0.079	0.006
ud	0.090	0.161	0.206	0.030	0.000	0.059	0.124	0.241	0.054	0.002	0.044	0.089	0.193	0.037	0.001
1wd	0.038	0.068	0.086	0.011	0.000	0.031	0.056	0.080	0.012	0.000	0.030	0.056	0.083	0.012	0.000
sh	0.011	0.019	0.033	0.004	0.000	0.015	0.021	0.027	0.003	0.000	0.010	0.019	0.027	0.004	0.000
h	0.245	0.355	0.413	0.037	0.001	0.189	0.284	0.400	0.076	0.005	0.157	0.228	0.366	0.056	0.003
ah	0.127	0.191	0.240	0.026	0.000	0.100	0.161	0.225	0.046	0.002	0.082	0.127	0.233	0.031	0.000
aw	0.066	0.179	0.300	0.045	0.002	0.098	0.167	0.253	0.046	0.002	0.072	0.125	0.220	0.036	0.001
1ww	0.025	0.052	0.433	0.068	0.047	0.018	0.032	0.046	0.007	0.000	0.022	0.033	0.052	0.007	0.000
2ww	0.033	0.051	0.104	0.012	0.000	0.029	0.055	0.073	0.078	0.006	0.026	0.043	0.080	0.011	0.000
Angular measurements															
MA	44.0	47.772	55	2.446	5.985	42	46.426	51	2.074	4.305	42	47.200	59	3.595	12.924
CA	15.0	20.803	28	3.107	9.655	10	19.970	29	4.344	18.877	10	17.942	25	4.129	17.055
SA	0.5	1.590	5	0.930	0.866	1	2.500	4	0.904	0.818	1	2.242	5	1.017	1.034
SS'	151.0	161.742	171	5.026	25.267	147	156.264	167	5.029	25.291	147	154.742	166	4.767	22.726
SPA	158.0	168.151	175	3.700	13.757	160	166.588	172	3.210	10.310	160	165.200	172	3.332	11.105
Ratios															
h/D	0.445	0.481	0.533	0.021	0.000	0.428	0.483	0.581	0.029	0.000	0.439	0.480	0.519	0.020	0.000
D/ud	3.718	4.656	5.500	0.418	0.175	3.282	5.075	7.000	0.858	0.737	3.266	5.657	7.900	0.985	0.970
ud/h	0.361	0.450	0.583	0.054	0.003	0.238	0.419	0.658	0.085	0.007	0.276	0.380	0.632	0.079	0.006
ah/aw	0.643	1.129	2.606	0.340	0.115	0.735	0.965	1.234	0.132	0.017	0.650	1.037	1.388	0.165	0.027
ah/d	0.300	0.396	0.612	0.062	0.003	0.346	0.394	0.471	0.030	0.000	0.034	0.380	0.504	0.071	0.005
1wd/D	0.057	0.094	0.172	0.020	0.000	0.063	0.100	0.152	0.025	0.000	0.069	0.122	0.219	0.035	0.001
D/NW	0.115	0.192	0.224	0.020	0.000	0.103	0.163	0.573	0.081	0.006	0.098	0.128	0.215	0.029	0.000
1ww/2ww	0.521	0.826	1.200	0.154	0.023	0.066	0.758	1.000	0.178	0.031	0.519	0.781	1.000	0.104	0.010
Number of whorls															
NW	2.909	3.823	4.413	0.275	0.076	3.380	3.908	4.472	0.287	0.082	2.908	3.671	4.188	0.322	0.103
TW	1.433	2.538	3.587	0.410	0.168	1.794	2.376	3.166	0.316	0.099	1.720	2.191	2.936	0.334	0.112
pW	0.118	1.284	2.844	0.588	0.239	0.950	1.531	2.105	0.287	0.082	0.955	1.499	2.119	0.307	0.094

Jararaca in nine morphological features, between Jararaca and Caxadaço samples in eight features, and between samples from Caxadaço and Parnaioca in seven features – i.e., samples from Caxadaço and Parnaioca were less dissimilar to each other than to the sample from Jararaca (Table III).

ANOVA revealed significant differences in the D/ud, ud/h, ah/aw, 1wd/D, and D/NW ratios across samples. Bonferroni's test showed Parnaioca and Jararaca samples to differ in four of these ratios (D/ud, ud/h, 1wd/D, and D/NW), Caxadaço and Parnaioca samples to differ in three (D/ud, 1wd/D, and D/NW), and Jararaca and Caxadaço samples to differ on two of these ratios (ah/aw and D/NW) – i.e., differences in measurement

ratios were most pronounced between samples collected from the Jararaca and Parnaioca trails, as were differences in linear measurements (Table III).

The mean total number of whorls (NW) differed significantly between the samples from Caxadaço (greater mean) and Parnaioca (Table III). The mean total number of teleoconch whorls (TW) differed significantly between the Jararaca and Parnaioca samples. The mean number of protoconch whorls (pW) differed significantly not only between samples from Jararaca and Caxadaço, but also between samples from Jararaca and Parnaioca (Table III).

Discriminant analysis

The discriminant analysis (Fig. 11) allowed the distinction of all three samples (Wilks's Lambda = 0.300, F = 6.689, df = 22, p = 0.000), particularly with respect to the sample from Jararaca, which differed the most from the others. The samples from Caxadaço and Parnaioca were more similar to each other than each was to the sample from Jararaca (Fig. 11). This analysis correctly classified 67% of the specimens (Fig. 11), with 34 out of 102 being incorrectly classified.

Table III. Results of ANOVA followed by Bonferroni's multiple comparison test, applied to linear and angular measurements, ratios, and number of whorls of *Happiella* cf. *insularis* specimens collected from the Jararaca (Jar), Caxadaço (Cax), and Parnaioca (Par) trails, Ilha Grande. Differences were considered statistically significant* at p ≤ 0.05. For abreviations see Table II.

	p	Jar x Cax	Jar x Par	Cax x Par
Linear measurements				
D	0.000*	0.000*	0.000*	0.001*
d	0.000*	0.000*	0.000*	0.002*
ud	0.000*	0.002*	0.000*	0.002*
1wd	0.000*	0.000*	0.000*	1.000
sh	0.036*	0.099	1.000	0.061*
h	0.000*	0.000*	0.000*	0.000*
ah	0.000*	0.002*	0.000*	0.000*
aw	0.000*	0.753	0.000*	0.000*
1ww	0.000*	0.000*	0.001*	1.000
2ww	0.008*	0.033*	0.012*	1.000
Angular measurements				
MA	0.470	1.000	1.000	0.751
CA	0.010*	1.000	0.010*	0.101
SA	0.000*	0.000*	0.002*	0.187
SS'	0.000*	0.000*	0.000*	0.611
SPA	0.003*	0.193	0.002*	0.285
Ratios				
h/D	0.947	1.000	1.000	1.000
D/ud	0.000*	0.106	0.000*	0.012*
ud/h	0.000*	0.212	0.000*	0.082
ah/aw	0.024*	0.020*	0.559	0.427
ah/d	0.490	1.000	0.757	1.000
1wd/D	0.000*	1.000	0.000*	0.004*
D/NW	0.000*	0.000*	0.000*	0.003*
1ww/2ww	0.175	0.238	0.450	1.000
Number of whorls				
NW	0.005*	0.807	0.107	0.004*
TW	0.001*	0.192	0.000*	0.099
pW	0.015*	0.022*	0.056*	0.929

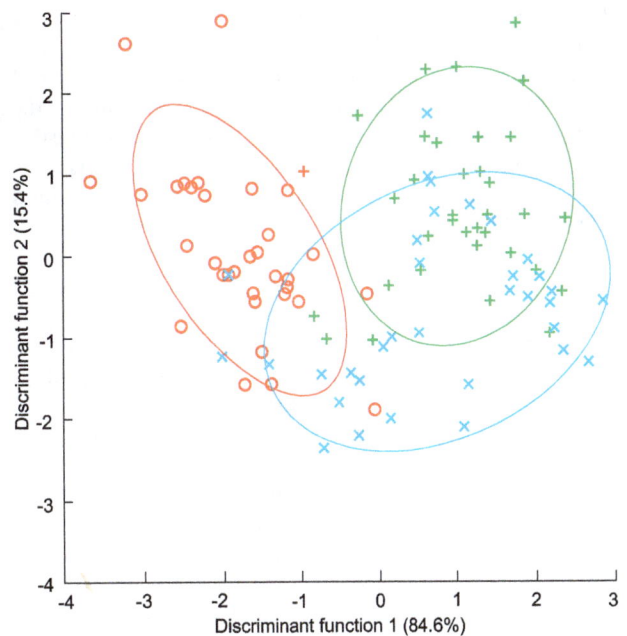

Figure 11. Graphic depiction of the discriminant analysis of morphometric variables of three *Happiella* cf. *insularis* samples collected from the Jararaca (○), Caxadaço (×), and Parnaioca (+) trail areas, Ilha Grande, RJ, Brazil.

The proportions of explanation were of 84.6% and 15.4% for the first and second discriminant functions, respectively. The coefficients also revealed the following variables to be major contributors to the degree of differentiation achieved with the first function: shell height (h), lower sutural angle (SS'), spire angle (SPA), aperture height (ah), number of protoconch whorls (pW) e number of teleoconch whorls (TW).

Discriminant function 1 = -2.263 h – 1.077 SS' + 0,964 SPA + 0.877 ah + 0.736 pW + 0.674 TW – 0.411 2ww + 0.296 aw + 0.294 + 0.113 1wd + 0.038 CA.

Discriminant function 2 = -1.046 NW + 0.992 SS' – 0.941 SPA + 0.900 TW – 0.870 ah + 0.799 pW – 0.780 aw + 0.465 h + 0.396 CA + 0.249 1wd + 0.035 2ww.

DISCUSSION

The morphology of the shell of *H.* cf. *insularis* fits the description for Systrophiidae perfectly: shell thin, translucent, polished, generally smooth, spire apex slightly prominent, discoid, simple peristomatic edge. According to Baker (1925), *Happiella* shells are characterized by a very low spire and an umbilicus normally reduced to a small perforation. In his original description of *H.* cf. *insularis*, Boëttger (1889) reported a maximum diameter of 5.25 mm, very close to the mean values found for the Caxadaço and Parnaioca samples, and shell height of 2 mm, very close to the mean obtained for the sample from Parnaioca. Aperture height and aperture width were originally described as measuring 2 mm and 2.25 mm, respectively, with the latter measurement falling within the confidence interval of the sample from Jararaca. The original description of the number of whorls (3.5) is also within the confidence intervals of both Jararaca and Parnaioca samples. According to the original description, the umbilicus size is one-fourth the maximum diameter of the shell, a similar ratio to that found in our specimens.

Our results revealed significant differences among the three populations of *H.* cf. *insularis* examined, which may be explained by differences in the degree of forest conservation in each area surveyed. The original vegetation in the Caxadaço and Parnaioca areas has been disturbed, more dramatically so in the latter, where the vegetation was entirely slashed down in some areas to make way for the now disabled Vila Dois Rios-Jararaca Dam road and for a number of plantations that served the now defunct Cândido Mendes Penal Colony. Along the Caxadaço Trail, inhabited until the construction of the penal colony by local native fishermen, the environmental changes were less pronounced. The Jararaca Trail region, by contrast, is better preserved. It has a relatively undisturbed secondary forest in lower-altitude areas and primary forest in higher areas. This translates into a deeper leaf litter layer, a more closed canopy, and lower temperature (Table I). As in other instances (SHIMEK 1930, BOYCOTT 1934, CAIN 1977, CLARKE *et al.* 1978, TILLIER 1981, EMBERTON 1995b, COOK 1997, WELTER-SCHULTES 2000, TESHIMA *et al.* 2003), better conservation certainly influences the environmental conditions overall, including leaf litter quality, structure, humidity, and depth, which in turn influence mollusk morphology (GOULD 1968, PEAKE 1978, CIPRIANI 2007). Investi-

gating *Ainohelix editha* (Adams, 1868) in Hokkaido, Japan, TESHIMA *et al.* (2003) demonstrated that shell size and growth rates are adaptations to the environmental conditions; CHIBA (2004), investigating the genus *Mandarina* Pilsbry, 1894 on the Bonin Islands, found that differentiation in shell shape and dimensions are accelerated in degraded environments.

In the present study, the smallest mean values were obtained for specimens collected from the Parnaioca Trail area (Table II), the most disturbed of the three areas (shallower leaf litter, more open canopy, higher temperature). The lower capacity of the local leaf litter to retain water is likely responsible for the smaller size of the snails. The smaller mean aperture height of the shell may represent a strategy to prevent excessive water loss (GOODFRIEND 1986). MACHIN (1967), PEAKE (1978), EMBERTON (1982), and GOODFRIEND (1986), along with other investigators, have reported that smaller specimens are found in terrestrial gastropod populations living in dry areas with a strong incidence of sunlight.

The mean shell diameter and number of whorls were greatest in snails collected from Jararaca (Table II), consistent with the hypothesis that higher humidity and lower temperatures promote increased rates of shell growth in terrestrial gastropods (GOULD 1984, GOODFRIEND 1986, EMBERTON 1994). BAUR (1988) concluded that the size of the shell of *Chondrina clienta* (Westerlund, 1883) increases in higher temperatures and lower population densities, as verified by ANDERSON *et al.* (2007) for *Oreohelix cooperi* (Binney, 1838). BAUR (1988) commented that the phenotypic plasticity found in *C. clienta* may be adaptive, as the genetic makeup of snails allows for different shell growth patterns under different environmental conditions.

The populations of the most disturbed areas – the Caxadaço and Parnaioca trails – are more similar to each other than to the population of the Jararaca Trail, which is the best-preserved area (Fig. 11). The greater similarity observed between the Parnaioca and Caxadaço samples can be explained by the intermediate degree of conservation of the Caxadaço region, which is closer to the degree of conservation of the Parnaioca area than to that of the Jararaca region.

The findings of this study corroborate investigations conducted in other countries showing that morphological differentiation is a result of the isolation of populations in areas that are distinct in vegetation cover, dominant plant species, maximum altitude, and soil type – i.e., areas that offer different microhabitats. As CHIBA (2004) pointed out, degraded environments accelerate differentiation by eliciting new ecological interactions and new habitat conditions, thus subjecting species to a number of selective pressures.

We believe that a similar process has occurred in the areas investigated in the present study, where the different degrees of forest degradation, added to different degrees of moisture, contributed to the morphological differentiation of the three studied populations of *H.* cf. *insularis*. However, as Brazilian species of Systrophiidae are not yet well defined, we have decided not to treat the three populations as separate species, and we recommend that they continue to be identified as *H.* cf. *insularis* until ongoing anatomical studies are concluded, and a decision on their taxonimical status is reached.

ACKNOWLEDGEMENTS

This study is one of the results of the Project "Fauna malacológica aquática e terrestre da Ilha Grande", supported by research grants from the Fundação de Amparo à Pesquisa do Estado do Rio de Janeiro (FAPERJ) to the second author (APQ1 E-26/110.402/2010 and E-26/110.362/2012). We would like to express our gratitude to PEIG/INEA (Parque Estadual da Ilha Grande/Instituto Estadual do Ambiente) for license 18/2007; to IBAMA/Sisbio (Instituto Brasileiro do Meio Ambiente e dos Recursos Naturais Renováveis/Sistema de Autorização e Informação em Biodiversidade) for license 19836-1 to ABB and 10812-1 to SBS; to Capes (Coordenação de Aperfeiçoamento do Pessoal de Ensino Superior for a PhD scholarship to ABB; to the reviewers who gave valuable suggestions to improve the manuscript; to Cientifica Consultoria for review of English and to CEADS (Centro de Estudos Ambientais e Desenvolvimento Sustentável da UERJ) for logistic support.

LITERATURE CITED

Alho, C.J.R.; M. Schneider & L.A. Vasconcellos. 2002. Degree of threat to the biological diversity in the Ilha Grande State Park (RJ) and guidelines for conservation. **Brazilian Journal of Biology 62** (3): 375-385. doi: 10.1590/S1519-69842002000300001.

Alves, S.L.; A.S. Zaú; R.R. Oliveira; D.F. Lima & C.J.R. Moura. 2005. Sucessão florestal e grupos ecológicos em Floresta Atlântica de encosta, Ilha Grande, Angra dos Reis/RJ. **Revista Universidade Rural: Série Ciências da Vida 25** (1): 26-32.

Anderson, T.K.; K.F. Weaver & R.P. Guralnick. 2007. Variation in adult shell morphology and life-history traits in the land snail *Oreohelix cooperi* in relation to biotic and abiotic factors. **Journal of Molluscan Studies 73**: 129-137. doi: 10.1093/mollus/eym006.

Baker, H.B. 1925. The Mollusca collected by the University of Michigan-Williamson expedition in Venezuela. Part III. **Occasional Papers of the Museum of Zoology 156**: 1-44.

Baur, B. 1988. Microgeographical variation in shell size of the land snail *Chondrina clienta*. **Biological Journal of the Linnean Society 35**: 247-259. doi: 10.1111/j.1095-8312.1988.tb00469.x.

Boëttger, O. 1889. Bemerkung uber ein paar brasilianische Landschneken, nebst Beschreibung drein neuer Hyalinien von dort. Nachrichtsblatt der deutschen **Malakozoologischen 20** (1-2): 27-30.

Boycott, A.E. 1934. The habitats of land mollusca in Britain. **Journal of Ecology 22**: 1-38.

Cain, A.J. 1977. Variation in the spire index of some coiled gastropods shells, and its evolutionary significance. **Philosophical Transactions of the Royal Society B277**: 377-428.

Callado, C.H.; A.A.M. Barros; L.A. Ribas; N. Albarello; R. Gagliardi & C.E.S. Jascone. 2009. Flora e cobertura vegetal, p. 91-162. *In*: M. Bastos & C.H. Callado (Eds). **O Ambiente da Ilha Grande**. Rio de Janeiro, UERJ/CEADS, 562p.

Chiba, S. 2004. Ecological and morphological patterns in communities of land snails of the genus *Mandarina* from the Bonin Islands. **Journal of Evolutionary Biology 17**: 131-143. doi: 10.1046/j.1420-9101.2004.00639.x.

Chiba, S. & A. Davison. 2007. Shell shape and habitat use in the North-west Pacific land snail *Mandarina polita* from Hahajima, Ogasawara: current adaptation or ghost of species past? **Biological Journal of the Linnean Society 91**: 149-159. doi: 10.1111/j.1095-8312.2007.00790.x.

Cipriani, R. 2007. Modelando las conchas de los moluscos, o la búsqueda de la espiral perfecta, p. 3-11. *In*: S.B. Santos; A.D. Pimenta; S.C. Thiengo; M.A. Fernandez & R.S. Absalão (Eds). **Tópicos em Malacologia – Ecos do XVIII Encontro Brasileiro de Malacologia**. Rio de Janeiro, Sociedade Brasileira de Malacologia, XIV+365p.

Clarke, B.; W. Arthur; D.T. Horsley & D.T. Parkin. 1978. Genetic variation and natural selection in pulmonate molluscs. 219-270. *In*: V. Fretter & J. Peake (Eds). **Pulmonates: Systematics, Evolution and Ecology**. Londres, Academy Press, 540p.

Cook, L.M. 1997. Geographic and ecological patterns in Turkish land snails. **Journal of Biogeography 24**: 409-418. doi: 10.1111/j.1365-2699.1997.00139.x.

Diver, C. 1931. A method to determining the number of whorls of a shell and its application to *Cepaea hortensis* Müll. **Proceedings of the Malacological Society of London 19**: 1931.

Emberton, K.C. 1982. Environment and shell shape in the Tahitian land snail *Partula otaheitana*. **Malacologia 23** (1): 23-35.

Emberton, K.C. 1994. Partitioning a morphology among its controlling factors. **Biological Journal of the Linnean Society 53**: 353-369.

Emberton, K.C. 1995a. Land-snail community morphologies of the highest-diversity sites of Madagascar, North America and New Zealand, with recommended alternatives to height-diameter plots. **Malacologia 36** (1-2): 43-66.

Emberton, K.C. 1995b. Sympatric convergence and environmental correlation between two land-snail species. **Evolution 3**: 469-475.

Engelman, K. 1997. SYSTAT 7.0. Chicago, SPSS Inc press, 421p.

Fonseca, A.L.M. & J.W. Thomé. 1994. Conquiliomorfologia e anatomia dos sistemas excretor e reprodutor de *Radiodiscus thomei* Weirauch, 1965 (Gastropoda, Stylommatophora, Charopidae). **Biociências 2** (1): 163-188.

Goodfriend, G.A. 1986. Variation in land-snail shell form and size and its causes: a review. **Systematic Zoology 2**: 204-223.

Gould, S.J. 1968. Ontogeny and the explanation of form: an allometric analysis. **Paleontological Society Memoirs 2**: 81-98.

GOULD, S.J. 1984. Covariance sets and ordered geographic variation in *Cerion* from from Aruba, Bonaire and Curaçao: a way of studying nonadaptation. **Systematic Zoology** 33 (2): 217-237.

KLECK, W. 1982. **Discriminant analysis**. Sage University Paper Series on Quantitative Applications in the Social Sciences, 07-0119. Beverly Hills, Sage Publications, 71p.

KREBS, J.C. 1998. **Ecological Methodology**. New York, Benjamin Cummings, XII + 620p.

LEVINE, D. M.; M.L. BERENSON & D. STEPHAN. 2000. **Estatística: teoria e aplicações**. Rio de Janeiro, LTC, 812 p.

MACHIN, J. 1967. Strutural adaptation for reducing water-loss in three species of terrestrial snail. **Journal of Zoology 152:** 55-65. doi: 10.1111/j.1469-7998.1967.tb01638.x.

MONTEIRO, D.P. & S.B. SANTOS. 2001. Conquiliomorfologia de *Tamayoa* (*Tamayops*) *banghaasi* (Thiele) (Gastropoda, Systrophiidae). **Revista Brasileira de Zoologia** 18 (4): 1049-1055. doi: 10.1590/S0101-81752001000400002.

MORRETES, F.L. 1949. Ensaio de catálogo dos moluscos do Brasil. **Arquivos do Museu Paranaense 7:** 1-216.

MYERS, N.; R.A. MITTERMEIER; C.G. MITTERMEIER; G.A.B. FONSECA & J. KENT. 2000. Biodiversity hotspots for conservation priorities. **Nature 403:** 853-858. doi: 10.1126/science.1067728.

OLIVEIRA, R.R. 2002. Ação antrópica e resultantes sobre a estrutura e composição da Mata Atlântica na Ilha Grande, RJ. **Rodriguésia** 53 (82): 33-58. doi: 10.1590/S1414-753X2007000200002.

PARMAKELIS, A.; E. SPANOS; G. PAPAGIANNAKIS; C. LOUIS & M. MYLONAS. 2003. Mitochondrial DNA phylogeny and morphological diversity in the genus *Mastus* (Beck, 1837): a study in recent (Holocene) island group (Koufonisi, south-east Crete). **Biological Journal of the Linnean Society 78:** 383-399. doi: 10.1046/j.1095-8312.2003.00152.x.

PARODIZ, J.J. 1951. Métodos de Conquiliometria. **Physis 20** (38): 241-248.

PEAKE, J.F. 1978. Distribution and ecology of Stylommatophora, p. 429-526. *In*: V. FRETTER & J. PEAKE (Eds). **Pulmonates. Systematics, Evolution and Ecology**. New York, Academic Press, vol. 2A, 540p.

ROCHA, C.F.D.; H.G. BERGALLO; M.A.S. ALVES & M.V. SLUYS. 2003. **A biodiversidade nos grandes remanescentes florestais do Estado do Rio de Janeiro e nas restingas da Mata Atlântica**. São Carlos, Editora Rima, 160p.

SANTOS, S.B.; A.B. BARBOSA; R.M.R.B. BRAGA; J.L. OLIVEIRA & R.F. XIMENES. 2010. Moluscos da Ilha das Flores, São Gonçalo, Rio de Janeiro. **Informativo SBMa 173:** 10-14.

SHIMEK, B. 1930. Land snails as indicators of ecological conditions. **Ecology 11** (4): 673-686. doi: 10.2307/1932328.

SIMONE, L.R.L. 2007. **Land and freshwater molluscs of Brazil**. São Paulo, EGB, Fapesp, 390p.

SLUYS, M.V.; R.V. MARRA; L. BOQUIMPANI-FREITAS; & C.F.D. ROCHA. 2012. Environmental factors affecting calling behavior of sympatric frog species at an Atlantic Rain Forest area, Southeastern Brazil. **Journal of Herpetology 46** (1): 41-46.

SOLEM, A. & F.M. CLIMO. 1985. Structure and habitat correlations of sympatric New Zealand land snail species. **Malacologia 26:** 1-30.

SPIEGEL, M.R. 1993. **Estatística**. São Paulo, Makron Books, Coleção Schaum, 643p.

TESHIMA, H.; A. DA VISON; Y. KUWAHARA; J. YOKOHAMA; S. CHIBA; T. FUKUDA; H. OGIMURA & M. KAWATA. 2003. The evolution of extreme shell shape variation in the land snail *Ainohelix editha*: a phylogeny and hybrid zone analysis. **Molecular Ecology 12:** 1869-1878. doi: 10.1046/j.1365-294X.2003.01862.x.

THIELE, J. 1927. Über einige brasilianische Landschnecken. **Abhandlungen der Senckenbergischen Naturforschenden Gesellschaft 40** (3): 307-329.

THIELE, J. 1931. **Handbuch der Systematischen Weichtierkunde**. Jena, Gustav Fischer, vol. 1, 778p.

TILLIER, S. 1981. Clines, convergence and character displacement in new Caledonian diplommatinids (land prosobranchs). **Malacologia 21** (1-2): 177-208.

VALORVITA, I. & VÄISÄNEN, R. A. 1986. Multivariate morphological discrimination between *Vitrea contracta* (Westerlund) and *V. crystallina* (Müller)(Gastropoda, Zonitidae). **Journal Molluscan Studies 52:** 62-67. doi: 10.1093/mollus/52.1.62

VERMEIJ, G.J. 1971. Gastropod evolution and morphological diversity in relation to shell geometry. **Journal of Zoology 163:** 15-23. doi: 10.1111/j.1469-7998.1971.tb04522.x.

VERA-Y-CONDE, C.F. & C.F.D. ROCHA. 2006. Habitat disturbance and small mammal richness and diversity in an atlantic rainforest area in southeastern Brazil. **Brazilian Journal of Biology 66** (4): 983-990. doi: 10.1590/S1519-69842006000600005.

WELTER-SCHULTES, F.W. 2000. Human-dispersed land snails in Crete, with special reference to *Albinaria* (Gastropoda: Clausiliidae). **Biologia Gallo-hellenica 24:** 83-106.

ZAR, J. H. 1999. **Biostatistical Analysis**. New Jersey, Prentice-Hall, 663p.

ZILCH, A. 1959. **Gastropoda: Euthyneura**. Berlim, Borträger, vol. 2, 834p.

10

The feeding habits of the eyespot skate *Atlantoraja cyclophora* (Elasmobranchii: Rajiformes) in southeastern Brazil

Alessandra da Fonseca Viana[1,2] & Marcelo Vianna[1]

[1] *Laboratório de Biologia e Tecnologia Pesqueira, Instituto de Biologia, Universidade Federal do Rio de Janeiro. Avenida Carlos Chagas Filho 373, Bloco A, 21941-902 Rio de Janeiro, RJ, Brazil.*
[2] *Corresponding author. E-mail: fviana.ale@gmail.com*

ABSTRACT. The stomach contents of the eyespot skate, *Atlantoraja cyclophora* (Regan, 1903), were examined with the goal to provide information about the diet of the species. Samples were collected off the southern coast of Rio de Janeiro, Brazil, near Ilha Grande, between January 2006 and August 2007, at a depth of about 60 m. The diet was analyzed by sex, maturity stages and quarterly to verify differences in the importance of food items. The latter were analyzed by: frequency of occurrence, percentage of weight and in the Alimentary Index. The trophic niche width was determined to assess the degree of specialization in the diet. Additionally, the degree of dietary overlap between males and females; juveniles and adults and periods of the year were defined. A total of 59 individuals of *A. cyclophora* were captured. Females and adults were more abundant. The quarters with the highest concentrations of individuals were in the summer of the Southern Hemisphere: Jan-Feb-Mar 06 and Jan-Feb-Mar 07. Prey items were classed into five main groups: Crustacea, Teleosts, Elasmobranchs, Polychaeta, and Nematoda. The most important groups in the diet of the eyespot skate were Crustacea and Teleosts. The crab *Achelous spinicarpus* (Stimpson, 1871) was the most important item. The value of the niche width was small, indicating that a few food items are important. The comparison of the diet between males and females and juveniles and adults indicates a significant overlap between the sexes and stages of maturity; and according to quarters, the importance of prey groups differed (crustaceans were more important in the quarters of the summer and teleost in Jul-Aug-Sep and Oct-Nov-Dec 06), indicating seasonal differences in diet composition. Three groups with similar diets were formed in the cluster analysis: (Jan-Feb-Mar 06 and 07); (Apr-May-Jun 06 and Jul-Aug-Sep 07); (Jul-Aug-Sep 06 and Oct-Nov-Dec 06).

KEY WORDS. Diet; elasmobranch; trophic niche; Rajidae; Rio de Janeiro.

Dietary analyses may be used to understand variations in the growth, reproduction, migration and the behavioral aspects of food capture. They may result in increased knowledge of resource sharing and competition between organisms (ROSECCHI & NOUAZE 1987), and a broader understanding of trophic ecology, which can be used in ecosystem management (ZAVALA-CAMIN 1996). Fish diet may differ ontogenetically, spatially and seasonally. For instance, juveniles and adults differ in the size of the food they consume, and differences in food availability may affect the diet of broadly distributed species (ZAVALA-CAMIN 1996).

Elasmobranchs play an important role in marine ecosystems, occupying high trophic levels (EBERT & BIZARRO 2007). They play an important role in the energy flow between the benthic and pelagic regions (AGUIAR & VALENTIN 2010). As a result of their place in the food chain, benthonic skates may accumulate contaminants such as mercury, and are therefore good indicators of pollution (LACERDA *et al.* 2000).

Benthonic skates have suffered increased fishing pressure in recent years (MASSA *et al.* 2006). They are also caught very frequently as bycatch. These two factors make fishery the main anthropogenic factor that affects elasmobranch populations. (VOOREN & KLIPPEL 2005). Overfishing of skates and rays can change the abundance and distribution of their populations, and result in proportional growth of populations of species in lower trophic levels (WALKER & HISLOP 1998).

Studies on the diet of elasmobranchs are less frequent than studies involving other marine fishes. According to AGUIAR & VALENTIN (2010), only 44 studies on the alimentary biology and ecology of elasmobranchs have been published in Brazilian journals, and only one of these analyzed the diet of *Atlantoraja cyclophora* (Regan, 1903). Elasmobranchs are carnivorous, eating less variety of prey than the herbivorous or omnivorous teleosts. EBERT & BIZARRO (2007) studied 60 species of Rajiformes and indicated that in the diet of skates, teleosts and decapods are dominant groups. Rajiformes are secondary or tertiary consumers and are classified as benthopelagic or epibenthic predators specialized in marine invertebrates or small crustaceans (AGUIAR & VALENTIN 2010).

The eyespot skate *A. cyclophora* belongs to Rajidae. This oviparous skate with demersal habits is found in the South Atlantic Ocean, from Rio de Janeiro (Brazil) to the south of the Mar del Plata (Argentina). The species is found from 30 to 300 m depth (mainly below 50 m) (GOMES *et al.* 2010), generally in cold waters of the continental shelf and upper continental slope, and does not seem to have a preference for any peculiar granulometry of the sediment. Normally, females of *A. cyclophora* are heavier and wider than males (ODDONE & VOOREN 2004). The species is caught as bycatch in coastal demersal fisheries, and has suffered growing fishing pressure. As a result, it is considered "vulnerable" in the Red List of the International Union for Conservation of Nature (CHEUNG *et al.* 2005). Given the scarcity of studies on *A. cyclophora*, it is the time for scientists to focus their attention on the species (MASSA *et al.* 2006).

The goal of this study was to generate information about the diet of *Atlantoraja cyclophora* in southeastern Brazil, by identifying the main food items to species; and to ascertain differences in the importance of food items according to sex, maturity stages and periods of the year.

MATERIAL AND METHODS

Between January 2006 and August 2007, 14 samples were hauled from the southern coast of Rio de Janeiro, Brazil, near Ilha Grande (Fig. 1). Each haul lasted one hour, and was conducted at about 60 m depth. The boat worked with a bottom-pair trawl. The net had a 20 mm mesh between opposing knots at the body and sleeves and 18 mm in codend.

The area of the study is located near the Ilha Grande Bay (22°50'-23°20'S, 44°00'-44°45'W), in the southern coast of Rio de Janeiro. The sediment is mainly characterized by medium/fine sand, but also silt and clay. In this region three water masses occur: Tropical Water, Coastal Water and South Atlantic Central Water. The Tropical Water (TW) occurs in the upper layer, with high temperatures and salinity (T > 20°C, S > 36.4). The Coastal Water (CW), with high temperature (24°C) and low salinity (34.9), normally occurs in the inner shelf. The South Atlantic Central Water (SACW) has low temperature and salinity (T < 20°C, 34 < S < 36.4) and can emerge when the wind conditions are favorable for upwelling. Therefore, in late spring and summer, this mass occupies the inner shelf, and in winter and autumn it moves offshore.

After collection, the material was chilled and later frozen until processing. Samples of *A. cyclophora* were identified according to GOMES *et al.* (2010), measured (disc width, cm), weighed (total weight, g) and sexed (voucher C.DBAV UERJ1256). Then, the skates were dissected and the maturity stage of each specimen (juvenile or adult) was determined according to a combination of internal and external characteristics (ODDONE & VOOREN 2005). The stomachs were removed, weighed, fixed in 10% formalin and conserved in 70% ethanol. The stomach contents were analyzed and prey items were

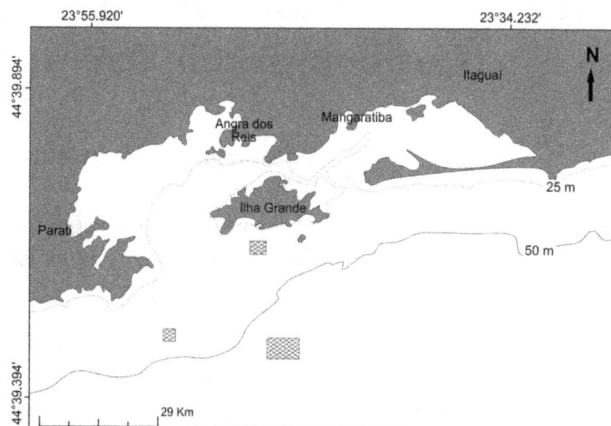

Figure 1. Location of samples taken, between January 2006 and August 2007, near Ilha Grande, on the southern coast of Rio de Janeiro, Brazil.

weighed and identified to the lowest possible taxonomic level. Identification was made with the help of specialists.

A cumulative prey curve (CORTES 1997, FERRY & CAILLIET 1996) was employed to determine whether the sample was large enough to precisely describe the diet. This method plots the cumulative number of randomly pooled stomachs against the cumulative number of prey types. This curve was constructed using the program EstimateSWin820 and fifty randomizations were performed.

Food items were analyzed by Frequency of Occurrence (%FO), expressed as the percentage of the stomachs analyzed containing the prey item; and Percentage Weight (%W), which is the percentage of the weight of the prey item with respect to the total weight of all prey items combined. These measurements were combined in the Alimentary Index (IAi), according to KAWAKAMI & VAZZOLER (1980), and modified to use the Percentage Weight (%W), according to the equation: $\%IAi_1 = ((\%FO_1 \times \%W_1) / \Sigma (\%FO_T \times \%W_T)) \times 100$, where $\%FO_1$ = percentage of stomachs in which determined item occur; $\%W_1$ = percentage of weight of determined item. The trophic niche width was determined according to the Standardized Levins Index (HURLBERT 1978) to assess the degree of diet specialization, according to the following equation: $Bi = [(\Sigma jPij^2)^{-1} - 1] (n-1)^{-1}$, where Pij = proportion of the prey j in the diet of the predator I; n = number of prey categories. In order to provide information about prey importance and feeding strategy, an adaptation of the COSTELLO's (1990) method by AMUNDSEN *et al.* (1996) was employed.

The diet was analyzed by sex, maturity stages of the skates and quarterly. The Trophic Niche Overlap Index by PIANKA (1973), was employed to determine the degree of dietary overlap between males and females and juveniles and adults, according to the following equation: $Oxy = \Sigma (Pxi \times Pyi) / \sqrt{\Sigma(\Sigma Pxi^2 \times \Sigma Pyi^2)}$, where Pxi = proportion of prey i in predator x; Pyi = proportion of prey i in predator y.

In the temporal analyses, six quarters were defined: 1 (Jan-Feb-Mar 06), 2(Apr-May-Jun 06), 3(Jul-Aug-Sep 06), 4(Oct-Nov-Dec 06), 5 (Jan-Feb-Mar 07), 6 (Jul-Aug-Sep/07). The diet was determined for each quarter and a cluster analysis was performed, employing Ward's Method, using the Past Statistic Program to define groups with similar diet.

RESULTS

We analyzed a total of 59 stomachs of *A. cyclophora*; of these, two were empty and were not included in the analyses. Of the remaining 57 stomachs, those from females and adults were more abundant (Table I), with disc width ranging from 14.8 to 47.8 cm (mean = 38.04 ± 8.25; median = 40.3). While Jan-Feb-Mar 06 were the quarters with the highest concentration of skates, Apr-May-Jun 06 and Jul-Aug-Sep 07 (Table I) had the lowest concentrations. The cumulative prey curve (Fig. 2) stabilized after about 40 stomachs, indicating that this number was representative for the sampling area.

Table I. Sex ratio according to maturity stages and quarters of the year of *Atlantoraja cyclophora*, collected in southern coast of Rio de Janeiro, Brazil.

	Females	Males	Total
Juveniles	10	7	17
Adults	24	16	40
Jan-Feb-Mar 06	14	10	24
Apr-May-Jun 06	0	1	1
Jul-Aug-Sep 06	5	2	7
Oct-Nov-Dec 06	7	6	13
Jan-Feb-Mar 07	8	3	11
Jul-Aug-Sep 07	0	1	1

Prey items were classified into five main groups (Crustacea, Teleosts, Elasmobranchs, Polychaeta, Nematoda) (Table II). The group Crustacea was the most important in the diet (79.4%IAi). Teleosts were also important, since this group has a significant FO%. The crab *Achelous spinicarpus* (Stimpson, 1871) (= *Portunus spinicarpus*) was the most important item overall (53.5%IAi), followed by Portunidae (23.6%IAi). Item fragments of teleosts were important (9.7%IAi), and *Dactylopterus volitans* (Linnaeus, 1758) was a very representative teleost (8.2%IAi). The trophic niche width of *A. cyclophora* was 0.1, indicating that a few food items are very important.

The prey importance and feeding strategy diagrams (Figs 3 and 4) show that no food item was dominant in the diet of *A. cyclophora*. We reached this conclusion because no items occur in area IV of the diagram (Fig. 3). However, two items were the most important, Brachyura and fragments of Teleosts. The other items were rare, because they occur in area III (Fig. 3). When whole groups were taken into consideration, Crustacea were dominant (presence in are IV), but Fishes were

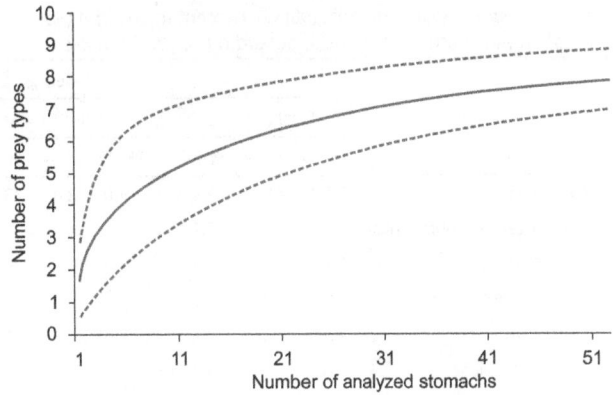

Figure 2. Cumulative prey curve in the diet, of *Atlantoraja cyclophora*, with a confidence interval of 95% upper and lower, in southeastern Brazil.

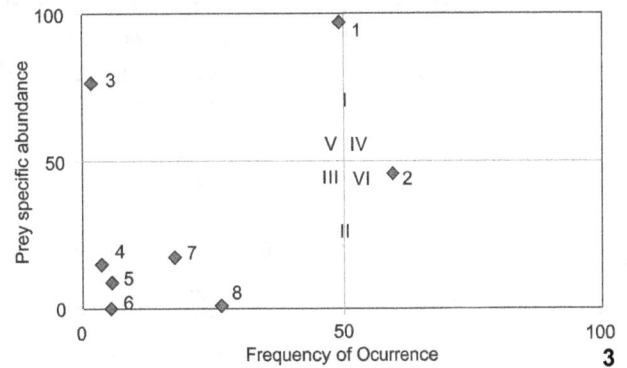

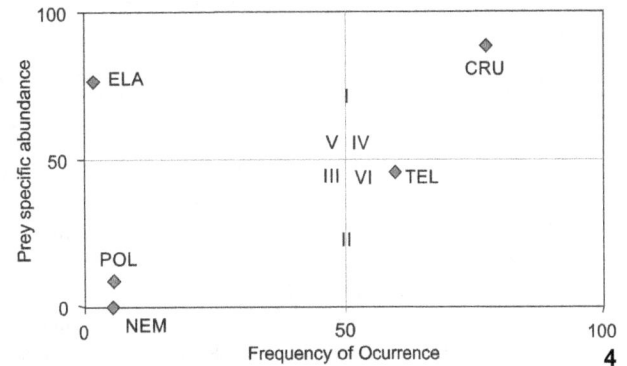

Figures 3-4. Diet of *Atlantoraja cyclophora* analyzed by Costello's method (1990), adapted by Amundsen et al. (1996), in southeastern Brazil. where: feeding strategy – specialist (I) or generalist (II); importance of prey – rare (III) or dominant (IV) and width of trophic niche – high between phenotype (V) or high within phenotype (VI) by food items (3) and by groups (4). Food items: 1) Brachyura, 2) Teleosts fragments, 3) Rajidae, 4) Stomatopoda, 5) Polychaeta, 6) Nematoda, 7) Caridea/Dendobranchiata, 8) Crustacean fragments. Groups: CRU) Crustacea, TEL) Teleost fishes, ELA) Elasmobranch, POL) Polychaeta, NEM) Nematoda.

Table II. Frequency of ocurrence (%FO), percentage weight (%W) and alimentary index (%IAi), of prey items in the total diet of *Atlantoraja cyclophora*, in southeastern Brazil, according to sex (females and males) and stages of maturity.

Prey items	Sex									Stages of maturity					
	Total (n = 57)			Females (n = 34)			Males (n = 23)			Juveniles (n = 17)			Adults (n = 40)		
	% FO	% W	% IAi	% FO	% W	% IAi	% FO	% W	% IAi	%FO	% W	%IAi	%FO	% W	%IAi
Crustacea	77.2	747.0	79.4	79.0	72.4	77.3	78.3	85.9	89.5	82.4	85.5	90.5	80.0	73.8	79.7
Caridea/Dendobranchiata	5.3	0.3	0.1	2.9	0.0	0.0	8.7	1.6	0.9	11.8	2.5	1.6	2.5	0.1	0.0
Caridea	10.5	2.1	1.3	11.8	1.5	0.9	8.7	4.6	2.7	5.9	2.4	0.8	12.5	2.0	1.2
Dendobranchiata	1.8	0.1	0.0				4.3	0.1	0.0				2.5	0.0	0.0
Brachyura	10.5	3.5	2.2	11.8	2.7	1.5	8.7	7.0	4.1	11.8	14.4	9.6	10.0	2.5	1.2
Leucosiidae	1.8	1.5	0.2				4.3	8.9	2.6				2.5	1.7	0.2
Parthenopidae	1.8	0.1	0.0				4.3	0.4	0.1				2.5	0.1	0.0
Portunidae	14.0	27.9	23.6	14.7	25.8	18.2	13.0	37.9	33.5				20.0	30.3	28.8
Achelous spinicarpus	22.8	38.9	53.5	29.4	42.0	59.2	17.4	24.4	28.8	11.8	64.8	43.1	30.0	36.8	52.3
Stomatopoda	3.5	0.1	0.0	5.9	0.1	0.0				5.9	0.1	0.0	2.5	0.1	0.0
Crustacean fragments	26.3	0.3	0.5	23.5	0.1	0.1	30.4	1.1	2.3	41.2	1.4	3.4	20.0	0.2	0.2
Teleosts	59.6	25.1	20.6	61.8	27.3	22.7	56.5	14.0	10.5	64.7	11.0	9.2	57.5	26.2	20.3
Dactylopterus volitans	8.8	15.5	8.2	14.7	18.7	13.2							12.5	16.8	10.0
Teleost fragments	50.9	3.2	9.7	50.0	2.5	6.0	52.2	6.4	22.6	64.7	11.0	40.4	45.0	2.5	5.3
Pleuronectiformes	1.8	1.2	0.1				4.3	7.3	2.1				2.5	1.3	0.2
Symphurus sp.	1.8	5.1	0.5	2.9	6.1	0.9							2.5	5.5	0.7
Polydactylus sp.	1.8	0.1	0.0				4.3	0.3	0.1				2.5	0.1	0.0
Elasmobranchs	1.8	0.2	0.0	2.9	0.3	0.0				5.9	3.0	0.2			
Rajidae	1.8	0.2	0.0	2.9	0.3	0.0				5.9	3.0	1.0			
Polychaeta	5.3	0.0	0.0	5.9	0.0	0.0	4.3	0.0	0.0	11.8	0.3	0.0	2.5	0.0	0.0
Euclymene sp.	1.8	0.0	0.0	2.9	0.0	0.0				5.9	0.1	0.0			
Sthenelais sp.	3.5	0.0	0.0	2.9	0.0	0.0	4.3	0.0	0.0	5.9	0.2	0.1	2.5	0.0	0.0
Nematoda	5.3	0.0	0.0	2.9	0.0	0.0	8.7	0.1	0.0	11.8	0.1	0.0	2.5	0.0	0.0
Nematods	5.3	0.0	0.0	2.9	0.0	0.0	8.7	0.1	0.0	11.8	0.1	0.1	2.5	0.0	0.0

also important (presence in area VI, with a high frequency of occurrence). Polychaeta and Nematoda were rare (Fig. 4).

The value of dietary overlap between females and males was high, 0.93. In both cases Crustaceans were more important (Table II) (77.3% IAi and 89.5% IAi respectively) followed by teleost, which were more heavily represented by females (22.7% IAi and 10.5% IAi respectively). Regarding the items, Portunidae (18.2% IAi for females and 33.5% IAi for males) and *A. spinicarpus* (59.2% IAi for females and 28.8% IAi for males) were the most important for both sexes. Other items like Caridea and crustacean fragments were more represented in the diet of males (2.7 and 2.3% IAi respectively).

The value of dietary overlap between juveniles and adults was also quite high, 0.99. In this case, the crustaceans were also more representative for the both stages of maturity (Table II) (90.5 and 79.7% IAi respectively), followed by teleost (9.2 and 20.3% IAi respectively). The polychaetes and nematodes

were less relevant, but more relevant to juveniles. In both groups *A. spinicarpus* was the most representative item (43.1% IAi for juveniles and 52.3%.IAi for adults).

The importance of prey groups differed according to season (Table III). The cluster analysis (Fig. 5) indicated that three groups were formed: Group 1 (Jan-Feb-Mar 06 and Jan-Feb-Mar 07); Group 2 (Apr-May-Jun 06 and Jul-Aug-Sep 07) and Group 3 (Jul-Aug-Sep 06 and Oct-Nov-Dec 06), indicating seasonal differences in diet composition. The quarters Jan-Feb-Mar 06 and Jan-Feb-Mar 07 had a great importance of crustaceans (97.3 and 98.9% IAi respectively), mainly Portunidae and *A. spinicarpus*. In the quarters of Jul-Aug-Sep 06 and Oct-Nov-Dec 06 teleost were the most representative (57.7 and 63.7% IAi respectively), especially *D. volitans*. The quarters, Apr-May-Jun 06 and Jul-Aug-Sep 07 had predominance of crustaceans (76.2 and 100% IAi respectively), but the number of individuals captured during this period was less representative.

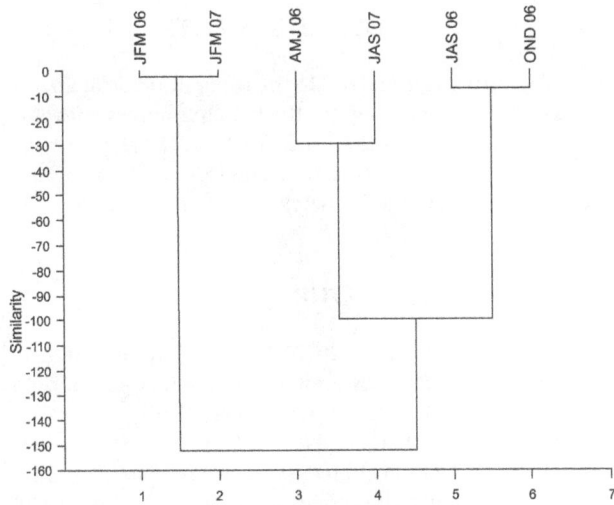

Figure 5. Dendrogram resulting from the cluster analysis, employing Ward's Method, according to quarter, in southeastern Brazil, where JFM correspond to Jan-Feb-Mar; AMJ to Apr-May-Jun, JAS to Jul-Aug-Sep and OND to Oct-Nov-Dec.

DISCUSSION

Females of *A. cyclophora* were larger at onset of sexual maturity than males, which is in agreement with results of other studies on Rajidae (ODDONE & VOOREN 2005, ODDONE *et al.* 2008). Females also reached larger size than males, as observed in other studies with *A. cyclophora* (ODDONE & VOOREN 2004, ODDONE *et al.* 2008). Skates were more abundant in the summer (Jan-Feb-Mar 06 and Jan-Feb-Mar 07). Summer is the season of greatest upwelling of the South Atlantic Central Water (SACW), which generates the oceanographic conditions favored by this species (ODDONE & VOOREN 2004).

Atlantoraja cyclophora feeds on benthic prey, mainly crustaceans and teleost. Crabs, mainly *A. spinicarpus*, were the most important item in the diet of *A.cyclophora*, probably because they are abundant in the area (PIRES 1992). The prey importance and feeding strategy diagrams confirmed these results, showing the dominance of crustaceans in the diet of the skate, followed by teleosts. Our results indicate that the diet of *A.cyclophora* is specialized. Although the Fishes did not have a high value of Alimentary Index (% IA_i) and was not a dominant item, this group

Table III. Frequency of Ocurrence (%FO), Percentage Weight (%W) and Alimentary Index (%IAi) of prey items in the diet of *Atlantoraja cyclophora* according to the quarters, in southeastern Brazil.

Prey items	Jan-Mar 06 (n = 24)			Apr-Jun 06 (n = 1)			Jul-Sep 06 (n = 7)			Oct-Dec 06 (n = 13)			Jan-Mar 07 (n = 11)			Jul-Sep 07 (n = 1)		
	% FO	%W	IAi	% FO	%W	IAi	% FO	%W	IAi	% FO	%W	IAi	% FO	%W	IAi	%FO	%W	IAi
Crustacea	83.3	95.6	97.3	100	76.2	76.2	85.7	37.9	42.3	46.2	51.1	36.3	100	97.2	98.9	100	100	100
Caridea/Dendobranchiata	4.2	0.6	0.1	100	76.2	76.2				7.7	0.0	0.0						
Caridea	4.2	0.6	0.1				57.1	6.8	17.9				9.1	0.7	0.2			
Dendobranchiata																100	66.7	66.7
Brachyura	16.7	10.4	5.5							7.7	1.1	0.4	9.1	0.3	0.1			
Leucosiidae										7.7	9.9	3.7						
Parthenopidae										7.7	0.4	0.1						
Portunidae	4.2	11.7	1.5				28.6	30.1	39.8	7.7	14.2	5.3	36.4	50.9	67.1			
Achelous spinicarpus	37.5	72.2	85.7				14.3	1.1	0.7	15.4	23.9	17.7	18.2	45.0	29.7			
Stomatopoda										15.4	0.6	0.5						
Crustacean fragments	29.2	0.2	0.1				14.3	0.0	0.0	7.7	1.0	0.4	45.5	0.3	0.5	100	33.3	33.3
Teleosts	50.0	4.4	2.7	100	23.8	23.8	71.4	62.1	57.7	84.6	48.8	63.7	54.5	1.9	1.0			
Dactylopterus volitans							14.3	36.0	23.8	23.1	40.5	45.1	9.1	0.8	0.2			
Teleost fragments	50.0	4.4	6.9	100	23.8	23.8	28.6	1.0	1.3	69.2	8.0	26.6	45.5	1.1	1.8			
Pleuronectiformes							14.3	4.9	3.3									
Symphurus sp.							14.3	20.2	13.3									
Polydactylus sp.										7.7	0.4	0.1						
Elasmobranchs													9.1	0.8	0.1			
Rajidae													9.1	0.8	0.3			
Polychaeta										7.7	0.0	0.0	18.2	0.1	0.0			
Euclymene sp.										7.7	0.0	0.0						
Sthenelais sp.													18.2	0.1	0.1			
Nematoda	4.2	0.0	0.0							7.7	0.0	0.0	9.1	0.0	0.0			
Nematods	4.2	0.0	0.0							7.7	0.0	0.0	9.1	0.0	0.0			

had a high Frequency of Occurrence (% FO), particularly in teleost fragments. These values are small because of the low% P and prey-specific abundance. Therefore we cannot estimate the true importance of this group. As previously observed by Zavala-Camin (1996), Fish are digested more rapidly than other groups that have a carapace, such as crustaceans.

A diet based on crustaceans and teleosts was also observed for a congeneric species, and for *A. cyclophora* by other authors (Soares *et al.* 1992, Bizarro *et al.* 2007). In a study of the diet of *A. cyclophora* in Ubatuba (23°20′S-24°00′S, 44°30′W-45°30′W), São Paulo, Brazil, the most important items were crustaceans and teleosts, and Brachyura was also relevant (Soares *et al.* 1992). In the same study, the feeding habits of *Raja castelnaui* (= *A. castelnaui*) were also analyzed and showed greater representation of crustaceans and teleosts. The diets of *Raja binoculata, R. inornata,* and *R. rhina* were studied in California (USA), and crustaceans and fish were the most important groups (Bizarro *et al.* 2007).

The diets of juveniles and adults overlapped significantly. However, teleost and *A. spinicarpus* are more important in the diet of adults and Caridea/Dendobranchiata, polychaetes and nematodes are more important in the diet of juveniles. The diets of both sexes were also very similar. Nevertheless, teleost and crabs are slightly more frequent in the diet of females, while males tend to eat more shrimp. These small differences can be attributed to differences in the method of catching food. It is difficult for smaller individuals to capture teleosts, which are more agile (Vianna *et al.* 2000).

The temporal analysis of feeding may have been compromised by the low abundance of individuals at certain times of the year, as for instance in Apr-May-Jun 06 and Jul-Aug-Sep 07. However, analysis of the other quarters indicated that there are two groups with similar diets. The water masses found in the region modify the physical and chemical parameters of the water during the seasons, which suggests a change in the structure and dynamics of the benthic fauna in the region (Pires 1992). Portunidae and *A. spinicarpus* were the most important groups consumed in Jan-Feb-Mar 06 and Jan-Feb-Mar 07, i.e., the summer of 2006 and the summer of 2007 (when the SACW approaches the coast), which may be explained by the increased availability of *A. spinicarpus* in this region at this time of year (Braga *et al.* 2005). At Ubatuba, *A. spinicarpus* is also important in the food of other rajids such as *R. agassizi* and *P. extenta,* especially in the summer, which is related to the local abundance of the species (Muto *et al.* 2001). Therefore, the period of greatest abundance of *A. spinicarpus* corresponds to the optimal oceanographic conditions for *A. cyclophora.* The upwelling of SACW may explain the importance of this crab in the diet of the skate. Seasonal changes in diet were observed in other studies of rajids (Muto *et al.* 2001, Bornatowski *et al.* 2010).

Although this study did not analyze a large number of stomachs, the data are important for interpreting the feeding biology of this skate, and to evaluate its responses to environmental conditions and fishing pressures.

ACKNOWLEDGEMENTS

The authors are grateful to the Laboratório de Biologia e Tecnologia Pesqueira group for the help in samples, measurements and dissection of the skates, to Paulo C. Paiva for helping with polychaete identification, and to Tereza C.G. da Silva and Karina A. Keunecke for helping with crustacean identification.

LITERATURE CITED

Aguiar, A.A. & J.L. Valentin. 2010. Biologia e Ecologia Alimentar de Elasmobrânquios no Brasil. **Oecologia Australis 14** (2): 464-489. doi:10.4257/oeco.2010.1402.09

Amundsen, P.A.; H.M. Gabler & F.J. Staldvik. 1996. A new approach to graphical analysis of feeding strategy from stomach contents data-modification of the Costello (1990) method. **Journal of Fish Biology 48**: 607-614. doi:10.1111/j.1095-8649.1996.tb01455.x

Bizarro, J.J.; H.J. Robinson; C.S. Rinewalt & D.A. Ebert. 2007. Comparative feeding ecology of four sympatric skate species off central California, USA. **Environmental Biology of Fishes 80** (2-3): 197-220. doi:10.1007/s10641-007-9241-6

Bornatowski, H.; M.C. Robert & L. Costa. 2010. Feeding of guitarfish *Rhinobatos percellens* (Walbaum, 1972) (Elasmobranchii, Rhinobatidae), the target of artisanal fishery in Southern Brazil. **Brazilian Journal of Oceanography 58** (1): 45-52. doi:10.1590/S1679-87592010000100005

Braga, A.A.; A. Fransozo; G. Bertini & P.B. Fumis. 2005. Composição e abundância dos caranguejos (Decapoda, Brachyura) nas regiões de Ubatuba e Caraguatatuba, litoral norte paulista, Brasil. **Biota Neotropica 5** (2). doi:10.1590/S1676-06032005000300004

Cheung, W.W.L.; T.J. Pitcher & D. Pauly. 2005. A fuzzy logic expert system to estimate intrinsic extinction vulnerabilities of marine fishes to fishing. **Biological Conservation 124**: 97-111. doi:10.1016/j.biocon.2005.01.017

Cortes, E. 1997. A critical review of methods of studying fish feeding based on analysis of stomach contents: application to elasmobranch fishes. **Canadian Journal of Fisheries and Aquatic Sciences 54**: 726-738. doi: 10.1139/f2012-051

Costello, M.J. 1990. Predator feeding strategy and prey importance: a new graphical analysis. **Journal of Fish Biology 36**: 261-263. doi:10.1111/j.1095-8649.1990.tb05601.x

Ebert, D.A & J.J Bizarro. 2007. Standardized diet compositions and trophic levels of skates (Chondrichthyes: Rajiformes: Rajoidei). **Environmental Biology of Fishes 80**: 221-237. doi:10.1007/s10641-007-9227-4

Ferry, L.A & G.M. Cailliet. 1996. Sample size and data analysis: are we characterizing and comparing diet properly?, p. 71-80. *In*: D. Mackinlay & K. Shearer (Eds). **Feeding Ecology and Nutrition in Fish.** São Francisco, American Fisheries Society.

GOMES, U.L.; C.N. SIGNORI; O.B.F. GADIG & H.R.S. SANTOS. 2010. **Guia para identificação de tubarões e raias do Rio de Janeiro.** Technical Books, Rio de Janeiro. 234p.

HURLBERT, S.H. 1978. The measurement of niche overlaps and some relatives. **Ecology 59:** 67-77. doi:10.2307/1936632

KAWAKAMI, E. & G. VAZZOLER. 1980. Método gráfico e estimativa de índice alimentar aplicado no estudo de alimentação de peixes. **Boletim do Instituto Oceanográfico 29** (2): 205-207. doi: 10.1590/S0373-55241980000200043

LACERDA, L.D.; H.H.M. PARAQUETTI; R.V. MARINS; C.E. REZENDE; I.R. ZALMON; M.P. GOMES & V. FARIAS. 2000. Mercury content in shark species from the south-eastern Brazilian coast. **Revista Brasileira de Biologia 60** (4): 571-576. doi: 10.1590/S0034-71082000000400005

MASSA, A.; N. HOZBOR & C.M. VOOREN. 2006. *Atlantoraja cyclophora.* In: IUCN (Ed.). **Red List of Threatened Species.** Version 2009.1. Available online at: http://www.iucnredlist.org [Accessed: 21 july 2009]. doi: 10.1007/978-1-4020-9703-4_8

MUTO, E.Y.; L.S.H. SOARES & R. GOITEIN. 2001. Food resource utilization of the skates *Rioraja agassizi* (Müller & Henle, 1841) and *Psammobatis extenta* (Garman, 1913) on the Continental shelf off Ubatuba, south-eastern Brazil. **Revista Brasileira de Biologia 61** (2): 217-238. doi: 10.1007/978-1-4020-9703-4_8

ODDONE, M.C & C.M. VOOREN. 2004. Distribution, abundance and morphometry of *Atlantoraja cyclophora* (Regan, 1903) (Elasmobranchii: Rajidae) in southern Brazil, Southwestern Atlantic. **Neotropical Ichthyology 2** (3): 137-144. doi: 10.1590/S1679-62252004000300005

ODDONE, M.C. & C.M. VOOREN. 2005. Reproductive biology of *Atlantoraja cyclophora* (Regan 1903) (Elasmobranchii: Rajidae) off southern Brazil. **ICES Journal of Marine Science 62:** 1095-1103.

ODDONE, M.C.; W. NORBIS; P.L. MANCINI & A.F. AMORIM. 2008. Sexual development and reproductive cycle of the Eyespot skate *Atlantoraja cyclophora* (Regan, 1903) (Condrichthyes: Rajidae: Arhynochobatinae), in southeastern Brazil. **Acta Adriatica 49** (1): 73-87. doi: 10.1016/j.icesjms.2005.05.002

PIANKA, E.R. 1973. The structure of lizard communities. **Annual Review of Ecology, Evolution and Systematics 4:** 53-74. doi: 10.1146/annurev.es.04.110173.000413

PIRES, A.M.S. 1992. Structure and dynamics of benthic megafauna on the continental shelf offshore of Ubatuba, southeastern Brazil. **Marine Ecology Progresses Series 86:** 63-76. doi: 10.3354/meps086063

ROSECCHI, E. & Y. NOUAZE. 1987. Comparison de cinq índices alimentaires utilises dans lánalyse des contenus stomacaux. **Revue des Travaux de L'institut des Pêches Maritimes 49** (4): 111-123.

SOARES, L.S.H.; C.L.B. ROSSI-WONGTSCHOWSKI; L.M.C. ALVARES; E.Y. MUTO & M. LOS ANGELES. 1992. Grupos tróficos de peixes demersais da plataforma continental interna de Ubatuba, Brasil. I. Condrichthyes. **Boletim do Instituto Oceanográfico 40** (1/2): doi: 10.1590/S0373-55241992000100006

VIANNA, M.; C.A. ARFELLI & A.F. AMORIM. 2000. Feeding of *Mustelus canis* (Elasmobranchii, Triakidae) caught off south-southeast coast of Brazil. **Boletim do Instituto de Pesca 26** (1): 79-84.

VOOREN, C.M. & S. KLIPPEL. 2005. Diretrizes para a conservação de espécies ameaçadas de elasmobrânquios, p. 213-228. In: C.M. VOOREN & S. KLIPPEL (Eds). **Ações para a conservação de tubarões e raias do Brasil.** Porto Alegre, Editora Igaré.

WALKER, P.A & G. HISLOP. 1998. Sensitive skates or resilient rays? Spatial and temporal shifts in ray species composition in the central and north-western North Sea between 1930 and the present day. **ICES Journal of Marine Science 55:** 392-402. doi:10.1006/jmsc.1997.0325

ZAVALA-CAMIN, L.A. 1996. **Introdução aos estudos sobre alimentação natural em peixes.** Maringá, EDUEM, 129p.

11

Evolution of bill size in relation to body size in toucans and hornbills (Aves: Piciformes and Bucerotiformes)

Austin L. Hughes

Department of Biological Sciences, University of South Carolina, Columbia SC 29205 USA. E-mail: austin@biol.sc.edu

ABSTRACT. Evidence that the bill of the Toco Toucan, *Ramphastos toco* Statius Muller, 1776, has a specialized role in heat dissipation suggests a new function for the large and light-weight bill of the toucan family (Piciformes: Ramphastidae). A prediction of this hypothesis is that bill length in toucans will increase with body mass at a rate greater than the isometric expectation. This hypothesis was tested in a phylogenetic context with measurements of skeletal elements in adult males of 21 toucan species. In these species, 64.3% of variance in relative skeletal measurements was accounted for by the contrast between bill and body size. Maxilla length and depth increased with body mass at a greater than isometric rate relative to both body mass and other linear skeletal measures. By contrast, no such trend was seen in a parallel analysis of 24 hornbill species (Bucerotiformes), sometimes considered ecological equivalents of toucans. The unique relationship between bill size and body mass in toucans supports the hypothesis that the evolution of a heat dissipation function has been a persistent theme of bill evolution in toucans.

KEY WORDS. Allometry; heat dissipation; Ramphastidae.

The adaptive significance of the large and remarkably light-weight bill of members of the Neotropical toucan family (Piciformes: Ramphastidae) has been the subject of much speculation (SHORT & HORNE 2001, TATTERSALL *et al.* 2009). Although VAN TYNE (1929: 39) suggested that the toucan's bill has no "especial adaptive function", a number of adaptive hypotheses have been proposed. BÜHLER (1995) proposed that the bill's large size and serrated edges originally evolved primarily as an adaptation for reaching and grasping fruit; later "tooth-like" markings on the bill may have evolved as adaptations to minimize mobbing by other birds when toucans prey on their nests (SICK 1993, BÜHLER 1995). SHORT & HORNE (2001) suggested a similar evolutionary sequence, while emphasizing the likely importance of species-specific bill markings in species recognition and courtship. Toucan bills are often brightly colored, and a few species show sexual dimorphism in bill coloration (SHORT & HORNE 2001). When bill color dimorphism occurs it is usually not very marked (SHORT & HORNE 2001), but its presence suggests that sexual selection may be another evolutionary force acting on toucan bills, at least in some species.

A further contribution to understanding the function of the toucan's bill was provided by evidence that the bill of the Toco Toucan, *Ramphastos toco* Statius Muller, 1776, serves as a key surface area for heat dissipation (TATTERSALL *et al.* 2009), which the bird can use to regulate body temperature by controlling blood flow. There is evidence that bills of a variety of avian taxa can function in heat dissipation (HAGAN & HEATH 1980, SCOTT *et al.* 2008, GREENBERG *et al.* 2012a, b, GREENBERG & DANNER 2013), suggesting that heat dissipation may be a

plesiomorphic function of the avian bill. In the Ramphastidae, it might be hypothesized that the ancestral heat-dissipation function has become elaborated by the evolution of a highly modifiable vascular radiator (TATTERSALL *et al.* 2009). On this hypothesis, the emergence of this vascular adaptation has been an additional factor favoring the evolution of large bill size in toucans, in conjunction with other selective pressures such as frugivory and signaling. Relatively little is known of toucans' thermal biology in nature, but the family is entirely Neotropical in distribution, and most species inhabit tropical lowland forests (SHORT & HORNE 2001), where high daily maximum temperatures occur year-round (GRUBB & WHITMORE 1996).

Consistent with a role for the toucan bill in heat-dissipation, TATTERSALL *et al.* (2009) presented evidence that bill length in juvenile and adult Toco Toucan increases as a function of body mass at a rate greater than the isometric expectation; i.e., greater than an exponent of 1/3 expected for a linear dimension (ALEXANDER 1971). Likewise, SYMONS & TATTERSALL (2010) provided evidence that across toucan species bill length increases as a function of body mass at a rate greater than linear expectation, using published data on 34 species of Ramphastidae. Such a relationship is expected if the bill plays a role in dissipating body heat, since metabolic rate increases with body mass with an exponent between 2/3 and 1.0, depending on activity level (GLAZIER 2008).

Here I analyze the evolution of bill size in relation both to the size of other major skeletal elements and to body mass across the family Ramphastidae in order to test the hypothesis of isometry against an alternative consistent with the bill's proposed role in heat dissipation, using statistical methods that control for phylogenetic relationships. A phylogenetic approach makes it possible to test the hypothesis that there has been a trend toward bill sizes greater than the isometric expectation throughout the evolution of this family. The hornbills (Bucerotiformes) are considered Old World ecological equivalents of the toucans, filling similar ecological niches in their respective ecosystems; most members of both families are cavity-nesting frugivores of tropical forests, and the two families have convergently evolved large slightly downcurved bills (KEMP 1995, KINNAIRD & O'BRIEN 2007). Because of these ecological parallels, I conduct a similar analysis with hornbills to compare the patterns of bill evolution relative to body size in the two groups of large-billed tropical birds. By comparing pattern of bill allometry in these two families, I test for a distinctive pattern of bill evolution in toucans, which would be consistent with the hypothesis that the toucan's bill plays an exceptionally highly developed role in heat dissipation.

MATERIAL AND METHODS

Measurements were made on complete skeletal specimens of adult males belonging to 21 species of toucans (Piciformes: Ramphastidae) and on complete skeletons of adult males of 24 species of hornbills (Bucerotiformes: Bucorvidae and Bucerotidae) from the U.S. National Museum of Natural History. The same measurements were also made on adult females of 16 of the toucan species and 14 of the hornbill species; since the patterns were similar for males and females, only the results for males are reported here. As an outgroup to root the phylogenetic tree of toucans, two species of New World barbets (Piciformes: Capitonidae) were used, *Capito aurovirens* (Cuvier, 1829) and *Semnornis ramphastinus* (Jardine, 1855) (Fig. 1). As an outgroup to root the phylogenetic tree of hornbills, the Eurasian Hoopoe, *Upupa epops* Linnaeus, 1758, (Upupiformes: Upupidae) was used (Fig. 2). Species were included based on available specimens but sampled all major lineages of both toucans and hornbills (Figs 1 and 2).

Because no comprehensive molecular phylogeny of toucans has been published, the phylogeny of toucans (Fig. 1) was derived from a combination of published DNA sequence-based phylogenies. The relationships among the ramphastid genera were based on NAHUM et al. (2003); see also PATANÉ et al. (2009). Relationships within the genus *Ramphastos* were based on PATANÉ et al. (2009); see also WECKSTEIN (2005). Relationships within *Pteroglossus* and *Baillonius* were based on EBERHARD & BERMINGHAM (2005) and PATEL et al. (2011). Relationships within *Aulacorhynchus* were based on BONACCORSO et al. (2011); and those within *Andigena* and *Selenidera* were based on LUTZ et al. (2013). The phylogeny of hornbills (Fig. 2) was based on the DNA sequence phylogeny of GONZALEZ et al. (2013). Most branching patterns indicated in Figs 1 and 2 were strongly supported in the original phylogenetic analyses by bootstrap prob-

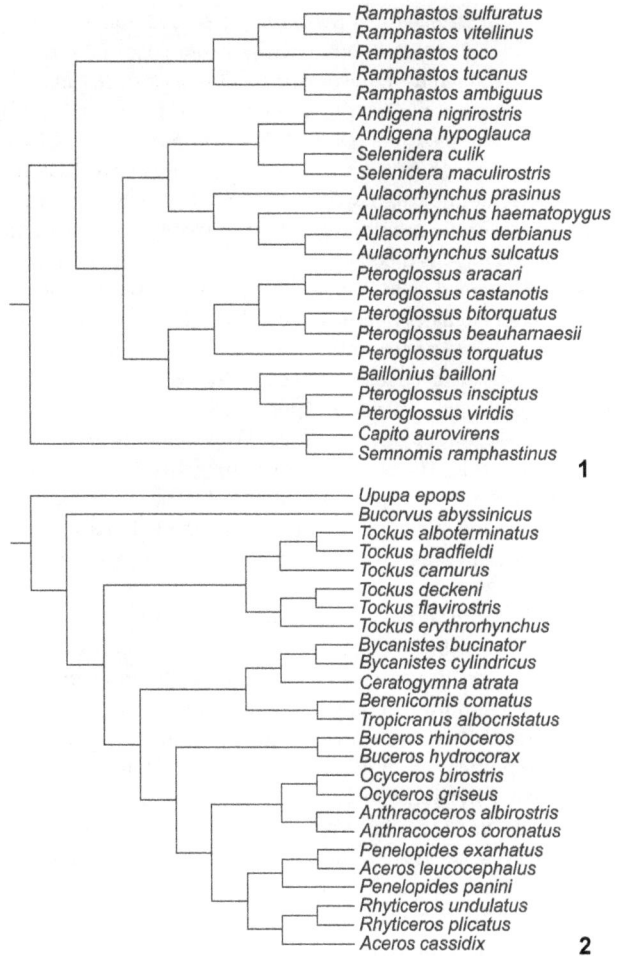

Figures 1-2. Phylogenies of species used in analyses: (1) phylogeny of toucans (Ramphastidae), rooted with two species of barbet, *Capito aurovirens* and *Semnornis ramphastinus*; (2) phylogeny of hornbills (Bucerotiformes), rooted with *Upupa epops*.

abilities, Bayesian posterior probabilities, or both. Preliminary analyses using the phylogeny of *Ramphastos* from HAFFER (1974, 1997) showed essentially identical results to those based on the phylogeny of PATANÉ et al. (2009); only the latter results are reported here.

In the case of toucans and barbets, the following nine skeletal measurements were made by digital caliper (BAUMEL 1993): 1) maxilla length, measured from the dorsal junction of the maxilla with the cranium to the tip of the bill (*Rostrum maxillare*); 2) maxilla depth, measured at the point of widest dorsal to ventral depth; 3) maxilla width, measured at the point of greatest lateral width; 4) cranium length, measured from the dorsal junction of the maxilla with the cranium to the posterior end (*Proeminentia cerebellaris*) of the cranium; 5) cranium width, measured at the point of greatest lateral width; 6) sternum, measured from *Apex carinae* to *Margo caudalis*; 7) synsacrum, measured from the anterior edge of *Ala preacetabularis* to the posterior edge of *Ala ischii*; 8) *femur*, measured from the proximal point of *Crista trochanteris* to the distal point of *Condylus lateralis*; and 9) tibiotarsus, measured from the proximal point of *Facies gastrocnemialis* to *Incisura intercondylaris*. In the case of the hornbill sample, Maxilla depth and cranium length were not included in analyses because the presence of the casque prevented comparable measurements in most species. Mean body mass values (in grams) for each species were obtained from Dunning (2008). In most species, values for males and females were given separately (Dunning 2008); and in those cases values for males were used. Data for male and female toucans are available in Appendix S1*, while male and female hornbills are available in Appendix S2*.

To test the sensitivity of these measurements to within-species variation, the same nine measurements were made on 10 adult males of *Ramphastos sulfuratus* Lesson, 1830 and 11 adult males of *R. toco*. Analysis of variance applied to log-transformed measurements was used to test for the relative magnitude of within-species and between-species components of variance in each of the nine measurements. In the case of all measurements, between species variance was significantly greater than within-species variance (p < 0.001 in every case except for cranium length, where p = 0.017, F-tests). The less pronounced between-species difference in cranium length than in the other measures was consistent with previous reports of low variance in similar measures (Höfling 1991).

Size-corrected transformations ("Mosimann transformations" – Mosimann 1970) were computed for the 9 skeletal measurements on toucans. Where the x_i are the individual measurements, let $z_i = \ln [x_i/G(x)]$, where G(x) is the geometric mean of the nine measurements within each species (Mosimann 1970). Principal components (PCs) were extracted from the correlation matrix of the z_is; the PC scores were used to provide size-independent indices of body shape for each toucan species (Darroch & Mosimann 1985, Jungers *et al.* 1995, Hughes 2013). The values used in these computations are shown in Appendix S1*. Principal components extracted Mosimann-transformed variables are preferable to principal components extracted from raw data, because the former are more effective in correctly identifying similarities in shape independent of body size (Jungers *et al.* 1995). Because maxilla depth and cranium length could not be accurately measured in the case of hornbills, in order to compare the two families, Mosimann transformations were computed for the remaining seven variables separately for each family; and principal components were extracted from these transformed variables.

To test hypotheses regarding isometric relationships among skeletal measures and between skeletal measures and body mass, all measurements were first log-transformed. The isometric expectation for the slope (b) of a log-log regression

(i.e., the allometric exponent) of any linear skeletal measure on any other linear skeletal measure is 1.0. The isometric expectation for b in a log-log regression of a linear measure on body mass (predicted to be proportional to body volume) is 1/3 (Alexander 1971). Because the toucan's maxilla is approximately triangular in cross-section (Short & Horne 2001), the external surface area of the bill consists largely of the area on the two lateral bill surfaces. Assuming that each of these surfaces has the approximate shape of an elongated triangle, the surface area of the maxilla can be roughly approximated by the product maxilla length times maxilla depth. The isometric expectation for b in a log-log regression of the product of two linear measures on body mass is 2/3 (Alexander 1971).

Isometric expectations were tested in two ways: 1) traditional analyses, in which phylogeny was not taken into account but rather each species was treated an independent unit of analysis; and 2) phylogenetically independent contrasts. In traditional analyses, the outgroup species were not included in the regressions. On the basis of the phylogenetic trees (Figs 1 and 2), phylogenetically independent contrasts were constructed using the PDAP (Garland *et al.* 1993) contrasts plug-in within Mesquite version 2.75 (Maddison & Maddison 2011). Regressions between phylogenetically independent contrasts were conducted without fitting an intercept (Garland *et al.* 1992). PCs extracted from the correlation matrix of the z_is were mapped on the toucan phylogeny by maximum parsimony using the "Map Continuous" function in Mesquite with default settings.

Following the recommendation of Smith (2009) for testing the null hypothesis of isometry, reduced major axis (RMA) was used rather than ordinary least squares (OLS) to estimate regression coefficients (Sokal & Rohlf 1995). The results with OLS (not shown) were very similar to those of RMA in the present case because correlations between variables were high. For all allometric regressions reported here (N = 58), the linear correlation coefficient ranged from 0.735 to 0.988 (mean = 0.887 ± 0.008 S.E., median = 0.893). OLS was used to estimate regression lines (Smith 2009). All reported significance levels are corrected for multiple testing by the Bonferroni method (Sokal & Rohlf 1995). Statistical analyses were conducted in Minitab (http://www.minitab.com).

RESULTS

Relative length of skeletal elements

The first principle component (PC1) extracted from the correlation matrix of size-corrected transformations of nine linear skeletal measures of toucans accounted for 64.3% of the variance and represented a contrast between two sets of variables: 1) maxilla length and maxilla depth; and 2) the other variables except for sternum (Table I). Thus PC1 could be interpreted as a size-corrected measure of the contrast between bill size and body

---

*Available as Online Supplementary Material accessed with the online version of the manuscript at http://www.scielo.br/zool

size. PC2, accounting for 17.7% of the variance, seemed to mainly consist of a contrast between sternum and maxilla width (Table I). In order to provide a visual image of how the contrast between bill and body size has evolved across the Ramphastidae, PC1 values were mapped across the phylogeny of toucans. The highest values (indicating greatest bill size relative to body size) were seen in *Ramphastos* (Fig. 3). The phylogeny also supported the hypothesis of a parallel increase in bill size relative to body size in the *Pteroglossus/Baillonius* lineage (Fig. 3).

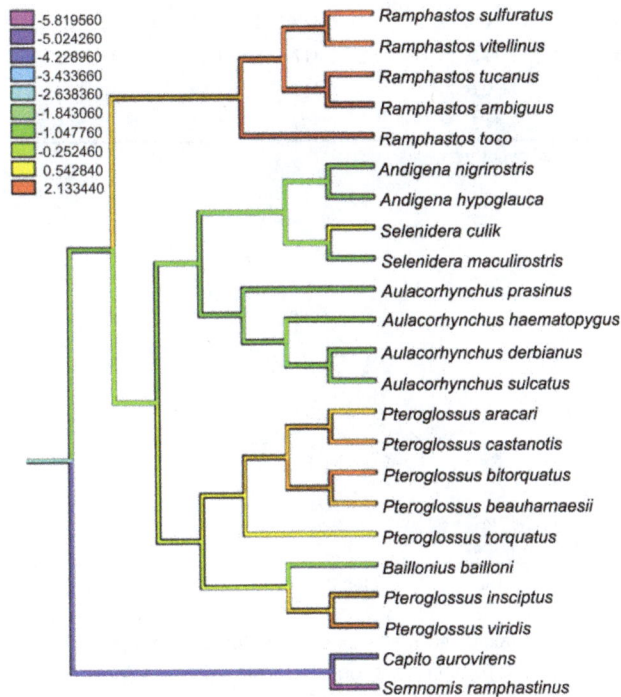

Figure 3. Parsimony-reconstructed PC1 scores across the phylogeny of toucans.

Table I. Variable loadings on the first two principal components (PC1 and PC2) derived from transformed skeletal measurements of 21 species of Ramphastidae.

Variable	PC1	PC2
Maxilla Length	0.407	-0.065
Maxilla Depth	0.391	0.069
Maxilla Width	-0.030	-0.691
Cranium Length	-0.375	-0.106
Cranium Width	-0.379	-0.059
Sternum	0.008	0.663
Synsacrum	-0.368	0.239
Femur	-0.327	0.003
Tibiotarsus	-0.391	0.002
% variance	64.300	17.700

Because maxilla depth and cranium length could not be accurately measured in the case of hornbills, principal were extracted from the correlation matrix of size-corrected transformations of remaining seven linear skeletal measures of in each family (Table II). Even excluding maxilla depth and cranium length, PC1 (accounting for 60.8% of the variance) in the toucan data again appeared mainly to represent a contrast between bill size and body size (Table II). By contrast, in hornbills, PC1 accounted for only 30.8% of the variance and appeared to reflect mainly a contrast between body size and the width of both bill and cranium (Table II). The loading of maxilla length on PC1 in hornbills (-0.063) differed strikingly from that in toucans (0.481, Table II). Thus hornbills appeared to differ from toucans in that bill length relative to body size was not a major factor in cross-species comparisons of major skeletal elements.

Table II. Variable loadings on the principal components (PC1) derived from transformed skeletal measurements of 21 species of Ramphastidae and 24 species of Bucerotidae.

Variable	Ramphastidae	Bucerotidae
Maxilla Length	0.481	-0.063
Maxilla Width	0.481	0.583
Cranium Width	0.039	0.342
Sternum	-0.421	-0.377
Synsacrum	0.001	-0.432
Femur	-0.451	-0.145
Tibiotarsus	-0.394	-0.436
% variance	60.800	38.000

Allometric relationships

In traditional analyses, not accounting for the phylogeny, the 9 log-transformed skeletal measures were regressed against log body mass (Table III). Likewise, phylogenetically independent contrasts in the same 9 log-transformed skeletal measures were regressed against phylogenetically independent contrasts in log body mass (Table III). The results were broadly similar in the two types of analysis (Table III). In both cases, the allometric exponent (b) for maxilla length and maxilla depth were significantly greater than the isometric expectation (1/3; Table III). In the case of phylogenetically independent contrasts, b for maxilla width was also significantly greater than the isometric expectation (Table III). In traditional analyses, but not in phylogenetically independent contrasts, b for femur was significantly greater than the isometric expectation (Table III). No other linear measure showed b significantly greater than the linear expectation in either type of analysis, but in the traditional analyses b for cranium length was significantly less than the isometric expectation (Table III). When the log of the product of maxilla length and maxilla depth,

Table III. Allometric exponents (b) of regression of skeletal measures on body mass of 21 species of Ramphastidae in traditional analyses and in phylogenetically independent contrasts.

Dependent variable	Null hypothesis[a]	Traditional analyses (non-phylogenetic)			Phylogenetically independent contrasts		
		b	t	p b	b	t	p[b]
Maxilla Length	1/3	0.647	3.44	< 0.05	0.884	3.55	< 0.05
Maxilla Depth	1/3	0.607	4.35	< 0.01	0.678	5.50	< 0.001
Maxilla Length x Maxilla Depth	2/3	1.239	3.96	< 0.01	1.527	4.88	< 0.001
Maxilla Width	1/3	0.339	0.28	N.S.	0.469	3.52	< 0.05
Cranium Length	1/3	0.189	-7.35	< 0.001	0.269	-2.51	N.S.
Cranium Width	1/3	0.305	-1.17	N.S.	0.348	0.41	N.S.
Sternum	1/3	0.430	1.58	N.S.	0.438	1.55	N.S.
Synsacrum	1/3	0.378	1.26	N.S.	0.417	1.30	N.S.
Femur	1/3	0.424	3.69	< 0.05	0.433	2.54	N.S.
Tibiotarsus	1/3	0.366	1.51	N.S.	0.371	1.26	N.S.

[a] The value shown is the isometric expectation of the exponent (b) under the relevant null hypothesis.
[b] All P-values shown have been corrected by the Bonferroni procedure.

was regressed against log body mass, in both types of analyses, b was significantly greater than the isometric expectation (2/3; Table III). These results imply that in the Ramphastidae both bill size and bill surface area increase with body mass at a greater rate than expected under isometry.

The hypothesis that maxilla length and maxilla depth show a distinctive pattern of evolution in the Ramphastidae was further tested by regressing logarithms of these measures on those of linear measures of non-maxillary structures (Table IV). In both traditional and phylogenetically based analyses, b exceeded the isometric expectation (1.0) in every case (Table IV). In both kinds of analyses, the regressions with maxilla length as the dependent variable, the b was significantly greater than the isometric expectation with cranium length, cranium width, and sternum as dependent variables (Table III). In both kinds of analyses, in regressions with maxilla depth as the dependent variable, b was significantly greater than the isometric expectation with cranium length, cranium width, sternum, synsacrum and tibiotarsus as dependent variables (Table IV). In the phylogenetically based analysis, maxilla length showed b greater than the isometric expectation when regressed on tibiotarsus, and maxilla depth also showed b greater than the isometric expectation when regressed on femur (Table IV).

When log-transformed skeletal measures of hornbills were regressed against log body mass, a very different pattern was seen from that seen in toucans (Table V). In hornbills, b for maxilla length did not differ significantly from the isometric expectation (Table V), resulting in distinct patterns in toucans and hornbills (Fig. 4). In traditional analyses, the only measure for which the slope significantly exceeded the isometric expectation was synsacrum, while the slope for cranium width was significantly less than the isometric expectation (Fig. 4). Likewise, in phylogenetically based analyses, the slope of the relationship for contrasts in log maxilla length did not

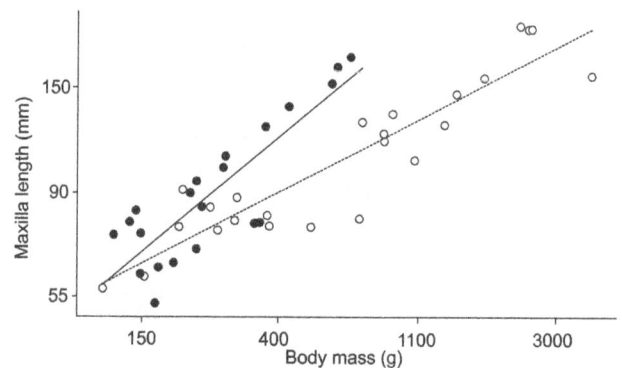

Figure 4. Maxilla length vs. body mass for toucans (solid circles) and hornbills (open circles) on a log scale, with OLS linear regression lines: for toucans, Y = 1.460 + 0.551X (solid line; adj. R^2 = 71.0%, p < 0.001) and for hornbills: Y = 2.445 + 0.342X (dotted line; adj. R^2 = 84.0%, p < 0.001).

differ significantly from the isometric expectation, and the only measure for which the slope exceeded the isometric expectation was synsacrum (Table V).

The Northern Ground Hornbill *Bucorvus abyssinicus* (Boddaert, 1783) (Fig. 2) had a relatively large synsacrum (129.9 mm) in comparison to the 23 other hornbill species of (mean = 62.6 ± 5.0 mm, range = 30.5 to 105.mm, Appendix S2*). A relatively large is consistent with the terrestrial habits and relatively large legs of the Northern Ground Hornbill (KEMP 1995). However, even when the Northern Ground Hornbill was excluded from the data set, a similar relationship was seen in the traditional analysis of the relationship between log synsacrum and log body mass (b = 0.409; test of equality to isometric expectation, p < 0.001). Likewise, in phylogenetically independent contrasts, when both the ancestral node and the node

Table IV. Allometric exponents (b) of regression of maxilla length and maxilla depth on other skeletal measures of 21 species of Ramphastidae in traditional analyses and phlogenetically independent contrasts.

| Independent Variable | Traditional analyses (non-phylogenetic) | | | | | | Phylogenetically Independent Contrasts | | | | | |
| | Maxilla Length | | | Maxilla Depth | | | Maxilla Length | | | Maxilla Depth | | |
	b	t	p [a]	b	t	p [a]	b	t	p [a]	b	t	p [a]
Cranium Length	3.424	3.62	< 0.05	3.209	4.24	< 0.01	3.597	3.53	< 0.05	2.782	4.63	< 0.001
Cranium Width	2.123	4.00	< 0.01	1.988	4.44	< 0.01	2.525	3.89	< 0.01	1.963	4.63	< 0.001
Sternum	1.503	3.58	< 0.05	1.408	3.64	< 0.01	2.024	3.79	<0.01	1.576	3.55	< 0.05
Synsacrum	1.709	2.74	N.S.	1.602	3.34	< 0.05	2.145	2.39	N.S.	1.650	2.96	< 0.05
Femur	1.528	2.20	N.S.	1.431	2.59	N.S.	2.012	2.40	N.S.	1.557	2.98	< 0.05
Tibiotarsus	1.769	2.75	N.S.	1.657	3.30	< 0.05	2.355	3.03	< 0.05	1.821	4.09	< 0.01

[a] All P-values shown have been corrected by the Bonferroni procedure. The null hypothesis in each case is that b = 1.0

Table V. Allometric exponents (b) of regression of skeletal measures on body mass of 24 species of Bucerotidae in traditional analyses and phlogenetically independent contrasts.

| Dependent variable | Null hypothesis [a] | Traditional analyses (non-phylogenetic) | | | Phylogenetically independent contrasts | | |
		b	t	P b	b	t	p b
Maxilla Length	1/3	0.372	1.15	N.S.	0.351	0.45	N.S.
Maxilla Width	1/3	0.317	-0.36	N.S.	0.333	-0.02	N.S.
Cranium Width	1/3	0.280	-3.17	< 0.05	0.302	-1.44	N.S.
Sternum	1/3	0.327	-0.43	N.S.	0.324	-0.56	N.S.
Synsacrum	1/3	0.407	5.39	< 0.001	0.392	3.18	< 0.05
Femur	1/3	0.325	-0.42	N.S	0.349	0.67	N.S.
Tibiotarsus	1/3	0.323	-0.44	N.S.	0.366	1.34	N.S.

[a] The value shown is the isometric expectation of the exponent (b) under the relevant null hypothesis.
[b] All P-values shown have been corrected by the Bonferroni procedure.

linking the Northern Ground Hornbill to the other hornbills (Fig. 2) were excluded, there was a similar relationship between contrasts in log synsacrum and contrasts in log body mass (b = 0.429, test of equality to isometric expectation, p < 0.01).

DISCUSSION

An examination of the relationship among linear measures of major skeletal measures and between those measures and body mass supported an unusual pattern of bill size evolution in the toucan family. Throughout the toucan family, the length and depth of the maxilla increased as a function of body mass at a rate greater than expected under isometry, implying disproportionately large bills per unit body mass in large-bodied toucan species, consistent with the hypothesis that heat dissipation has been an important factor in the evolution of the large bills of toucans (TATTERSALL et al. 2009). Since the capacity for radiation of heat from the bill is a function of surface area, it is further expected that bill surface area will increase with body mass at a rate greater than expected under isometry. The present analyses supported this prediction, since the results showed that product of toucan bill length and depth increases with body mass at a rate greater than the isometric expectation.

In spite of the ecological parallels between toucans and hornbills (KEMP 1995, KINNAIRD & O'BRIEN 2007), the present analyses provided no evidence of a greater than isometric increase in hornbill maxilla length as a function of body mass. These results are consistent with the hypothesis that the bill of hornbills does not play a role in heat dissipation analogous to that of toucans. This hypothesis will require further testing through physiological study of hornbills. It is of interest, however, that hornbills appear to make use of alternative heat-dissipation mechanisms from those seen in toucans; for instance, evaporative water loss from the bare skin under the wings, which is exposed by the hornbills' unique lack of underwing-coverts (KEMP 1995).

In contrast to the maxilla, in hornbills synsacrum length increased with body mass at a greater rate than expected under isometry. This pattern was seen even when the terrestrial Northern Ground Hornbill, in which the synsacrum was unusually large, was excluded from the analysis. The increase in the length of the synsacrum with body mass may reflect an enhanced need for weight support in the larger hornbills. That no similar trend is seen in toucans may reflect their substantially smaller body masses, as well as the fact that even arboreal hornbills spend more time on the ground than toucans

(KEMP 1995), with a consequent requirement to support the body weight on the pelvic girdle.

All phylogenies represent hypotheses, which are subject to revision in the light of additional data (GARLAND et al. 2005). In the present case, the fact that traditional and phylogenetic analyses yielded very similar results suggests that the conclusions are likely to be robust to phylogenetic revision. In addition, the phylogenetic perspective provided evidence that bill size increased relative to body size independently in different toucan lineages. In the toucans, 64.3% of variance in size-adjusted skeletal measures was accounted for by a composite variable (PC1) that could be interpreted as reflecting the contrast between bill and body size. PC1 increased markedly the genus *Ramphastos* and the *Pteroglossus/Baillonius* lineage (Fig. 3). Thus, the relationship between bill dimensions and body mass was a recurring feature of evolution across the phylogeny of toucans.

A fuller understanding of the evolution of the bill in toucans and hornbills will require investigation of the thermal biology of these species in a natural setting. At present little is known about the temperature regimes encountered by these birds in nature and the variety of behavioral and physiological strategies which they employ to cope with temperature extremes. Additional studies of morphological evolution, combining data on both within-species and between-species variation, can provide further insights into the selective forces acting on bill morphology. In particular, comparative study of the evolution of bill morphology in males and females will help to elucidate the potential role of sexual selection as a factor in shaping the evolution of the bill in these families.

Typically biological structures are multi-functional; thus, support for the heat-dissipation hypothesis precludes neither the hypothesis that reaching for and grasping fruit played a key role in the origin of the toucan's large bill, nor the hypothesis that the bill has secondarily evolved roles in aposematic and intraspecific signaling, including a role in sexual selection (BÜHLER 1995, SHORT & HORNE 2001). The apparent convergence between the bills of toucans and hornbills lends plausibility to the hypothesis that the original selective pressure favoring large bills in the toucan lineage arose from frugivory (BÜHLER 1995). At the same time, some role in heat dissipation is likely to be a plesiomorphic character of the bills of birds (HAGAN & HEATH 1980, SCOTT et al. 2008; GREENBERG et al. al. 2012a, b, GREENBERG & DANNER 2013). Thus, the relatively elaborate mechanisms of heat-dissipation seen in toucans may have arisen as an exaptation (GOULD & VRBA 1982); that is, the co-option of an existing structure for a new function. The present results, because they reveal that the relationship between bill dimensions and body mass has persisted across the toucan phylogeny, suggest that the co-option of the toucan bill for heat dissipation may represent an ancient feature within this family, which has acted in concert with other selective factors favoring large bill size.

ACKNOWLEDGMENTS

I am grateful to the staff of the U.S. National Museum, Bird Division, especially Chris Milensky, for access to specimens. Comments by Glenn J. Tattersall and an anonymous reviewer greatly improved the manuscript.

LITERATURE CITED

ALEXANDER, R.M. 1971. **Size and Shape.** London, Edward Arnold.

BAUMEL, J.J. 1993. **Handbook of Avian Anatomy: Nomina Anatomica Avium.** Cambridge, Nuttall Ornithological Club, 2nd ed.

BONACCORSO, E.; J.M. GUAYASAMIN; A.T. PETERSON & A.G. NAVARRO-SIGÜENZA. 2011. Molecular phylogeny and systematics of Neotropical toucanets in the genus *Aulacorhynchus* (Aves, Ramphastidae). **Zoologica Scripta 40**: 336-349. doi: 10.1111/j.1463-6409.2011.00475.x

BÜHLER, P. 1995. Grösse, Form und Färbung des Tukanschnabels – Grundlage für den evolutiven Erflog der Ramphastiden? **Journal für Ornithologie 136**: 187-193.

DARROCH, J.N.& J.E. MOSIMANN. 1985. Canonical and principal components of shape. **Biometrika 72**: 241-252.

DUNNING, J.C. 2008. CRC handbook of Avian Body Masses. Boca Raton, CRC Press, 2nd ed.

EBERHARD, J.R. & E. BERMONGHAM. 2005. Phylogeny and comparative biogeography of *Pionopsitta* parrots and *Pteroglossus* toucans. Molecular Phylogenetics and Evolution 36: 288-304. doi: 10.1016/j.ympev.2005.01.022

GARLAND JR, T.; P.H. HARVEY & A.R. IVES. 1992. Procedures for the analysis of comparative data using phylogenetically independent contrasts. **Systematic Biology 41**: 18-32. doi: 10.1093/sysbio/42.3.265

GARLAND JR, T.; A.W. DICKERMAN; C.M. JANIS & J.A. JONES. 1993. Phylogenetic analysis of covariance by computer simulation. **Systematic Biology 42**: 265-292.

GARLAND JR, T.; A.F. BENNET & E.L. REZENDE. 2005. Phylogenetic approaches in comparative physiology. **Journal of Experimental Biology 208**: 3015-3035. doi: 10.1242/jeb.01745

GLAZIER, D.S. 2008. Effects of metabolic level on the body size scaling of metabolic rate in birds and mammals. **Proceedings of the Royal Society B 275**: 1405-1410. doi:10.1098/rspb.2008.0118

GONZALEZ, J.-C.; B.C. SHELDON; N.J. COLLAR & J.A. TOBIAS. 2013. A comprehensive molecular phylogeny for the hornbills (Aves: Bucerotidae). **Molecular Phylogenetics and Evolution 67**: 468-483. doi: 10.1016/j.ympev.2013.02.012

GOULD, S.J. & E.S. VRBA. 1982. Exaptation – a missing term in the science of form. **Paleobiology 8**: 4-15.

GREENBERG, R. & R.M. DANNER. 2013. Climate, ecological release and bill dimorphism in an island songbird. **Biology Letters 9**: 20130118. doi:10.1098/rsbl.2013.0118

GREENBERG, R.; V. CADENA; R.M. DANNER & G. TATTERSALL. 2012a. Heat loss may explain bill size differences between birds occupying different habitats. **Plos One 7** (7): e40933. doi: 10.1371/journal.pone.0040933

GREENBERG, R.; R. DANNER; B. OLSEN & D. LUTER. 2012b. High summer temperature explains bill size variation in salt marsh sparrows. **Ecography 35**: 146-152. doi: 10.1111/j.1600-0587.2011.07002.x

GRUBB, P.J. & T.C. WHITMORE. 1966. A comparison of montane and lowland rain forest in Ecuador: II. The climate and its effects on the distribution and physiognomy of the forests. **Journal of Ecology 54**: 303-333.

HAFFER, J. 1974. **Avian Speciation in Tropical South America.** Cambridge, Nuttal Ornithological Club.

HAFFER, J. 1997. Foreword: species concepts and species limits in ornithology, p. 11-24. *In*: J. DEL HOYO; A. ELLIOTT & J. SARGATAL (Eds). **Handbook of the Birds of the World.** Barcelona, Lynx Ediciones, vol. 4.

HAGAN, A.A. & J.E. HEATH. 1980. Regulation of heat loss in the duck by vasomotion in the bill. **Journal of Thermal Biology 5**: 95-101. doi: 10.1016/0306-4565(80)90006-6

HÖFLING, E. 1991. Étude comparative du crâne chez les Ramphastidae (Aves, Piciformes). **Bonner Zoologische Beiträger 42**: 55-65.

HUGHES, A.L. 2013. Indices of Anseriform body shape based on relative size of major skeletal elements and the relationship to reproductive effort. **Ibis 155**: 835-846. doi: 10.1111/ibi.12087

JUNGERS, W.L.; A.B. FALSETTI & C.E. WALL. 1995. Shape, relative size, and size-adjustments in morphometrics. **American Journal of Physical Anthropology 38**: 137-161. doi: 10.1002/ajpa.1330380608

KEMP, A. 1995. **The Hornbills.** Oxford, Oxford University Press.

KINNAIRD, M.F. & T.G. O'BRIEN. 2007. **The Ecology and Conservation of Asian Hornbills: Farmers of the Forest.** Chicago, University of Chicago Press.

LUTZ, H.L.; J.D. WECKSTEIN; J.S. PATANÉ; J.M. BATES & A. ALEIXO. 2013. Biogeography and spatio-temporal diversification of *Selenidera* and *Andigena* toucans (Aves: Ramphastidae). **Molecular Phylogenetics and Evolution 69**: 873-883. doi: 10.1016/j.ympev.2013.06.017

MADDISON, W.P. & D.R. MADDISON. 2011. Mesquite: a modular system for evolutionary analysis. Version 2.75, available online at: http://mesquiteproject.org [Accessed: 30/IX/2011].

MOSIMANN, J. 1970. Size allometry: size and shape variables with characterizations of the lognormal and generalized gamma distributions. **Journal of the American Statistical Association 65**: 930-945.

NAHUM, L.A.; S.L. PEREIRA; F.M. FERNANDES; S.R. MATIOLI & A.WAJNTAL. 2003. Diversification of Ramphastinae (Aves, Ramphastidae) prior to the Cretaceous/Tertiary boundary as shown by molecular clock of mtDNA sequences. **Genetics and Molecular Biology 26**: 411-418. doi: 10.1590/S1415-47572003000400003.

PATANÉ, J.S.; J.D. WECKSTEIN; A. ALEIXO & J.M. BATES. 2009. Evolutionary history of *Ramphastos* toucans: molecular phylogenetics, temporal diversification, and biogeography. **Molecular Phylogenetics and Evolution 53**: 923-934. doi: 10.1016/j.ympev.2009.08.017

PATEL, S.; J.D. WECKSTEIN; J.S. PATANÉ; J.M. BATES & A. ALEIXO. 2011. Temporal and spatial diversification of *Pteroglossus* araçaris (Aves: Ramphastidae) in the neotropics: constant rate of diversification does not support an increase in radiation during the Pleistocene. **Molecular Phylogenetics and Evolution 58**: 105-115. doi: 10.1016/j.ympev.2010.10.016

SCOTT, G.R.; V. CADENA; G.R. TATTERSALL & W.K. MILSOM. 2008. Body temperature depression and peripheral heat loss accompany the metabolic and ventilatory responses to hypoxia in low and high altitude birds. **Journal of Experimental Biology 211**: 1326-1335. doi: 10.1242/ jeb.015958

SHORT, L.L. & J.F. HORNE. 2001. **Toucans, Barbets and Honeyguides.** Oxford, Oxford University Press.

SICK, H. 1993. **Birds in Brazil.** Princeton, Princeton University Press.

SMITH, R.J. 2009. Use and misuse of the reduced major axis for line-fitting. **American Journal of Physical Anthropology 140**: 476-486. doi: 10.1002/ajpa.21090

SOKAL, R.R. & F.J. ROHLF. 1995. **Biometry.** San Francisco, W.H. Freeman, 3rd ed.

SYMONS, R.E. & G.J. TATTERSALL. 2010. Geographical variation in bill size across bird species provides evidence for Allen's Rule. **The American Naturalist 176**: 188-197. doi: 10.1086/ 653666

TATTERSALL, G.J.; D.V. ANDRADE & A.S. ABE. 2009. Heat exchange from the toucan bill reveals a controllable vascular thermal radiator. **Science 325**: 468-470. doi: 10.1126/science.1175553

VAN TYNE, J. 1929. The life history of the toucan *Ramphastos brevicarinatus*. **University of Michigan Museum of Zoology, Miscellaneous Publications 19**: 1-43.

WECKSTEIN, J.D. 2005. Molecular phylogenetics of the Ramphastos toucans: implications for the evolution of morphology, vocalizations, and coloration. **Auk 122**: 1191-1209. doi: doi: 10.1642/0004-8038(2005)122[1191:MPOTRT]2.0.CO;2

12

Alpaida (Araneae: Araneidae) from the Amazon Basin and Ecuador: new species, new records and complementary descriptions

Regiane Saturnino[1,*], Bruno V.B. Rodrigues[1] & Alexandre B. Bonaldo[1]

[1]*Laboratório de Aracnologia, Coordenação de Zoologia, Museu Paraense Emílio Goeldi. Avenida Perimetral 1901, Terra Firme, 66077-830 Belém, Pará, Brazil.*
Corresponding author. E-mail: sf.regiane@gmail.com

ABSTRACT. Two new species of *Alpaida*, *A. levii* and *A. yanayacu*, the male of *A. iquitos* Levi, 1988 and the female of *A. gurupi* Levi, 1988 are described and illustrated for the first time. *Alpaida levii*, described from the states of Pará and Amazonas, is closely related to *A. delicata* (Keyserling, 1892), but differs in that males have a curved and distally pointed terminal apophysis, and females have the epigynum longer than wide and a drop-shaped median lobe. *Alpaida yanayacu* is only known from Ecuador and is characterized by long and rounded lateral lobes in ventral view and median lobe wide at base. A brief discussion about the morphological similarity among *A. levii*, *A. delicata* and *A. truncata* (Keyserling, 1865) is presented. Based on the information provided, new diagnoses are proposed for *A. delicata* and *A. truncata*. New records of *A. antonio* Levi, 1988, *A. bicornuta* (Taczanowski, 1878), *A. boa* Levi, 1988, *A. deborae* Levi, 1988, *A. delicata*, *A. erythrothorax* (Taczanowski, 1873), *A. guimaraes* Levi, 1988, *A. guto* Abrahim & Bonaldo, 2008, *A. gurupi*, *A. iquitos*, *A. leucogramma* (White, 1841), *A. murtinho* Levi, 1988, *A. negro* Levi, 1988, *A. rossi* Levi, 1988, *A. septemmammata* (O. Pickard-Cambridge, 1889), *A. simla* Levi, 1988, *A. tayos* Levi, 1988, *A. truncata*, *A. urucuca* Levi, 1988, *A. utiariti* Levi, 1988 and *A. veniliae* Levi, 1988 are presented.

KEY WORDS. Arachnida, distribution, Neotropical Region, spiders, taxonomy.

Species of *Alpaida* O. Pickard-Cambridge, 1889 are diurnal orb-weaving spiders occurring only in the Neotropical region. Currently, the genus is composed of 148 species (WORLD SPIDER CATALOG 2014), although it is estimated to contain about 200-300 species (LEVI 1988). Species of this genus are characterized by the glabrous body, orange to red carapace, abdomen and carapace without bristles; male palp with radix, embolus and terminal apophysis fused into one sclerite; mushroom-shaped paramedian apophysis connected to conductor; epigynum usually represented by a transverse sclerotized structure, with posterior lips, a median scape and copulatory openings located on each side between plate and lips (LEVI 1988).

Alpaida was revised by LEVI (1988), who dealt with 134 species, among which 94 were described as new. Many of the 40 previously known species had been erroneously transferred to *Alpaida* from different genera. Since LEVI's (1988) contribution only 11 species have been added to the genus: *A. guto* Abrahim & Bonaldo, 2008 described from the Brazilian Amazon; *A. itacolomi* Santos & Santos, 2010, *A. tonze* Santos & Santos, 2010, *A. caramba* Buckup & Rodrigues, 2011 and *A. arvoredo* Buckup & Rodrigues, 2011 described from the Atlantic Forest; *A. teresinha* Braga-Pereira & Santos, 2013 and *A. toninho* Braga-Pereira & Santos, 2013 from coastal forest areas in Brazil; *A. monzon audiberti* Dierkens, 2014 and *A. oyapockensis* Dierkens, 2014 from French Guiana; *A. losamigos* Deza & Andía, 2014 and *A. penca* Deza & Andía, 2014 from Peru.

As discussed by BRAGA-PEREIRA & SANTOS (2013), even though *Alpaida* is amongst the largest Neotropical spider genera, much of its diversity is still to be described (see LEVI 1988). Since 52 species are known only from females (including the two subspecies of *A. truncata* and *A. monzon audiberti*) and 17 species only from males, there is still much work to be done in the taxonomy of the genus. Since the revision of LEVI (1988), previously unknown opposite sexes were described for only seven species: the male of *A. scriba* (Mello-Leitão, 1940) by BUCKUP & MEYER (1993), the males of *A. citrina* (Keyserling, 1892) and *A. octolobata* Levi, 1988 by RODRIGUES & MENDONÇA (2011), the males of *A. hoffmanni* Levi, 1988, *A. kochalkai* Levi, 1988, *A. lomba* Levi, 1988 and the female of *A. arvoredo* by BUCKUP & RODRIGUES (2011).

Several species of the genus have large distribution ranges. Considering that only 53% of the described species are known from both sexes, it may be questionable to describe a new species in the absence of strong evidence for matching it with the opposite sex of a known species.

As documented in a growing number of faunistic inventories (e.g., HÖFER & BRESCOVIT 2001, BONALDO et al. 2009, CAFOFO et al. 2013), the diversity of *Alpaida* species in the Amazon Basin is high. In some particularly well-collected sites, up to 17 sympatric species have been found (unpublished data). In view of this, we have endeavored to examine all specimens of *Alpaida* deposited at the Museu Paraense Emílio Goeldi, which contain expressive material from the Amazon region. As a result, two new species of *Alpaida* are described, one based on both sexes from Brazil, the other on a female from Ecuador. One of the new species is morphologically related to *A. delicata* (Keyserling, 1892), which was compared to *A. truncata* (Keyserling, 1865) by Levi (1988). A discussion about the morphological similarity among the three species is presented and new diagnoses are proposed for *A. delicata* and *A. truncata*. Additionally, the male of *A. iquitos* Levi, 1988 and the female of *A. gurupi* Levi, 1988 are described for the first time and new records for twenty one species of *Alpaida* are documented: *A. antonio* Levi, 1988, *A. bicornuta* (Taczanowski, 1878), *A. boa* Levi, 1988, *A. deborae* Levi, 1988, *A. delicata*, *A. erythrothorax* (Taczanowski, 1873), *A. guimaraes* Levi, 1988, *A. guto* Abrahim & Bonaldo, 2008, *A. gurupi*, *A. iquitos*, *A. leucogramma* (White, 1841), *A. murtinho* Levi, 1988, *A. negro* Levi, 1988, *A. rossi* Levi, 1988, *A. septemmammata* (O. Pickard-Cambridge, 1889), *A. simla* Levi, 1988, *A. tayos* Levi, 1988, *A. truncata*, *A. urucuca* Levi, 1988, *A. utiariti* Levi, 1988 and *A. veniliae* Levi, 1988.

MATERIAL AND METHODS

The specimens examined were deposited in the Museu Paraense Emílio Goeldi, Belém, Brazil (MPEG, curator: Alexandre Bonaldo), Instituto Nacional de Pesquisas da Amazônia, Manaus, Brazil (INPA, curator: Célio Magalhães) and Museo de Zoología, Sección de Invertebrados, Pontificia Universidad Catolica, Quito, Ecuador (QCAZ, curator: Clifford Keil). The description format and palpal terminology follows Levi (1988). Male palps were expanded by alternated immersion in 10% KOH (potassium hydroxide) solution in distilled water for a few minutes. All measurements are in millimeters. The specimens were photographed and measured using a Leica M205A, with LAS automontage software. Additional information not originally inserted in labels is included between [brackets]. The following abbreviations are used: (A) terminal apophysis, (ALE) anterior lateral eyes, (AME) anterior median eyes, (C) conductor, (E) embolus, (L) distal lobe, (LL) lateral lobes, (MA) median apophysis, (N) notch, (PLE) posterior lateral eyes, (PME) posterior median eyes, (PM) paramedian apophysis, (PMP) posterior median plate, (R) radix, (S) median lobe of the scape. The distribution maps were made using the program QGIS (QGIS Development Team 2012) and include records only from the specimens listed here.

TAXONOMY

Alpaida levii sp. nov.
Figs. 1-9, 29

Type material. Male holotype from Brazil, *Pará*: Juruti (Mutum), 02°36'10.6"S 56°12'25.8"W, 14.IX.2002, D. Guimarães leg., deposited in MPEG (24370). Paratypes: Brazil, *Amazonas*: Manaus (Reserva Adolpho Ducke), 02°59'05.97"S

59"55'42.58"W, 1 female, 01.VIII.2008, J. ten Caten leg. (INPA); 1 male, 14.VIII.2008, R. Saturnino leg. (INPA); Canutama, 08°39'14.8"S 64°21'34.3"W, 1 male and 1 female, 30.IV.2007, R. Saturnino leg. (INPA); 1 female (INPA); 08°39'11.6"S 64°21'34.6"W, 1 male and 1 female, 06.V.2007, R. Saturnino leg. (INPA); Pará: Juruti (Capiranga), 02°28'0.6"S 56°12'42.2"W, 1 female, 15.IX.2002, A.B. Bonaldo leg. (MPEG 24371); 1 male (MPEG 24375); 02°29'19"S 56°06'34"W, 1 female, 15.IV.2008, B.V.B. Rodrigues leg. (MPEG 24374); 02°30'25.4"S 56°11'04.8"W, 1 male, 09.II.2007, J.A.P. Barreiros leg. (MPEG 24376); (Mutum), 02°36'10.6"S 56°12'25.8"W, 1 male and 1 female, 10.IX.2002, D. Guimarães leg. (MPEG 24372); Portel (Floresta Nacional de Caxiuanã), 01°57'38.9"S 51°36'45.3"W, 1 female, 12.V.2005, D.F. Candiani leg. (MPEG 24373).

Diagnosis. Males of *A. levii* resemble those of *A. delicata* by the dorsal abdominal coloration pattern (Figs. 1, 11), by the finger-shaped distal lobe (L, Fig. 3) and by the strong spines on tibiae I and II (Fig. 2). They differ by the distally pointed and curved retrolateral apical sector of the terminal apophysis (A, Fig. 3) (spoon-shaped in *A. delicata*; Fig. 12) – see Levi (1988: Figs. 476-478) for comparison – and by the reduced cymbial prolateral projection (Fig. 3) (well-developed in *A. delicata*), see Levi (1988: Fig. 477) for comparison. Females of *A. levii* resemble those of *A. delicata* by the two black humps on posterior end of abdomen (Figs. 7, 10), but differ by the drop-shaped median lobe of the scape (Fig. 8), epigynum longer than wide (Fig. 8) (wider than long in *A. delicata*), posterior median plate medially constricted (Fig. 9) (wide in *A. delicata*); lateral lobes of epigynum diamond-shaped in posterior view (Figs. 8, 9), see Levi (1988: figs. 472-474) for comparison.

Description. Male (MPEG 24370). Total length 6.1. Carapace length 2.6, width 2.0, height 0.6. Clypeus height 0.05. Sternum length 1.3, width 0.9. Abdomen length 3.6, width 1.8, height 1.5. Leg formula I/IV/II/III. Leg lengths: femur, I 3.3, II 2.7, III 1.9, IV 3.0; patella, I 1.1, II 1.0, III 0.8, IV 0.9; tibia, I 2.8, II 2.2, III 1.4, IV 2.5; 1metatarsus, I 3.0, II 2.5, III 1.4, IV 2.8; tarsus, I 1.1, II 1.0, III 0.7, IV 0.9. Eye diameters and interdistances: AME 0.1, ALE 0.1, PME 0.11, PLE 0.1; AME-PME 0.08, AME-ALE 0.34, PME-PLE 0.4, AME-AME 0.14, PME-PME 0.07. Carapace pale yellow, with two gray diagonal stripes on each side of the carapace (Fig. 1). Sternum pale yellow with brown margins, and a central black stripe. Endites and labium pale brown with white apices and brown margins. Chelicerae and legs yellow. Apices of femur, patella, tibia and metatarsus gray. Abdomen longer than wide, rectangular, with rounded borders. Dorsal side pale gray, with many white dots with different sizes (Fig.1). Two black humps and four black spots posteriorly. Two dark spots anteriorly. Venter pale gray with dark gray stripes, epigastric area slightly darker. Palp: palpal patella with two long macrosetae; median apophysis triangular-shaped in mesal view; paramedian apophysis formed by a distally expanded branch (Figs. 3, 4); embolus longer than terminal apophysis basal prong (Figs. 3-5); distal lobe of the terminal

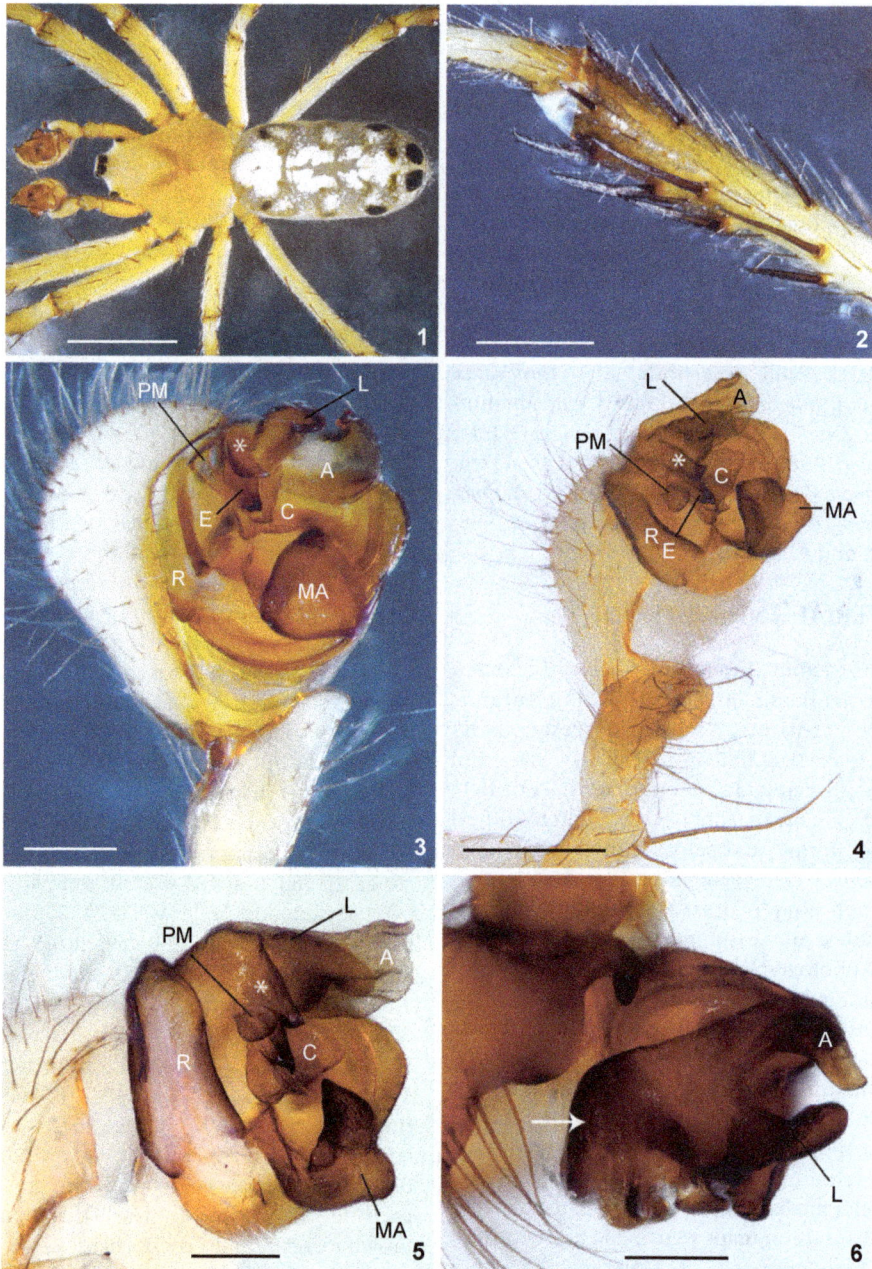

Figures 1-6. *Alpaida levii* **sp. nov.**, male: (1) habitus dorsal; (2) spines of the tibia I; (3) palpus, mesal view; expanded palpus (MPEG 24376): (4) mesal view; (5) detail of the mesal view; (6) detail of the terminal apophysis. (A) Terminal apophysis, (C) conductor, (E) embolus, (L) distal lobe, (MA) median apophysis, (PM) paramedian apophysis, (*) basal prong of the terminal apophysis, (R) radix. Scale bars: (1) = 2 mm, (2, 4) = 0.5 mm, (3, 5, 6) = 0.2 mm.

apophysis finger-shaped in mesal view (Figs. 3, 6); terminal apophysis thin and curved (Fig. 3).

Female (MPEG 24374). Total length 7.8. Carapace length 3.2, width 2.5, height 0.7. Clypeus height 0.13. Sternum length 1.3, width 1.1. Abdomen length 6.0, width 2.6, height 2.2. Leg formula I/IV/II/III. Leg lengths: femur, I 3.5, II 3.0, III 2.0, IV 3.6; patella, I 1.3, II 1.1, III 0.8, IV 1.1; tibia, I 3.1, II 2.5, III 1.4, IV 2.8; metatarsus, I 3.3, II 2.6, III 1.5, IV 2.9; tarsus, I 1.1, II 1.1, III 0.7, IV 1.0. Eye diameters and interdistances: AME 0.15, ALE 0.13, PME 0.15, PLE 0.14; AME-PME 0.1, AME-ALE 0.6,

PME-PLE 0.6, AME-AME 0.16, PME-PME 0.13. Carapace pale yellow, with two lateral gray, diagonal stripes (Fig. 4). Sternum pale yellow with brown margins, and a central gray stripe. Endites and labium pale brown with white apices and brown margins. Chelicerae and legs yellow. Apices of femur, patella, tibia and metatarsus gray. Abdomen longer than wide, rectangular, with rounded borders. Dorsum pale gray, almost entirely covered by white and yellow spots and differently sized stripes (Fig.4). Posteriorly with two black humps and four black spots. Venter pale gray, epigastric area slightly darker. Epigynum longer than wide, with inconspicuous notch (Fig. 8); posterior median plate narrow (Fig. 9); lateral lobes diamond-shaped (Figs. 8, 9); scape drop-shaped.

Additional material examined. BRAZIL, *Amazonas*: Manaus (Reserva Adolpho Ducke), 02°59'11.33"S 59°56'14.69"W, 1 female, 30.VII.2008, J. ten Caten leg. (INPA); Coari (Base de Operações Geólogo Pedro de Moura, Porto Urucu), 04°52'07.6"S 65°15'53.6"W, 1 male, 22.VII.2003, A.B. Bonaldo leg. (MPEG 19874); 04°50'32"S 65°04'80"W, 1 female, 05.IX.2006, S. C. Dias leg. (MPEG 13762); Manicoré, 04°54.705'S 61°06.788'W, 1 female, 14.VII.2007, L.T. Miglio leg. (INPA). *Pará*: Melgaço (Floresta Nacional de Caxiuanã, TEAM 2), 01°43'43.2"S 51°29'00.6"W, 1 female, 26.IV.2006, J.A.P. Barreiros leg. (MPEG 24348); 1 female, R.B. Lopes leg. (MPEG 24369); (Floresta Nacional de Caxiuanã, TEAM 3), 01°43'59.2"S 51°30'38.6"W, 1 female, 04.X.2005, N. Abrahim leg. (MPEG 24339); 1 female, 17.IV.2006, C.B. Lopes leg. (MPEG 24353); (Floresta Nacional de Caxiuanã, TEAM 6), 01°44'18.02"S 51°27'48.01"W, 1 female (MPEG 24358); 1 female, 15.IV.2006, R.B. Lopes leg. (MPEG 24360); 1 female, 02.VIII.2011, Equipe MPEG (MPEG 24350); 1 female (MPEG 24366); (Floresta Nacional de Caxiuanã, Estação Científica Ferreira Penna), 01°44'15.5"S 51°26'42.0"W, 1 female, 10.VII.2002, J.P. Sifuerte leg. (MPEG 24356); (Floresta Nacional de Caxiuanã, TEAM 4), 01°45'12.8"S 51°31'14.7"W, 1 female, 12.X.2005, N. Abrahim leg. (MPEG 24351); 1 female, 23.IV.2006, E.J. Sales leg. (MPEG 24346); Juruti (Capiranga), 02°28'0.6"S 56°12'42.2"W, 2 females, 07.IX.2002, D. Guimarães leg. (MPEG 24357); 1 female, 12.IX.2002, D. Guimarães leg. (MPEG 24368); 1 female, A.B. Bonaldo leg. (MPEG 24365); 1 female, 15.IX.2002, A.B. Bonaldo leg. (MPEG 24363); 02°28'22.1"S 56°12'29.4"W, 1 female, 11.VIII.2008, N.C. Bastos leg. (MPEG 24359); 1 female, 10.III.2006, S.C. Dias leg. (MPEG 9133); (Barroso), 02°27'0.11"S 56°00'60"W, 1 female, 09.II.2009, B.V.B. Rodrigues leg. (MPEG 24354); 02°27'45.5"S 56°00'51"W, 1 female, 12.IV.2008, N.C. Bastos leg. (MPEG 24361); 1 male, 12.VIII.2008, N.F. Lo Man Hung leg. (MPEG 24340); 02°27'51.4"S 56°00'08.6"W, 1 female, 22.V.2009, N. Abrahim leg. (MPEG 24333); (Beneficiamento), 02°30'25.4"S 56°11'04.8"W, 1 female, 09.II.2007, N.F. Lo Man Hung leg. (MPEG 24364); 02°30'27.4"S 56°10'39.5"W, 1 female, 10.VIII.2010, B.V.B. Rodrigues leg. (MPEG 24336); 1 female, N.C. Bastos leg. (MPEG 24343); 1 female, 20.II.2011, N.C. Bastos leg. (MPEG 24334); (Mutum), 02°33'06.9"S 56°13'29.0"W, 1 female, 11.VIII.2010, B.V.B. Rodrigues leg. (MPEG 24335); 1 female,

(MPEG 24349); 1 female, N.C. Bastos leg. (MPEG 24347); 1 female, N. Abrahim leg. (MPEG 24352); 02°33'07.2"S 56°13'06.2"W, 1 female, 07.IX.2002, D. Guimarães leg. (MPEG 24337); 02°36'11.2"S 56°12'36.3"W, 1 female, 04.VIII.2004, D.R. Santos-Souza leg. (MPEG 24341); 1 female (MPEG 24342); 02°36'44.7"S 56°11'39.2"W, 1 female, 27.V.2009, N. Abrahim leg. (MPEG 24338); 1 female, N.F. Lo Man Hung leg. (MPEG 24355); 1 female, 19.VIII.2011, R. Saturnino leg. (MPEG 24345); Novo Progresso (Serra do Cachimbo), 09°22'02.9"S 55°01'11.9"W, 1 male, 06.IV.2004, D. Guimarães leg. (MPEG 6358).

Distribution. Brazil, states of Pará and Amazonas.

Etymology. The specific name is a patronym to honor the late arachnologist Herbert W. Levi, and to recognize his immense contribution to spider taxonomy. His seminal work inspired and will continue to inspire generations of arachnologists.

Variation in paratypes. Six males, total length. 4.8 to 7.4; carapace: 2.3 to 3.0; number of spines on tibiae I and II: 5 to 10. Seven females, total length: 7.4 to 9.0; carapace: 2.0 to 3.2.

Alpaida delicata (Keyserling, 1892)
Figs. 10-13, 31

Epeira delicata Keyserling, 1892: 183, pl. 9, fig. 135, 6 (females and 4 males syntypes from Espírito Santo, Brazil, deposited in Natural History Museum (BMNH), London, not examined).
Araneus taczanowskii Simon, 1897: 473 (female holotype from Tefé, Est. Amazonas, Brazil, deposited in MNHN, not examined); Bonnet, 1955: 609. Synonimized by Levi, 1988.
Alpaida delicata: Levi, 1988: 458, figs. 472-478; Dierkens, 2014: 17, figs. 7, 8, 30, 41.

Diagnosis. Males of *A. delicata* resemble those of *A. levii* by the dorsal abdominal coloration pattern (Figs. 1, 11), by the finger-shaped distal lobe (L, Figs. 3-6) and by the strong spines on tibiae I and II (Fig. 2). They differ by the spoon-shaped retrolateral apical sector of the terminal apophysis (Figs. 12, 13) (distally pointed and curved in *A. levii*; Figs. 3-5) and by the well-developed cymbial prolateral projection (reduced in *A. levii*), see LEVI (1988: figs. 476-478) for comparison. Females of *A. delicata* resemble those of *A. levii* by the two black humps on posterior end of abdomen (Figs. 7, 10) and by the epigynum with the notch not well demarcated, but differs by the sinuous lips in ventral view, epigynum wider than long (longer than wide in *A. levii*; Figs. 8, 9); posterior median plate medially wide (medially constricted in *A. levii*; Fig. 9); lateral lobes rounded in posterior view (diamond-shaped in *A. levii*; Figs. 8, 9), see LEVI (1988: figs. 472-474) for comparison.

Material examined. BRAZIL, *Amazonas*: Presidente Figueiredo (Reserva Biológica de Uatumã), 01°49'05.55"S 59°14'34.23"W, 1 female (INPA); 01°49'37.24"S 59°14'31.81"W, 1 male (INPA); Manaus (Reserva Adolpho Ducke), 02°57'30.23"S 59°55'57.45"W, 1 female (INPA); 02°59'21.44"S 59°57'18.62"W, 1 female (INPA); Autazes, 04°09'26.3"S 60°07'53"W, 1 male and 4 females (INPA); 04°09'55.4"S 60°07'53.9"W, 1 male (INPA);

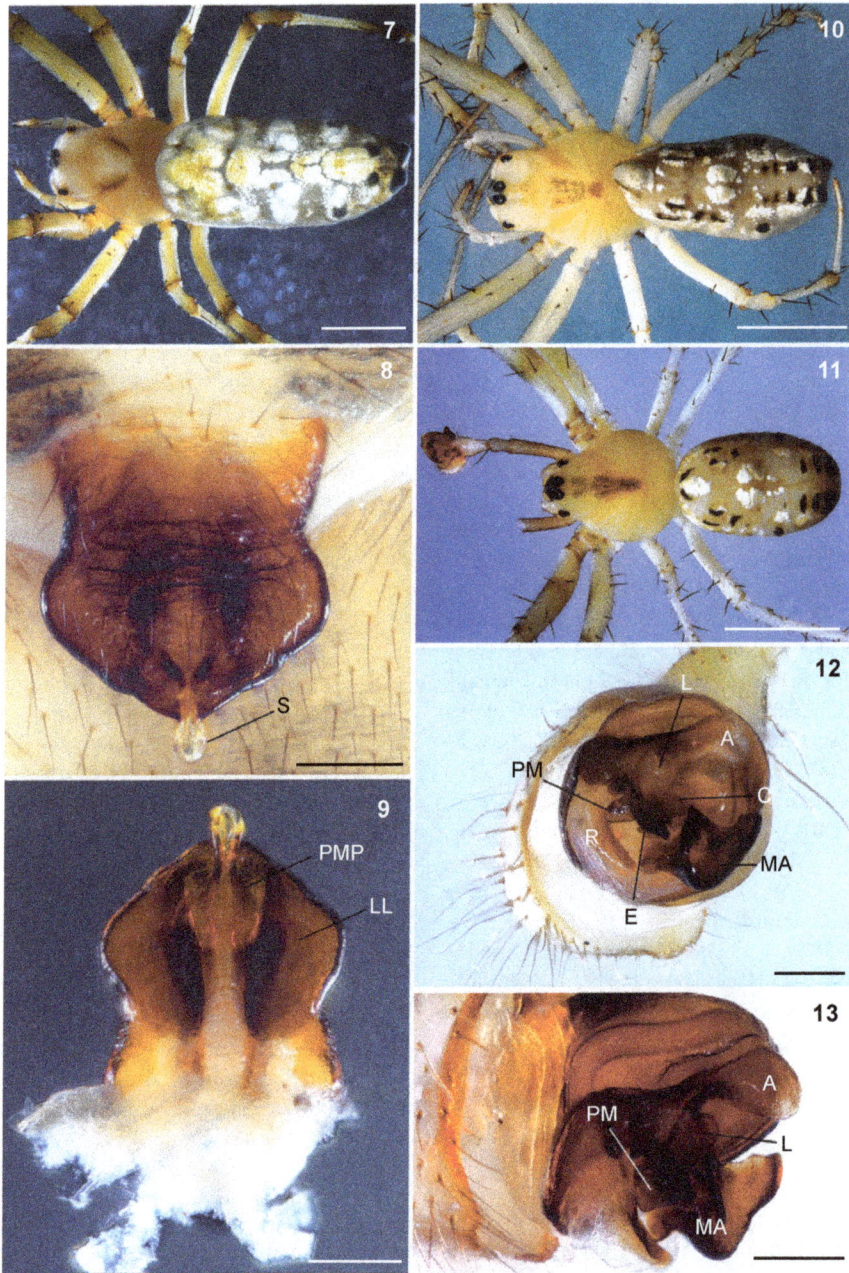

Figures 7-13. (7-9) *Alpaida levii* **sp. nov.**, female: (7) habitus dorsal; (8) epigynum ventral view; (9) epigynum posterior view. (10-13) *Alpaida delicata*: (10) female (MPEG 24408), habitus dorsal; (11) male (MPEG 9136), habitus dorsal; (12-13) expanded palpus: (12) mesal view; (13) detail of the terminal apophysis, median apophysis and paramedian apophysis. (A) Terminal apophysis, (C) conductor, (E) embolus, (L) distal lobe, (LL) lateral lobes, (MA) median apophysis, (PM) paramedian apophysis, (PMP) posterior median plate, (R) radix, (S) median lobe of the scape, (*) basal prong of the terminal apophysis. Scale bars: (7, 10, 11) = 2 mm, (8, 9, 12, 13) = 0.2 mm.

Canutama, 08°38'49"S 64°22'05.5"W, 1 female (INPA); 1 male (INPA); 08°39'05.8"S 64°22'05.6"W, 2 females (INPA); 08°39'06.8"S 64°22'05.8"W, 1 female (INPA); 1 male (INPA); 08°39'11.6"S 64°21'34.6"W, 1 male and 1 female (INPA); 1male (INPA); 1 female (INPA); 2 females (INPA); 08°39'14.8"S 64°21'34.3"W, 1 male (INPA); 1 female (INPA). *Pará*: Melgaço (Floresta Nacional de Caxiuanã, TEAM 1), 01°42'24.00"S 51°27'34.30"W, 1 male (MPEG 24539); (Floresta Nacional de

Caxiuanã, TEAM 2), 01°43'43.20"S 51°29'0.70"W, 1 male (MPEG 24541); 1 male (MPEG 24540); 1 male (MPEG 24522); 1 female (MPEG 24528); (Floresta Nacional de Caxiuanã, TEAM 3), 01°43'59.20"S 51°30'38.60"W, 1 female (MPEG 24542); 1 female (MPEG 24545); 1 male (MPEG 24543); 1 male (MPEG 24544); 1 female (MPEG 24523); 1 female (MPEG 24527); 1 female (MPEG 24526); 1 male (MPEG 24524); 1 male and 1 female (MPEG 24525); (Castanhal do Jacaré), 01°44'13.5"S 51°25'32.8"W, 1 male (MPEG 24560); (Floresta Nacional de Caxiuanã, Estação Científica Ferreira Penna), 01°44'15.5"S 51°26'42.0"W, 1 female (MPEG 24561); (Floresta Nacional de Caxiuanã, TEAM 4), 01°45'12.80"S 51°31'14.70"W, 1 female (MPEG 24529); 2 females (MPEG 24531); 1 female (MPEG 24532); 1 male (MPEG 24530); 1 female (MPEG 24535); 1 female (MPEG 24536); 1 female (MPEG 24546); 1 female (MPEG 24547); 1 male (MPEG 24548); (Floresta Nacional de Caxiuanã, Caiçara), 01°46'41.4"S 51°25'28.7"W, 1 female (MPEG 24558); (Floresta Nacional de Caxiuanã, TEAM 5), 01°47'23.66"S 51°34'52.18"W, 1 female (MPEG 24533); 1 male (MPEG 24534); 1 female (MPEG 24537); 1 female (MPEG 24538); 1 female (MPEG 24550); 1 male (MPEG 24552); 1 female (MPEG 24551); 1 male and 1 female (MPEG 24556); 1 female (MPEG 24557); 1 male and 1 female (MPEG 24555); 1 female (MPEG 24554); 1 female (MPEG 24549); 2 females (MPEG 24553); (Terra Preta), 01°51'19.30"S 51°25' 57.50"W, 1 male (MPEG 24559); (Portel, Plote PPBio), 01°57'38.9"S 51°36'45.3"W, 1 male and 1 female (MPEG 24562); 1 male (MPEG 24563); 1 male (MPEG 24564); (Cametá, Curuçambaba, Área de Floresta), 02°06'27.2"S 49°18'33.1"W, 1 male (MPEG 24577); 1 male (MPEG 24575); 02°07'27.6"S 49°18'52.7"W, 1 male (MPEG 24574); 02°06'39.4"S 49°18' 40.7"W, 1 female (MPEG 24578); 1 female (MPEG 24580); (Curuçambaba, Área de Praia), 02°06'31.4"S 49°18'55.8"W, 1 female (MPEG 24579); 1 female (MPEG 24576); 1 male and 1 female (MPEG 24581); 2 females (MPEG 24573); (Moju, Campo experimental da Embrapa), 02°11'12.44"S 48°47'34.31"W, 1 male (MPEG 24321); (Juruti, Capiranga), 02°28'0.60"S 56°12'42.20"W, 2 females (MPEG 24412); 1 female (MPEG 24380); 02°28'22.1"S 56°12'29.4"W, 1 female (MPEG 24430); 1 female (MPEG 24421); 1 female (MPEG 24414); 1 male (MPEG 9203); 1 male (MPEG 9132); (Barroso), 02°27'41.7"S 56°00'11.6"W, 1 female (MPEG 24420); 02°27'51.4"S 56°00'08.6"W, 1 female (MPEG 24422); 1 female (MPEG 24426); 1 female (MPEG 24403); 1 female (MPEG 24390); 1 female (MPEG 24404); (Ferrovia Km 23), 02°29'19"S 56°06'34"W, 1 male. (MPEG 24401); (Beneficiamento), 02°30'25.4"S 56°11'04.8"W, 1 male (MPEG 24416); 02°30'27.4"S 56°10'39.5"W, 1 male (MPEG 24378); 1 female (MPEG 24387); 1 female (MPEG 24393); (Mutum), 02°33'06.9"S 56°13'29.0"W, 1 female (MPEG 24398); 1 male and 1 female (MPEG 24399); 1 female (MPEG 24417); 1 female (MPEG 24394); 1 male (MPEG 24383); 02°33'18.0"S 56°13'22.4"W, 1 male (MPEG 24384); 1 female (MPEG 24418); 02°36'10.6"S 56°12'25.8"W, 1 female (MPEG 24396); 1 male (MPEG 24410); 02°36'11.2"S 56°12' 36.3"W, 2 females (MPEG 24405); 1 male (MPEG 24411); 2 fe-

males (MPEG 24391); 1 female (MPEG 24392); 02°36'44.7"S 56°11'39.2"W, 1 male (MPEG 24386); 1 male (MPEG 9136); 1 female (MPEG 9137); 1 female (MPEG 24379); 1 female (MPEG 24423); 1 male (MPEG 24419); 1 male (MPEG 24427); 1 female (MPEG 24425); 1 female (MPEG 24429); 1 male (MPEG 24382); 1 female (MPEG 24400); 2 males (MPEG 24406); 1 male (MPEG 24409); 1 male (MPEG 24388); 1 female (MPEG 24385); 1 male (MPEG 24377); 1 female (MPEG 24389); 1 female (MPEG 24381); 1 male (MPEG 24397); 1 female (MPEG 24402); 02°36'45.2"S 56°11'27.5"W, 1 male (MPEG 24413); 1 female (MPEG 24407); 1 female (MPEG 24408); 1 male and 1 female (MPEG 24395); 1 female (MPEG 24415); 02°36'45.7"S 56°11'38.2"W, 1 female (MPEG 24424); 1 male (MPEG 24428).

Distribution. Previously known from Colombia, Peru, Bolivia, French Guiana and Brazil (Amazonas: Tefé; Pará: Melgaço; Espírito Santo). Recorded here also from Presidente Figueiredo, Manaus, Autazes and Canutama, state of Amazonas; Moju and Juruti, state of Pará, Brazil.

Alpaida truncata (Keyserling, 1865)
Figs. 14-17, 32

Epeira truncata Keyserling, 1865: 807, pl. 19, figs. 21-22 (female from Uruguay, deposited in BMNH, not examined).

Alpaida truncata Levi, 1988: 472, figs. 570-578; Levi, 2002: 538, figs. 75-77, 260-261; Dierkens, 2014: 22, figs. 19-20, 34, 47.

Diagnosis. Males of *A. truncata* resemble those of *A. levii* and *A. delicata* by having the posteriorly hump-shaped abdomen (Figs. 7, 10, 14) and *A. queremal* by the extremely long and distally pointed median apophysis (Fig. 16), see Levi (1988: fig. 569) for comparison. They differ from other species of *Alpaida* by the c-shaped and digitiform paramedian apophysis (Figs. 16, 17), and a modified second tibia, flattened and wide, bearing two macrosetae; differ from *A. queremal* by a notch in the base of the median apophysis – modified from Levi (1988: see figs. 569, 577-578 for comparison). Females also resemble those of *A. levii*, *A. delicata* and *A. queremal* by the posteriorly hump-shaped abdomen, but differ from *A. levii* by the epigynum wider than longer, from *A. delicata* by the median lobe not well-developed and from *A. queremal* by the lack of the lateral lobes of the epigynum – modified from Levi (1988: see figs. 472-473, 564-565, 570-571 for comparison).

Description. Male and female. See Levi (1988): 472, figs. 570-578.

Material examined. BRAZIL, *Pará*: Santa Bárbara, 01°13' 37.32"S 48°17'45.59"W, 2 females (MPEG 24519); Benevides, 01°21'43.87"S 48°14'37.79"W, 1 male (MPEG 2972); Belém (Jardim Botânico Rodrigues Alves), 01°25'49.0"S 48°27'22.3"W, 1 female (MPEG 24516); 1 female (MPEG 24517); (Museu Paraense Emílio Goeldi, Campus de Pesquisa), 01°27'03.03"S 48°26'40.2"W, 1 male (MPEG 4907); 1 male (MPEG 24512); (Parque Zoobotânico Emílio Goeldi), 01°27'12"S 48°28'35"W, 1 male (MPEG 24513); 1 male (MPEG 24514); 1 female (MPEG

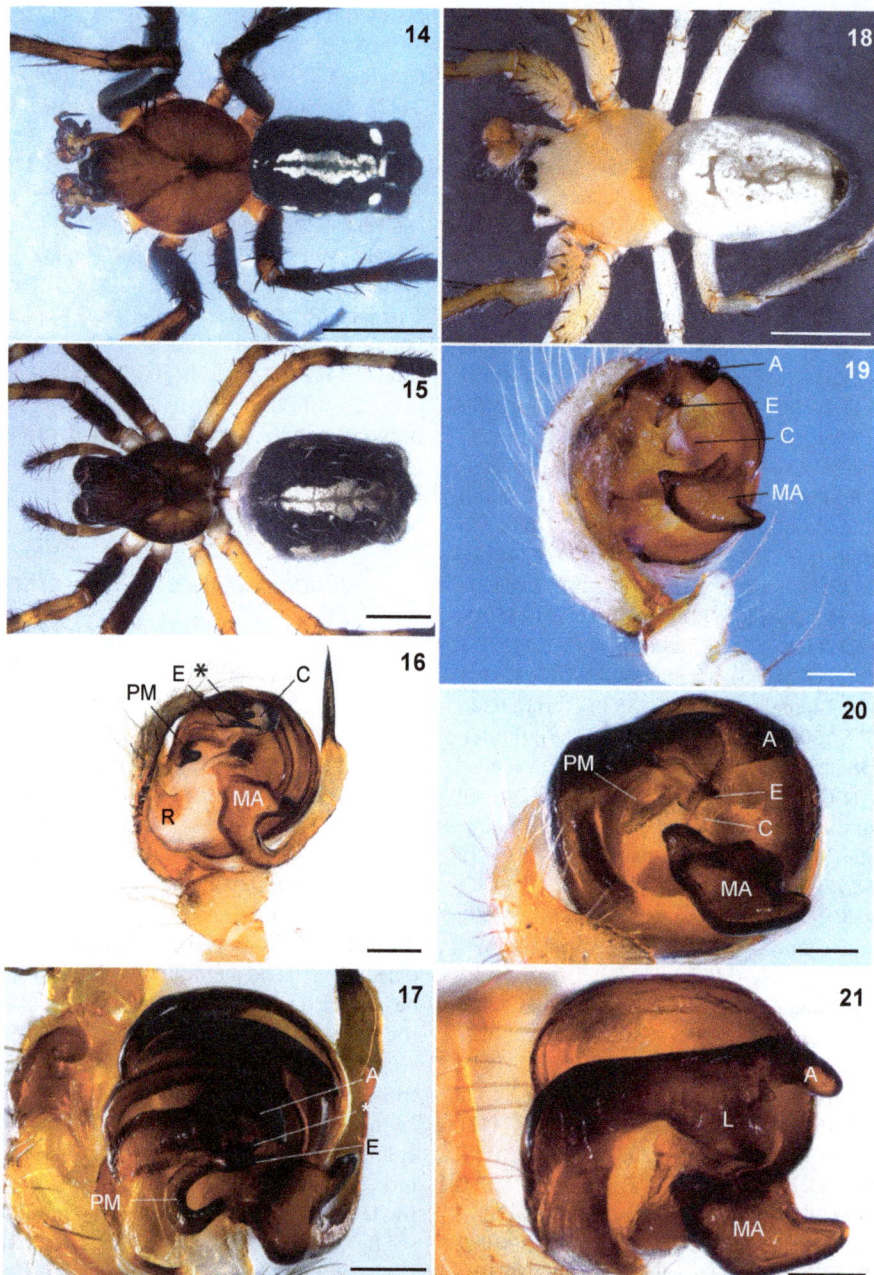

Figures 14-21. (14-17) *Alpaida truncata*: (14, 16, 17) male (MPEG 24508): (14) habitus dorsal; (16) palpus, mesal view; (17) expanded palpus, detail of the terminal apophysis, basal prong, embolus, median and paramedian apophyses; (15) female (MPEG 24503), habitus dorsal. (18-21) *Alpaida iquitos*, male: (18) habitus dorsal; (19) palpus, mesal view; (20-21) expanded palpus: (20) mesal view; (21) detail of the terminal apophysis and median apophysis. (A) Terminal apophysis, (C) conductor, (E) embolus, (L) distal lobe, (MA) median apophysis, (PM) paramedian apophysis, (R) radix, (*) basal prong of the terminal apophysis. Scale bars: (14, 15) = 2 mm, (16, 17) = 0.2 mm, (18) = 1 mm, (19-21) = 0.1 mm.

24515); (Mata do Betina, Universidade Federal do Pará), 01°28′02″S 48°26′33″W, 1 female (MPEG 24518); Melgaço (Floresta Nacional de Caxiuanã, Estação Científica Ferreira Penna), 01°44′15.5″S 51°26′42″W, 1 female (MPEG 8010); 1 female (MPEG 8011); 1 female (MPEG 8009); 01°44′18.02″S 51°27′48.01″W, 1 female (MPEG 22499); 01°47′32.7″S

51°25'59.2"W, 1 male (MPEG 24520); Juruti (Barroso), 02°27'51.4"S 56°00'08.6"W, 1 female (MPEG 24497); 1 male (MPEG 24500); 1 female (MPEG 24504); 1 male (MPEG 24505); (Capiranga), 02°28'22.1"S 56°12'29.4"W, 1 male (MPEG 24499); 1 female (MPEG 24501); 02°29'57.8"S 56°12'60"W, 1 male (MPEG 24502); (Beneficiamento), 02°30'04.9"S 56°09'46.6"W, 1 male (MPEG 24507); 02°30'25.4"S 56°11'04.8"W, 1 male and 1 female (MPEG 24498); 1 female (MPEG 24503); 02°30'27.4"S 56°10'39.5"W (MPEG 24506); 1 male (MPEG 24508); 1 female (MPEG 24511); 2 females (MPEG 24510); 1 female (MPEG 24509); (Mutum), 02°36'45.2"S 56°11'27.5"W, 1 female (MPEG 24596); Marabá (Serra Norte), 06°0'23.1"S 50°17'50.3"W, 1 male (MPEG 4243); 1 male (MPEG 4194); Novo Progresso, 07°08'07"S 55°24'51"W, 1 male (MPEG 4495); Serra do Cachimbo, 09°16'18.6"S 54°56'22.9"W (MPEG 6330).

Distribution. Previously known from Mexico to Argentina. Recorded here also from Benevides, Juruti and Marabá, state of Pará, Brazil.

Alpaida iquitos Levi, 1988
Figs. 18-21, 30

Alpaida iquitos Levi, 1988: 416, figs. 194-197 (female holotype and female paratype from Iquitos, Peru, V.1920, deposited in Museum of Comparative Zoology (MCZ), Harvard University, not examined).

Note. The male, described here, is identified as belonging to *A. iquitos* based on the morphological similarity and also in the sympatric distribution with the female previously described by Levi (1988).

Diagnosis. Males of *A. iquitos* resemble those of *A. bicornuta* by the wider than long median apophysis (MA, Figs. 19-21) and by the flattened distal lobe (L, Fig. 21), but differs from this and all other species of *Alpaida* by the sharply pointed, opposed proximal ends of median apophysis (MA, Figs. 19-21) and by the distally rounded retrolateral apical sector of the terminal apophysis (A, Figs. 19-21) – see Levi (1988: Figs. 17-18) for comparison. As diagnosed by Levi (1988), females differ from *A. variabilis* and *A. kochalkai* by having the posterior plate of the epigynum constricted in the middle and from *A. variabilis* by having the epyginum longer (Levi 1988: see figs. 194-197).

Description. Male (MPEG 24164). Total length 3.3. Carapace length 1.7, width 1.3, height 0.6. Clypeus height 0.05. Sternum length 0.7, width 0.6. Abdomen length 2.0, width 1.2, height 1.0. Leg formula I/II/IV/III. Leg lengths: femur, I 1.6, II 1.4, III 1.1, IV 1.3; patella, I 0.6, II 0.6, III 0.4, IV 0.4; tibia, I 1.3, II 1.0, III 0.6, IV 1.1; metatarsus, I 1.2, II 1.1, III 0.6, IV 1.0; tarsus, I 0.6, II 0.5, III 0.4, IV 0.4. Eyes diameters and interdistances: AME 0.12, ALE 0.09, PME 0.11, PLE 0.09; AME-PME 0.09, AME-ALE 0.21, PME-PLE 0.24, AME-AME 0.09, PME-PME 0.07. Carapace pale yellow. Sternum, endites and labium pale yellow with brown margins. Chelicerae pale yellow. Legs pale yellow; tibia, metatarsus and tarsus of legs I-II darker. Abdomen longer than wide, oval. Laterals of dorsal side, pale yellow, center white, with

a dark gray stripe posteriorly (Fig. 18). Venter pale yellow. Palp: palpal patella and palpal tibiae with one long macrosetae; median apophysis wider than long, sharply pointed, opposed proximal ends of median apophysis two-pointed (Figs. 19-21); paramedian apophysis distally expanded (Fig. 20); embolus short, covered by terminal apophysis (Figs. 19-20); basal prong of the terminal apophysis absent; distal lobe of the terminal apophysis well-developed, visible only with the expanded palpus in apical view of the terminal apophysis (Fig. 21); retrolateral portion of the terminal apophysis rounded (Figs. 19-21).

Female. See Levi (1988: 416, figs. 194-197).

Material examined. BRAZIL, *Pará*: Portel (Floresta Nacional de Caxiuanã, Plote PPBio), 01°57'38.9"S 51°36'45.3"W, 1 female, 10.V.2005, C.A. Lopes (MPEG 24184); 1 female, 12.V.2005, D.F. Candiani (MPEG 24179); Melgaço (Floresta Nacional de Caxiuanã, TEAM 1), 01°42'24.0"S 51°27'34.3"W, 1 male, 26.IX.2005, J.A.P. Barreiros leg. (MPEG 24167); (Floresta Nacional de Caxiuanã, TEAM 5), 01°43'21.6"S 51°25'51.2"W, 1 female, 08.X.2005, J.A.P. Barreiros leg. (MPEG 24175); 1 female (MPEG 24177); 1 male, 13.X.2005, Robinho leg. (MPEG 24164); (Floresta Nacional de Caxiuanã), TEAM 2, 01°43'43.2"S 51°29'00.6"W, 1 male, 03.X.2005, J.A.P. Barreiros leg. (MPEG 24168); 1 female, 28.IX.2005, B.B. Santos leg. (MPEG 24170); 1 female, 05.X.2005, J.A.P. Barreiros leg. (MPEG 24171); 3 females, N. Abrahim leg. (MPEG 24169); (Floresta Nacional de Caxiuanã, TEAM 3), 01°43'59.2"S 51°30'38.6"W, 1 female, 29.IX.2005, B.B. Santos leg. (MPEG 24173); 1 female, J.A.P. Barreiros leg. (MPEG 24176); 1 female, 04.X.2005, J.A.P. Barreiros leg. (MPEG 24174); 1 female, 17.IV.2006, C.A. Souza leg. (MPEG 24163); 1 female, R.B. Lopes leg. (MPEG 24162); (Floresta Nacional de Caxiuanã, Estação Científica Ferreira Penna), 01°44'15.5"S 51°26'42.0"W, 1 female, 16.XI.2001, Aires leg. (MPEG 24180); 1 female, 07-13.II.2002, A.B. Bonaldo leg. (MPEG 24181); 1 female, 17.IV.2002, M. Andrade leg. (MPEG 24182); 1 female, 05.VI.2004, A.B. Bonaldo leg. (MPEG 24183); 01°44'18.02"S 51°27'48.01"W, 1 female (MPEG 24185); 1 female (MPEG 24187); 1 female, 25.III.2002, N. Abrahim leg. (MPEG 24178); 1 female, 05.VI.2004, C. Trinca leg. (MPEG 24186); (Floresta Nacional de Caxiuanã, TEAM 4), 01°45'12.8"S 51°31'14.7"W, 1 female, 07.X.2005, Robinho leg. (MPEG 24172); 1 female, 12.X.2005, J.A.P. Barreiros leg. (MPEG 24165); 1 male, N. Abrahim leg. (MPEG 24166).

Distribution. Previously known from Brazil (Pará: Melgaço and Canindé [Paragominas]; Mato Grosso: Chapada dos Guimarães), Ecuador, French Guiana and Peru. Recorded here also from Portel, state of Pará, Brazil.

Alpaida gurupi Levi, 1988
Figs. 22-24, 30

Alpaida gurupi Levi, 1988: 429, figs. 278-279 (male holotype from Canindé, Rio Gurupi, Pará, Brazil, 27-28.II.1966, deposited in American Museum of Natural History (AMNH), not examined).

Note. The females, described here, were identified as belonging to *A. gurupi* based on the morphological similarity and

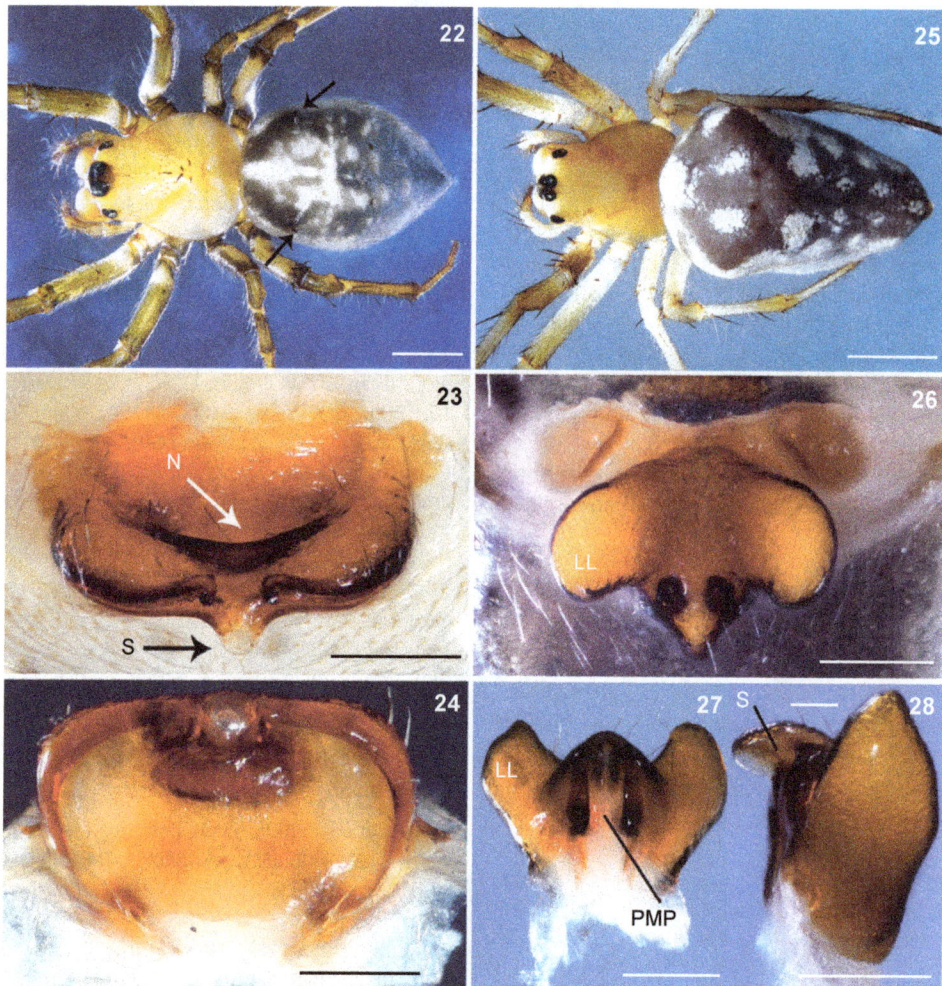

Figures 22-28. (22-24) *Alpaida gurupi*, female: (22) habitus dorsal, arrow = spines; (23-24) epigynum: (23) ventral view; (24) posterior view. (25-28) *Alpaida yanayacu* new species, female: (25) habitus dorsal; (26-28) epigynum: (26) ventral view; (27) lateral view; (28) posterior view. (LL) Lateral lobes, (N) notch, (PMP) posterior median plate, (S) median lobe of the scape. Scale bars: (22, 25) = 1 mm, (23-28) = 0.2 mm.

also in the sympatric distribution with the male previously described by Levi (1988).

Diagnosis. Females of *A. gurupi* resemble those of *A. amambay* by the epigynum wider than long, by the well demarcated notch and by short and rounded scape median lobe (Fig. 23), but differs by the notch occupying more than half of the epigynum width (Fig. 23), posterior median plate wide, without sinuous borders (Fig. 24), see Levi (1988: Figs. 274-275) for comparison. As diagnosed by Levi (1988), males differ from other species of *Alpaida* by the gently curved median apophysis and the hooded appearance of the terminal apophysis see Levi (1988: figs. 278-279).

Description. Female (MPEG 24190). Total length 5.4. Carapace length 2.2, width 1.8, height 0.8. Clypeus height 0.08. Sternum length 0.9, width 0.9. Abdomen length 3.3, width 2.0, height 1.8. Leg formula IV/I/II/III. Leg lengths: femur, I 2.0, II 1.8, III 1.4, IV 2.1; patella, I 0.8, II 0.8, III 0.6, IV 0.8; tibia, I 1.6, II 1.3, III 1.0, IV 1.8; metatarsus, I 1.5, II 1.4, III 1.0, IV 1.6; tarsus, I 0.8, II 0.7, III 0.6, IV 0.8. Eyes diameters and interdistances: AME 0.1, ALE 0.12, PME 0.17, PLE 0.14; AME-ALE 0.14, PME-PLE 0.3, AME-AME 0.12, PME-PME 0.2. Carapace pale yellow, with darker yellow stripes on cephalic and thoracic region of the carapace (Fig. 9). Sternum pale yellow with orange margins. Endites and labium pale orange with white apices and brown margins. Chelicerae yellow. Legs orange-brownish. Femur, patella, tibia and metatarsus gray. Abdomen longer than wide, oval, posteriorly pointed with rounded borders; anteriorly with two spines, one on each side (see arrows Fig. 22). Dorsal side dark gray, many white and pale gray dots with different sizes (Fig. 22). Ventral side dark gray with light

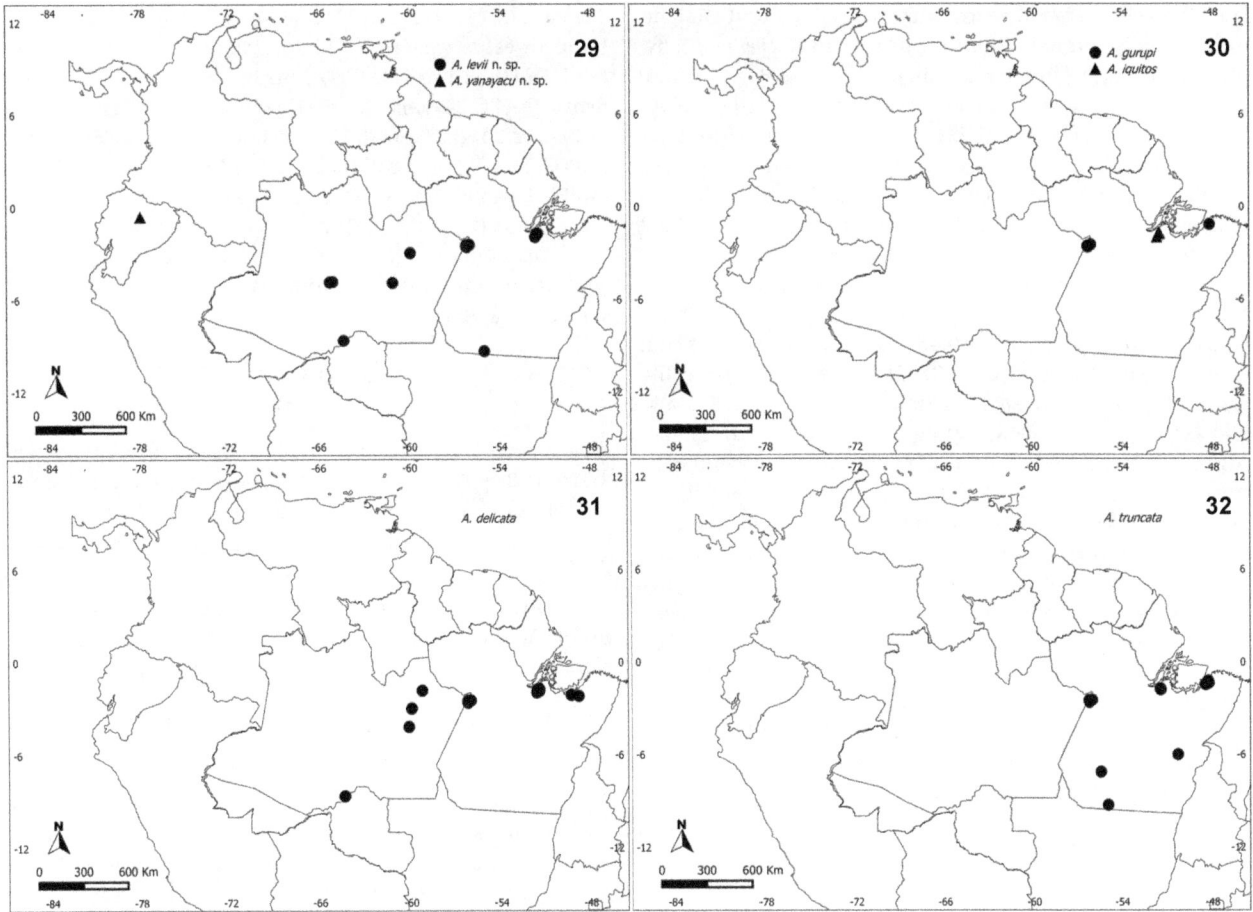

Figures 29-32. Distribution of *Alpaida* species in the Amazon region, North Brazil, and Ecuador: (29) *Alpaida levii* **sp. nov.** and *A. yanayacu* **sp. nov.**; (30) *A. gurupi* and *A. iquitos*; (31) *A. delicata*; (32) *A. truncata*.

gray stripes and many white dots. Epigynum wider than long, with evident notch; (Fig. 23); posterior median plate wide (Fig. 24); median lobe of scape short, drop-shaped (Fig. 23).

Male. See Levi (1988): 429, figs. 278-279.

Material examined. Brazil, *Pará*: Santo Antônio do Pará 01°09'09"S 48°07'45"W, 1 male, 07.IV.1975, R.F. da Silva leg. (MPEG 2973); Juruti (Barroso), 02°27'51.4"S 56°00'08.6"W, 1 female, 08.II.2007, N.F. Lo Man Hung leg. (MPEG 24199); 1 female, 16.XI.2007, N.F. Lo Man Hung leg. (MPEG 24197); 1 female (MPEG 24201); 1 male and 1 female, 08.VIII.2008, N.F. Lo Man Hung leg. (MPEG 24189); (Mutum), 02°33'06.9"S 56°13'29.0"W, 2 females, 12.VIII.2010, N.C. Bastos leg. (MPEG 24195); 02°33'18"S 56°13'22.4"W, 1 female, 05.V.2010, N.C. Bastos leg. (MPEG 24191); 02°36'11.2"S 56°12'36.3"W, 1 female, 04.VIII.2004, D.F. Candiani and D.R. Santos-Souza leg. (MPEG 24193); 1 female, 09.VIII.2004, D.F. Candiani leg. (MPEG 24196); 1 male, 09.VIII.2004, D.R. Santos-Souza leg. (MPEG 24194); 02°36'44.7"S 56°11'39.2"W, 1 female, 12.II.2007, N.F. Lo Man Hung leg. (MPEG 24198); 1 female (MPEG 24200); 1

female, 08.VIII.2008, N.C. Bastos leg. (MPEG 24188); 1 female, N.F. Lo Man Hung leg. (MPEG 24190); 1 female, L.T. Miglio leg. (MPEG 24192).

Distribution. Previously known from Colombia and Brazil (Pará: Canindé [Paragominas]). Recorded here also from Santo Antônio do Pará and Juruti, state of Pará, Brazil.

Variation. Six females, total length. 5.4 to 7.2; carapace: 2.4 to 2.9.

Alpaida yanayacu **sp. nov.**
Figs. 25-28, 29

Type material. Female holotype from Ecuador: Napo (Yanayacu Biological Station), 0°36'29.76"S 77°52'56.82"W, 25.XI.2009, A.B. Bonaldo leg., deposited in QCAZ.

Diagnosis. Females of *A. yanayacu* resemble those of *A. machala* Levi, 1988 and *A. eberhardi* Levi, 1988 by having the lateral lobes long and median lobe wide at its base, which is proportionally closer to *A. eberhardi*, since the median lobe is larger in both species than in *A. machala*; they differ by the lateral lobes rounded in ventral view (Fig. 26) (almost straight in *A. machala* and not visible in this view in *A. eberhardi*); by the borders of the posterior median plate parallel (Fig. 27); and by the less curved median lobe scape (Fig. 28) than in *A. machala*, see LEVI (1988: 555-556, 559-560) for comparison.

Description. Female (QCAZ). Total length 4.4. Carapace length 1.7, width 1.35, height 0.5. Clypeus height 0.05. Sternum length 0.75, width 0.67. Abdomen length 3.25, width 1.9, height 1.8. Leg formula I/II/IV/III. Leg lengths: femur, I 1.55, II 1.5, III 1.1, IV 1.6; patella, I 0.7, II 0.57, III 0.45, IV 0.47; tibia, I 1.47, II 1.17, III 0.75, IV 1.17; metatarsus, I 1.3, II 1.15, III 0.62, IV 1.15; tarsus, I 0.62, II 6.0, III 0.45, IV 0.52. Eyes diameters and interdistances: AME 0.1, ALE 0.07, PME 0.08, PLE 0.07; AME-PME 0.075, AME-ALE 0.26, PME-PLE 0.3, AME-AME 0.08, PME-PME 0.11. Carapace yellow, with two gray stripes on the lateral edges. Sternum black. Endites and labium dark brown with pale apices. Chelicerae yellow with distal third brown. Legs yellow, except the tarsus brown. Abdomen longer than wide, rounded anteriorly and pointed posteriorly. Dorsal side gray with white pigments of different sizes. Two pairs of white lateral stripes. Venter dusky gray from epigastric area to behind spinnerets. Epigynum wider than long, with notch not demarcated (Fig. 26); posterior median plate narrow with parallel borders (Fig. 27); scape relatively long and curved (Fig. 28).

Distribution. Known only from the type locality.

Etymology. The specific name is a noun in apposition taken from the type locality.

Alpaida antonio Levi, 1988
Fig. 33

Alpaida antonio Levi, 1988: 446, figs. 392-397 (female holotype from Fazenda Santo Antônio, Uruçuca, Bahia, Brazil, 27.XI.1977, deposited in Museu de Ciências Naturais (MCN), Fundação Zoobotânica do Rio Grande do Sul, not examined); Dierkens 2014: 15, figs. 1, 38.

Material examined. BRAZIL, *Amazonas*: Manicoré, 04°54'57"S 61°06'45.4"W, 1 male (INPA); 1 male (INPA); 1 male (INPA).

Distribution. Previously known from Guyana, French Guiana and Brazil (Pará: Melgaço, Canindé [Paragominas]; Bahia: Uruçuca, Camacã; Espírito Santo: Rio São José). Recorded here also from Manicoré, state of Amazonas, Brazil.

Alpaida bicornuta (Taczanowski, 1878)
Fig. 34

Epeira bicornuta Taczanowski, 1878: 168, pl. 2, fig. 18 (female lectotype and paralectotypes designated by Levi, 1988 from Pumamarca and Amable María, Junín, Peru, deposited in Polska Akademia Nauk (PAN), not examined).
Alpaida bicornuta: Levi, 1988: 387, figs. 11-18; Dierkens, 2014: 16, figs. 3, 4, 28, 39.

Material examined. BRAZIL, *Pará*: Bragança (Ilha das Canelas), 0°47'8.08"S 46°43'20.88"W, 1 female (MPEG 4980); 5 males and

2 females (MPEG 11187); 01°3'S 46°46'W, 1 female (MPEG 2971); Juruti (Área de várzea), 02°12'36.1"S 56°07'20.7"W, 1 female (MPEG 24493); 1 male (MPEG 24490); 1 male (MPEG 24491); 1 female (MPEG 24492); 1 female (MPEG 24494); 1 female (MPEG 24495); 02°24'33.2"S 56°26'10.6"W, 1 male (MPEG 24486); 1 male (MPEG 24487); (Barroso), 02°28'28.9"S 55°59'58.8"W, 1 female (MPEG 24489); 1 female (MPEG 24488); Marabá (Serra Norte, Fofoca), 05°58'13.81"S 50°21'28.16"W, 1 female (MPEG 4262).

Distribution. Previously known from Costa Rica to Argentina. Recorded here also from Bragança, Juruti and Marabá, state of Pará, Brazil.

Alpaida boa Levi, 1988
Fig. 35

Alpaida boa Levi, 1988: 447, figs. 408-409 (male holotype from Fonte Boa, Amazonas, Brazil, IX.1975, deposited in AMNH, not examined); Dierkens 2014: 16, figs. 5, 29.

Material examined. BRAZIL, *Pará*: Almerim (Jari), 0°53'16.53"S 52°50'41.59"W, 1 male (MPEG 7604).

Distribution. Previously known from French Guyana and Brazil (Amazonas: Fonte Boa). Recorded here also from Almerim, state of Pará, Brazil.

Alpaida deborae Levi, 1988
Fig. 36

Alpaida deborae Levi, 1988: 442, figs. 364-366 (female holotype from Browns Berg, 05°N, 55°27'W, Brokopondo Prov., Surinam, 20.II.1982, deposited in MCZ, not examined); Dierkens 2014: 16, figs. 6, 40.

Material examined. BRAZIL, *Pará*: Juruti (Área de Várzea), 02°24'33.2"S 56°26'10.6"W, 1 female (MPEG 24313).

Distribution. Previously known from Surinam, French Guiana and Brazil (Pará: Belém). Recorded here also from Juruti, state of Pará, Brazil.

Alpaida erythrothorax (Taczanowski, 1873)
Fig. 35

Singa erythrothorax Taczanowski, 1873: 126 (female lectotype, 2 males and 1 juvenile paralectotypes from Cayenne, French Guiana, deposited in PAN, not examined).
Alpaida erythrothorax: Levi, 1988: 444, figs. 376-378; Dierkens, 2014: 22.

Material examined. BRAZIL, *Pará*: Melgaço (Floresta Nacional de Caxiuanã), 01°44'18.02"S 51°27'48.01"W, 1 female (MPEG 24330); 1 female (MPEG 24332); 1 female (MPEG 24331); 01°44'15.5"S 51°26'42.0"W, 1 female (MPEG 24329); 1 female (MPEG 24328); Novo Progresso (Serra do Cachimbo), 09°16'18.6"S 54°56'22.9"W, 1 female (MPEG 6182); 1 female (MPEG 6363).

Distribution. Previously known from French Guiana and Brazil (Pará: Melgaço). Recorded here also from Novo Progresso, state of Pará, Brazil.

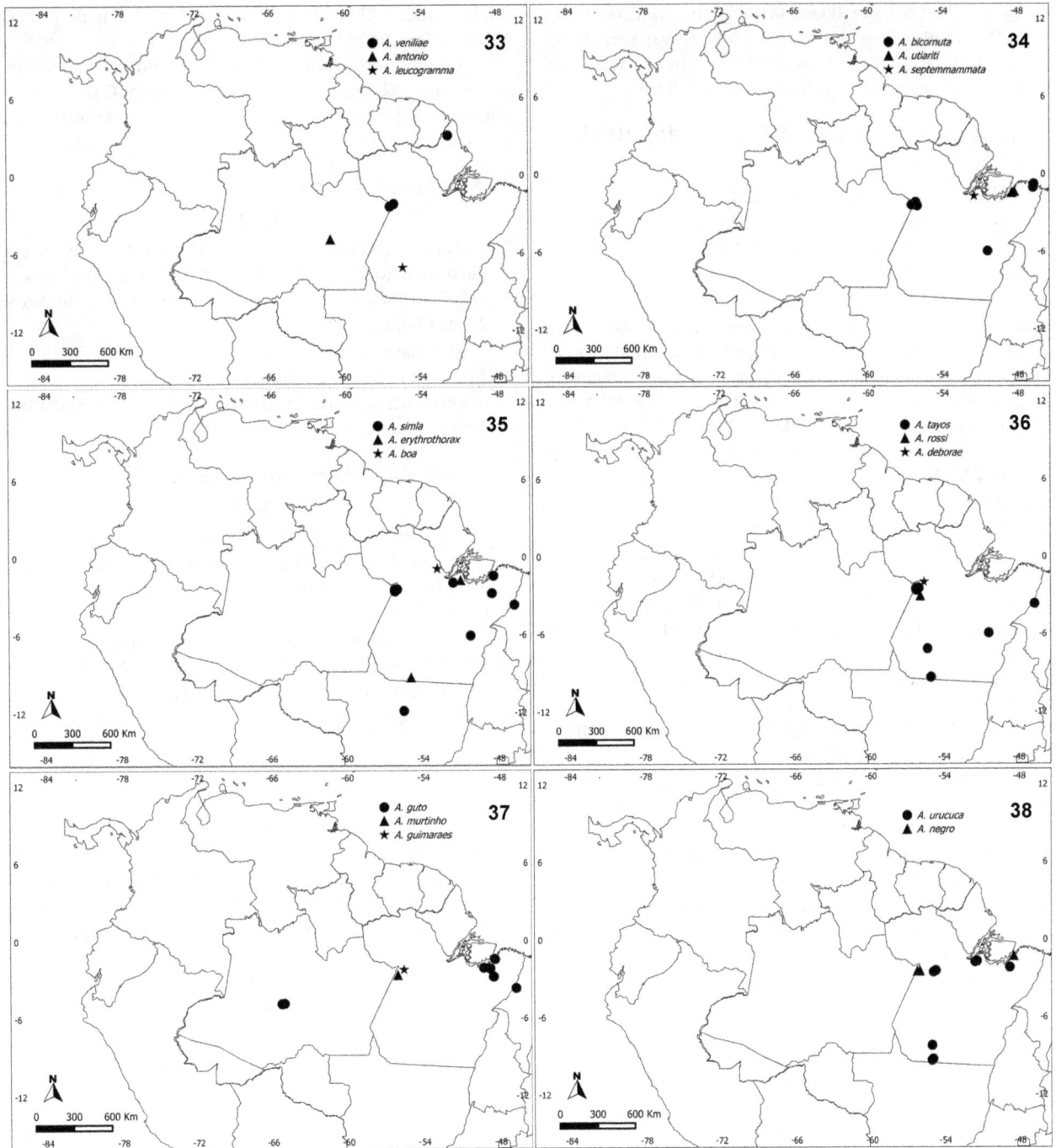

Figures 33-38. Distribution of new records of *Alpaida* species in the Amazon region, North Brazil: (33) *Alpaida antonio*, *A. leucogramma* and *A. veniliae*; (34) *A. bicornuta*, *A. septemmammata* and *A. utiariti*; (35) *A. boa*, *A. erythrothorax* and *A. simla*; (36) *A. deborae*, *A. rossi* and *A. tayos*; (37) *A. guimaraes*, *A. guto* and *A. murtinho*; (38) *A. negro* and *A. urucuca*.

Alpaida guimaraes Levi, 1988
Fig. 37

Alpaida guimaraes Levi, 1988: 390, figs. 19-24 (female holotype

from Chapada dos Guimarães, Mato Grosso, Brazil, 01.XII.1983, deposited in MCN, not examined); Dierkens, 2014: 23.

Material examined. Brazil, *Pará*: Juruti (Área de Várzea), 02°12′36.1″S 56°07′20.7″W, 1 female (MPEG 24317); 1 female

(MPEG 24319); 1 female (MPEG 24318); 1 female (MPEG 24320).

Distribution. Previously known from Guyana and Brazil (Pará: Jacareacanga; Bahia: Uruçuca; Mato Grosso: Xavantina). Recorded here also from Juruti, state of Pará, Brazil.

Alpaida guto Abrahim & Bonaldo, 2008
Fig. 37

Alpaida guto Abrahim & Bonaldo, 2008: 398, figs. 1-4 (male holotype from Floresta Nacional de Caxiuanã, Melgaço, Pará, Brazil, 09.V.2005, deposited in MPEG 5241, examined); Dierkens, 2014: 14.

Material examined. Brazil, *Amazonas*: Coari (Base de Operações Geólogo Pedro de Moura, Porto Urucu), 04°50′01″S 65°03′53″W, 1 female (MPEG 13776); 04°52′06″S 65°15′52″W, 1 female (MPEG 13736). *Pará*: Belém (Parque Estadual do Utinga), 01°25′18.8″S 48°25′48.3″W, 1 male and 3 females (MPEG 24203); 1 male (MPEG 24217); 3 females (MPEG 24218); 1 female (MPEG 24210); 1 female (MPEG 24222); 1 male (MPEG 24212); 1 female (MPEG 24202); 1 male and 4 females (MPEG 24204); 2 females (MPEG 24205); 1 male (MPEG 24206); 1 female (MPEG 24207); 1 female (MPEG 24208); 1 male (MPEG 24209); 1 female (MPEG 24211); 1 male (MPEG 24213); 1 female (MPEG 24214); 2 females (MPEG 24215); 1 male (MPEG 24216); 2 males and 5 females (MPEG 24219); 1 female (MPEG 24220); 1 male (MPEG 24221); 1 male and 6 females (MPEG 24223); (Reserva Mocambo), 01°26′48″S 48°25′1″W, 1 female (MPEG 24246); 2 males and 1 female (MPEG 24254); 1 female (MPEG 24245); 1 female (MPEG 24253); 1 male and 1 female (MPEG 24259); 1 female (MPEG 24255); 3 females (MPEG 24264); 1 female (MPEG 24252); 1 male (MPEG 24239); 1 female (MPEG 24251); 2 females (MPEG 24238); 1 male and 3 females (MPEG 24260); 1 male and 1 female (MPEG 24249); 1 male and 1 female (MPEG 24256); 3 males (MPEG 24240); 1 male (MPEG 24247); 1 male (MPEG 24248); 1 female (MPEG 24261); 1 male and 1 female (MPEG 24241); 2 males (MPEG 24242); 1 male and 1 female (MPEG 24243); 1 female (MPEG 24244); 1 male (MPEG 24250); 1 female (MPEG 24258); 1 female (MPEG 24262); 1 male and 1 female (MPEG 24257); 1 male (MPEG 24263); Cametá (Curuçambaba, Área de Floresta), 02°06′27.2″S 49°18′33.1″W, 4 females (MPEG 24586); 2 males (MPEG 24590); 1 male (MPEG 24588); 2 females (MPEG 24582); 1 male and 1 female (MPEG 24587); 1 male (MPEG 24589); 02°07′27.6″S 49°18′52.7″W, 2 females (MPEG 24585); 1 female (MPEG 24584); 1 female (MPEG 24583); Moju (Campo experimental da Embrapa), 02°09′38.9″S 48°47′50.64″W, 1 male (MPEG 24234); 02°10′41.52″S 48°47′37.13″W, 1 male (MPEG 24232); 02°11′44.37″S 48°47′38.79″W, 1 male (MPEG 24235); Tailândia (Fazenda Marupiara), 02°47′44.9″S 48°32′39.2″W, 1 male (MPEG 24230); 1 male (MPEG 24233); 02°48′43.7″S 48°30′44″W, 1 male (MPEG 24237); 1 male (MPEG 24231); 1 male (MPEG 24236). *Maranhão*: Centro Novo do Maranhão (Reserva Biológica do Gurupi), 03°41′21″S 46°45′16.5″W, 1 female (MPEG 24227); 2 males (MPEG 24228); 1 male (MPEG 24229); 03°41′33.84″S 46°44′46.62″W, 1 female

(MPEG 24226); 03°41′47.22″S 46°44′17.4″W, 1 male (MPEG 24224); 1 female (MPEG 24225).

Distribution. Previously known from Brazil (Pará: Melgaço and Santa Bárbara). Recorded here also from Coari, state of Amazonas; Belém, Moju and Tailândia, state of Pará; and Centro Novo do Maranhão, state of Maranhão, Brazil.

Alpaida leucogramma (White, 1841)
Fig. 33

Epeira (*Singa*) *leucogramma* White, 1841: 474 (female holotype from Rio de Janeiro, Brazil, deposited in BMNH, not examined).
Alpaida leucogramma: Levi, 1988: 391, figs. 32-38; Dierkens, 2014: 17, figs. 10, 31.

Material examined. Brazil, *Pará*: Novo Progresso, 07°08′07″S 55°24′51″W, 1 male (MPEG 4478).

Distribution. Previously known from Panama to Argentina. Recorded here also from Novo Progresso, state of Pará, Brazil.

Alpaida murtinho Levi, 1988
Fig. 37

Alpaida murtinho Levi, 1988: 399, figs. 84-85 (male holotype from Vila Murtinho, Rondônia, Brazil, 03.IV.1922, ex MCZ, deposited in Museu de Zoologia da Universidade de São Paulo (MZSP), not examined).

Material examined. Brazil, *Pará*: Juruti (Beneficiamento), 02°30′08.8″S 56°09′48.87″W, 12 males (MPEG 24314).

Distribution. Previously known from Brazil (Rondônia). Recorded here also from Juruti, state of Pará, Brazil.

Alpaida negro Levi, 1988
Fig. 38

Alpaida negro Levi, 1988: 448, figs. 410-414 (female holotype from Rio Negro, Paraná, Brazil, deposited in MZSP, not examined).

Material Examined. Brazil, *Pará*: Belém (Icoaraci), 01°17′59.5″ S 48°28′42.1″W, 1 male (MPEG 3348); Juruti (Barroso), 02°27′51.4″S 56°00′08.6″W, 1 female (MPEG 24485); (Capiranga), 02°28′22.1″S 56°12′29.4″W, 1 female (MPEG 24484).

Distribution. Previously known from Brazil (Mato Grosso and Paraná). Recorded here also from Belém and Juruti, state of Pará, Brazil.

Alpaida rossi Levi, 1988
Fig. 36

Alpaida rossi Levi, 1988: 447, figs. 404-407 (female holotype from Monzón Valley, Tingo María, Dpto. Huánuco, Peru, 10.XI.1954, deposited in California Academy of Sciences (CAS), not examined).

Material examined. Brazil, *Pará*: Juruti (Mutum), 02°33′18.0″S 56°13′22.4″W, 1 female (MPEG 24315); 02°33′13.8″S 56°13′22.1″W, 1 female (MPEG 24316).

Distribution. Previously known from Peru. Recorded here also from Juruti, state of Pará, Brazil.

Alpaida septemmammata (O. Pickard-Cambridge, 1889)
Fig. 34

Epeira septemmammata O. Pickard-Cambridge, 1889: 42, pl. 7, fig. 6 (fifteen females specimens from Teapa, Mexico, deposited in BMNH, type not located, both material not examined); Keyserling, 1892: 89. Pl. 4, fig. 67.

Alpaida septemmammata: Levi, 1988: 452, figs. 427-434; Dierkens, 2014: 21, figs. 15, 32.

Material examined. Brazil, *Pará*: Melgaço (Floresta Nacional de Caxiuanã, Estação Científica Ferreira Penna), 01°43'21.6"S 51°25'51.2"W, 1 male (MPEG 24482); 1 male (MPEG 24483).

Distribution. Previously known from Mexico to Argentina. Recorded here also from Melgaço, state of Pará, Brazil.

Alpaida simla Levi, 1988
Fig. 35

Alpaida simla Levi, 1988: 430, figs. 289-293 (female holotype, male and 6 immature paratypes from Simla, Trinidad, Lesser Antilles, IV.1964, deposited in MCZ, not examined); Bonaldo et al., 2009; Cafofo et al., 2013.

Material examined. Brazil, *Pará*: Belém (Reserva Mocambo), 01°26'48"S 48°25'1"W, 1 male (MPEG 24289); 1 male (MPEG 24290); Portel (Floresta Nacional de Caxiuanã, Plote PPBio), 01°57'38.9"S 51°36'45.3"W, 1 female (MPEG 13399); Tailândia (Fazenda Marupiara), 02°47'44.9"S 48°32'39.2"W, 1 female (MPEG 24288); Juruti (Capiranga), 02°28'0.6"S 56°12'42.2"W, 1 female (MPEG 24282); 02°28'22.1"S 56°12'29.4"W, 2 males (MPEG 8141); 1 female (MPEG 8188); 2 males (MPEG 8162); 1 male (MPEG 24265); 1 male (MPEG 24283); 1 female (MPEG 24286); (Barroso), 02°27'41.7"S 56°00'11.6"W, 1 male (MPEG 8148); 1 male (MPEG 8153); 1 male (MPEG 8180); 1 male (MPEG 8192); (Beneficiamento), 02°30'08.8"S 56°09'48.87"W, 2 females (MPEG 24270); 02°30'27.4"S 56°10'39.5"W, 1 female (MPEG 24275); (Mutum), 02°33'04.8"S 56°13'32.5"W, 1 female (MPEG 24267); 1 female (MPEG 24271); 1 female (MPEG 24276); 1 female (MPEG 24284); 1 male (MPEG 24280); 02°33'06.9"S 56°13'29.0"W, 1 male (MPEG 24277); 1 female (MPEG 24266); 1 male (MPEG 24269); 1 male (MPEG 24268); 1 male (MPEG 24272); 02°33'13.8"S 56°13'22.1"W, 1 male (MPEG 24285); 1 male (MPEG 24273); 02°33'18.0"S 56°13'22.4"W, 1 female (MPEG 24278); 1 female (MPEG 24281); 1 female (MPEG 24274); 02°36'44.7"S 56°11'39.2"W, 1 female (MPEG 24279); Marabá (Serra Norte), 06°4'22.10"S 50°14' 47.27"W, 2 males (MPEG 4196). *Maranhão*: Centro Novo do Maranhão (Reserva Biológica do Gurupi), 03°41'07.9"S 46°45' 46"W, 1 female (MPEG 24287). *Mato Grosso*: Sinop, 11°51'38.73"S 55°30'34.85"W, 1 male (MPEG 3350).

Distribution. Previously known from Trinidad & Tobago and Brazil (Pará: Melgaço and Portel). Recorded here also from Belém, Tailândia, Juruti and Marabá, Pará; Centro Novo do Maranhão, state of Maranhão; and Sinop, state of Mato Grosso, Brazil.

Alpaida tayos Levi, 1988
Fig. 36

Alpaida tayos Levi, 1988: 456, figs. 458-467 (female holotype from Los Tayos-Santiago, banana plantation, 03°04'S, 78°02'W, Prov. Morona-Santiago, Ecuador, 03.VIII.1976, deposited in MCZ, not examined).

Material examined. Brazil, *Pará*: Juruti (Capiranga), 02°28'0.6"S 56°12'42.2"W, 1 female (MPEG 24158); 02°28'22.1"S 56°12'29.4"W, 1 male (MPEG 24142); (Barroso), 02°27'51.4"S 56°00'08.6"W, 1 female (MPEG 24145); 1 male (MPEG 24161); 1 male (MPEG 24139); (Beneficiamento), 02°30'25.4"S 56°11'04.8"W, 1 male (MPEG 24140); 02°30'27.4"S 56°10'39.5"W 1 male (MPEG 24136); 1 female (MPEG 24132); (Mutum), 02°33'06.9"S 56°13'29.0"W, 1 female (MPEG 24133); 1 female (MPEG 24147); 1 male (MPEG 24134); 1 female (MPEG 24137); 1 female, (MPEG 24148); 1 female (MPEG 24135); 1 male (MPEG 24150); 1 female (MPEG 24146); 02°33'13.8"S56°13'22.1"W, 1 female (MPEG 24159); 1 male and 2 females (MPEG 24141); 02°33'18.0"S 56°13'22.4"W, 1 female (MPEG 24156); 02°36'11.2"S 56°12'36.3"W, 1 male (MPEG 24155); 02°36'44.7"S 56°11'39.2"W, 1 female (MPEG 9204); 1 female (MPEG 9134); 1 male (MPEG 163); 1 male (MPEG 24138); 1 female (MPEG 24143); 1 female (MPEG 24149); 1 female (MPEG 24151); 1 male (MPEG 24153); 2 females (MPEG 24152); 1 male (MPEG 24160); 02°36'45.2"S 56°11'27.5"W, 1 male (MPEG 24154); 1 female. (MPEG 24157); 02°36'45.7"S 56°11'38.2"W, 1 female (MPEG 24144); Marabá (Serra Norte), 05°57'48.56"S 50°24'1.7"W, 1 male (MPEG 4234); Novo Progresso, 07°09'53"S 55°18'53"W, 1 female (MPEG 4499); (Serra do Cachimbo), 09°21'59"S 55°02'01"W, 1 male (MPEG 6169). *Maranhão*: Centro Novo do Maranhão (Reserva Biológica do Gurupi), 03°41'07.92"S 46°45'46.08"W, 1 male (MPEG 24131).

Distribution. Guyana, Ecuador, Peru, French Guiana and Brazil (Pará: Ananindeua and Canindé [Paragominas]). Recorded here also from Juruti, Marabá, Novo Progresso, state of Pará; and Centro Novo do Maranhão, state of Maranhão, Brazil.

Alpaida urucuca Levi, 1988
Fig. 38

Alpaida urucuca Levi, 1988: 454, figs. 440-445 (female holotype from Fazenda Antonio, Uruçuca, Bahia, Brazil, 24.X.1979, deposited in MCN, not examined).

Material examined. Brazil, *Pará*: Melgaço (Floresta Nacional de Caxiuanã, 01°42'24"S 51°27'34.3"W, 1 female (MPEG 24325); 01°43'21.6"S 51°25'51.2"W, 1 female (MPEG 24326); 01°45'12.8"S 51°31'14.7"W, 1 female (MPEG 24323); 1 female (MPEG 24324); 01°46'36.00"S 51°35'12.21"W, 1 female (MPEG 24322); Moju (Campo experimental da Embrapa), 02°11'44.37"S 48°47'38.79"W, 1 female (MPEG 24327); Santarém (Alter-do-Chão), 02°26'33.18"S 54°43'8.70"W, 1 female (MPEG 16153); 1 female (MPEG 16154); 1 female (MPEG 16151); 02°32'59.02"S 54°54'05.20"W, 1 female (MPEG 16152); 1 female (MPEG 16408); 1 female (MPEG 16150); Altamira (Castelo dos Sonhos),

08°13'03"S 55°00'57"W, 1 female (MPEG 4488); Novo Progresso (Serra do Cachimbo), 09°16'18.6"S 54°56'22.9"W, 1 female (MPEG 6125); 09°22'02.9"S 55°01' 11.9"W, 1 female (MPEG 6086).

Distribution. Previously known from Brazil (Pará: Melgaço; Bahia: Uruçuca). Recorded here also from Moju, Santarém, Altamira and Novo Progresso, state of Pará, Brazil.

Alpaida utiariti Levi, 1988
Fig. 34

Alpaida utiariti Levi, 1988: 466, figs. 523-524 (male holotype from Utiariti, Mato Grosso, Brazil, 30.VII.1961, deposited in MZSP, not examined).

Material examined. BRAZIL, *Pará*: Belém (MPEG, Campus de Pesquisa), 01°27'03.0"S 48°26'40.2"W, 1 male (MPEG 22467); 1 male (MPEG 24521); Benevides, [01°21'43.87"S 48°14'37.79"W], 1 male (MPEG 4665).

Distribution. Previously known from Brazil (Mato Grosso: Utiariti). Recorded here also from Benevides, state of Pará, Brazil.

Alpaida veniliae Levi, 1988
Fig. 33

Epeira veniliae Keyserling, 1865: 817, pl. 19, fig. 23 (seven females and one male syntypes from New Granada, deposited in BMNH, not examined); Keyserling, 1893: 256, pl. 13, fig. 191.

Epeira pantherina Taczanowski, 1872: 132. Male lectoype designated by LEVI (1988) from Uaça, Amapá, Brazil (PAN), not examined. Synonymyzed by LEVI (1988).

Alpaida veniliae: Levi, 1988: 402, figs. 103-109; Dierkens, 2014: 22, figs. 21, 22, 35, 48.

Material examined. BRAZIL, *Amapá*: Oiapoque, 03°4'25.63"S 51°51'8.18"W, 2 males and 5 females (MPEG 5000). *Pará*: Juruti (Área de Várzea), 02°12'36.1"S 56°07'20.7"W, 1 male and 1 female (MPEG 24308); 1 female (MPEG 24297); 1 male (MPEG 24305) 1 female (MPEG 24309); 1 female (MPEG 24301); 1 female (MPEG 24311); 2 females (MPEG 24299); 2 males (MPEG 24304); 1 male (MPEG 24303); 1 female (MPEG 24312); 02°24'33.2"S 56°26'10.6"W, 3 females (MPEG 24291); 2 males and 1 female (MPEG 24293); 4 females (MPEG 24294); 2 females (MPEG 24295); 2 females (MPEG 24298); 1 male and 1 female (MPEG 24302); 1 female (MPEG 24306); 1 female (MPEG 24307); 2 females (MPEG 24310); 6 males and 11 females (MPEG 24292); 3 males and 4 females (MPEG 24296).

Distribution. Previously known from Panama to Argentina. Recorded here also from Oiapoque, state of Amapá; and Juruti, state of Pará, Brazil.

DISCUSSION

Alpaida levii new species shares somatic and genitalic characters with *A. delicata*: abdomen hump-shaped, cymbium prolaterally expanded and strong spines on tibiae I and II, sug-

gesting that *A. levii* and *A. delicata* may be sister species. Due to these similarities, records of *A. delicata* in the faunistic literature may not be accurate, as observed for at least some of the specimens of *A. levii* examined by us, which were previously determined as *A. delicata*.

LEVI (1988) compared *A. delicata* with *A. truncata* for diagnostic purposes, based on the hump-shaped abdomen, which is shared by both species. However, considering the new information provided here, we propose a new diagnosis for *A. delicata*, which we compare with *A. levii*. The median apophysis of *A. truncata* is extremely long, with a distally pointed tip (Fig. 16), more similar to that of by *A. queremal* Levi, 1988, while the median apophysis of *A. delicata* is short, medially excavated and quadrangular, similar to *A. levii*. *Alpaida delicata* and *A. levii* also share strong spines on tibiae I and II. The terminal apophysis of *Alpaida* species can present a distal lobe, a basal prong and a retrolateral apical sector, recognized here for the first time. The retrolateral apical sector can be very developed in some species, such as *A. delicata* (A, Fig. 12) and *A. levii* (A, Figs. 3-5) or reduced, as in *A. truncata* (A, Fig. 17). The distal lobe and the basal prong of the terminal apophysis are absent in some *Alpaida* species and the distribution of those characters may be important in a phylogenetic context.

Due to the high complexity of the palps of *Alpaida*, especially with regards to the terminal apophysis, and giving LEVI's (1988) choice to document the palps only in mesal view, it is difficult to identify all sclerites of all species and the identification of some species may be uncertain, especially when there is intraespecific variation. For this reason we document the expanded palp and details of some sclerites for the new species and for the males of *A. iquitos* and *A. truncata*. This refined information will facilitate the recognition of these taxa and an eventual phylogenetic analysis to clarify the relationship among species of *Alpaida*.

ACKNOWLEDGEMENTS

The authors would like to thank the reviewers and editor for their comments that help improve the manuscript. The authors are supported by grants: ABB – PQ grant #304965/2012-0, BVBR# 302358/2013-7 – CNPq, RS #3362 – CELPA/FADESP Monitoramento (REF.: 061/2013). Project 3362 is supported by Centrais Elétricas do Pará S.A. – CELPA and involves monitoring studies in the Marajó Archipelago, entitled "Monitoramento dos possíveis impactos da linha de transmissão do Marajó sobre a fauna".

LITERATURE CITED

ABRAHIM N, BONALDO AB (2008) A new species of *Alpaida* (Araneae, Araneidae) from Caxiuanã National Forest, Oriental Amazonia, Brazil. **Iheringia, Série Zoologia, 98**(3): 397-399. doi: 10.1590/S0073-47212008000300015

BONALDO AB, CARVALHO LS, PINTO-DA-ROCHA R, TOURINHO A, MIGLIO LT, CANDIANNI DF, LO-MAN-HUNG NF, ABRAHIM N, RODRIGUES BVB, BRESCOVIT AD, SATURNINO R, BASTOS NC, DIAS SC, SILVA BJF, PEREIRA-FILHO JMB, RHEIMS CA, LUCAS SM, POLOTOW D, RUIZ G, INDICATTI R (2009) Inventário e história natural dos aracnídeos da Floresta Nacional de Caxiuanã, p. 577-621. In: LISBOA PLB (Org.). **Caxiuanã: desafios para a conservação de uma Floresta Nacional na Amazônia.** Belém, Museu Paraense Emílio Goeldi.

BONNET P (1955) Bibliographia araneorum. **Toulouse 2**(1): 1-918.

BRAGA-PEREIRA GF, SANTOS AJ (2013) Two new species of the spider genus *Alpaida* (Araneae: Araneidae) from restinga areas in Brazil. **Revista Brasileira de Zoologia 30**(3): 324-328. doi: 10.1590/S1984-46702013000300010

BUCKUP EH, MEYER AC (1993) Sobre o macho de *Alpaida scriba* (Araneae, Araneidae). **Revista Brasileira de Entomologia 37**(2): 353-354.

BUCKUP EH, RODRIGUES ENL (2011) Espécies novas de *Alpaida* (Araneae: Araneidae), descrições complementares e nota taxonômica. **Iheringia, Série Zoologia, 101**(3): 262-267. doi: 10.1590/S0073-47212011000200013

CAFOFO EG, SATURNINO R, SANTOS AJ, BONALDO AB (2013) Riqueza e composição em espécies de aranhas da Floresta Nacional de Caxiuanã, p. 539-562. In: PLB LISBOA (Ed.). **Caxiuanã: Paraíso ainda preservado.** Belém, Museu Paraense Emílio Goeldi.

DIERKENS M (2014) Contribution à l'étude des Araneidae de Guyane française. V – Les genres *Alpaida* et *Ocrepeira*. **Bulletin Mensuel de la Societe Linneenne de Lyon 83**(1-2): 14-30.

HÖFER H, BRESCOVIT AD (2001) Species and guild structure of a Neotropical spider assemblage (Araneae; Reserva Ducke, Amazonas, Brazil). **Andrias 15**: 99-120.

KEYSERLING E (1865) Die Beiträge zur Kenntniss der Orbitelae Latr. **Verhandlungen der Kaiserlich-Königlichen Zoologisch-Botanischen Gesellschaft in Wien 15**: 799-856.

KEYSERLING E (1892) Die Spinnen Amerikas, Epeiridae. **Nürnberg 4**: 1-208.

KEYSERLING E (1893) Die Spinnen Amerikas. Epeiridae. **Nürnberg 4**: 209-377.

LEVI HW (1988) The neotropical orb-weaving spiders of the genus *Alpaida* (Araneae: Araneidae). **Bulletin of the Museum of Comparative Zoology 151**: 365-487.

LEVI HW (2002) Keys to the genera of araneid orbweavers (Araneae, Araneidae) of the Americas. **Journal of Arachnology 30**(3): 527-562.

PICKARD-CAMBRIDGE O (1889) Arachnida. Araneida. **Biologia Centrali-Americana, Zoology 1**: 1-56.

QGIS DEVELOPMENT TEAM (2012) **Quantum GIS Geographic Information System.** Open Source Geospatial Foundation Project, v. 1.8.0. Available online at: http://qgis.osgeo.org [Accessed: 28/11/2015]

RODRIGUES ENL, MENDONÇA JR MS (2011) Araneid orb-weavers (Araneae, Araneidae) associated with riparian forests in southern Brazil: a new species, complementary descriptions and new records. **Zootaxa 2759**: 60-68.

SIMON E (1897) Etudes arachnologiques. 27e Mémoire. XLII. Descriptions d'espèces nouvelles de l'ordre des Araneae. **Annales de la Société Entomologique de France 65**: 465-510.

TACZANOWSKI L (1872) Les aranéides de la Guyane française. **Horae Societatis Entomologicae Rossicae 8**: 32-132.

TACZANOWSKI L (1873) Les aranéides de la Guyane française. **Horae Societatis Entomologicae Rossicae 9**: 64-150.

TACZANOWSKI L (1878) Les Aranéides du Pérou central. **Horae Societatis Entomologicae Rossicae 14**: 140-175.

WHITE A (1841) Description of new or little known Arachnida. **Annals and Magazine of Natural History 7**: 471-477.

WORLD SPIDER CATALOG (2014) **World Spider Catalog.** Bern, Natural History Museum Bern, version 15.5. Available online at: http://wsc.nmbe.ch [Accessed: 7/10/2014].

13

Intra- and inter-annual variations in Chironomidae (Insecta: Diptera) communities in subtropical streams

Diane Nava[1], Rozane M. Restello[1] & Luiz U. Hepp[1,*]

[1]Programa de Pós-graduação em Ecologia, Universidade Regional Integrada do Alto Uruguai e das Missões. Avenida Sete de Setembro 1621, 99709-910 Erechim, RS, Brazil.
*Corresponding author. E-mail: luizuhepp@gmail.com

ABSTRACT. The structure and composition of stream benthic communities are strongly influenced by spatial and temporal factors. This study evaluated the intra and inter-annual variations in Chironomidae communities in subtropical streams. The organisms were sampled from 10 small-order streams during the summer and winter of 2010-2012. The number of chironomid specimens sampled was 7,568, distributed in 49 genera. Chironomid abundance and richness varied intra and inter-annually and community composition varied intra-annually (2010 and 2011). Water temperature, total organic carbon, nitrogen, and rainfall were correlated with chironomid community composition. The intra-annual variation of the community was dependent on climatic variations (temperature and rainfall) and changes caused by intensive agricultural use. We conclude that the temporal variation observed in the Chironomidae community correlates with climatic variations (rainfall) and changes in the total organic carbon and total nitrogen, caused by intensive agricultural land use.

KEY WORDS. Agriculture impacts; bioindicators; macroinvertebrates; rainfall.

The structure and composition of aquatic communities are influenced by spatial and temporal factors (SUAREZ 2008). Knowing how these factors act on biological communities facilitates the understanding of how local and regional factors influence species occurrence (POFF et al. 2006, SUAREZ 2008). The distribution of benthic macroinvertebrates is affected by factors such as type of substrate (HEPP et al. 2012), habitat characteristics (GALDEAN et al. 2000, BUSS et al. 2004), land use (HEPP et al. 2010), and climatic variations over a timescale (SCHEFFER & VAN NES 2007). Climatic variations have decisive effects on the distribution of benthic organisms (SMITH et al. 2003) and can occur at different timescales, both intra-annual (seasons) and inter-annual (between years).

Chironomidae (Insecta: Diptera) occur in great abundance and high diversity in most aquatic ecosystems in all continents (EPLER 2001, FERRINGTON 2008). They play an important role in the food web of aquatic communities, establishing links between producers and consumers, as well as participating in nutrient cycles (HENRIQUES-OLIVEIRA et al. 2003). Chironomids are tolerant to various changes in the environment (ROSIN & TAKEDA 2007, RESTELLO et al. 2012), and depending on the species, they may display negative or positive responses to human impacts (FERRINGTON 2008).

Chironomid communities can be effected by the integrity of the riparian zone. The state and the extent of the riparian vegetation correlates with differences in the abundance, richnness, and composition of chironomid communities in streams (SENSOLO et al. 2012). For instance, suppression of the riparian vegetation results in decreased overall diversity and increased numbers of tolerant taxa (AL-SHAMI et al. 2010a). Different chironomids inhabit different habitats and substrates (SANSEVERINO & NESSIMIAN 2001) although they are most frequent in heterogeneous and stable environments, where they attain high diversity (ROSA et al. 2011, 2013).Temporal variations in biological communities are mainly linked to climate-related changes (e.g., temperature and rainfall). Climate affects ecological processes such as competition, predation and recruitment (GRESENS et al. 2007). In addition to climatic factors, temporal variations in the structure and composition of chironomid communities may reflect the biological characteristics of the species that compose these communities (HEINIS & DAVIDS 1993, SIQUEIRA et al. 2008) or temporal changes in physical and chemical characteristics (AL-SHAMI et al. 2010b). Natural disturbances, such as spates caused by increased rainfall in human-impacted areas may carry chemicals from adjacent areas to the streams, thus affecting chironomid communities (GRESENS et al. 2007).

In this study, the intra- and inter-annual variation of Chironomidae communities in subtropical streams was assessed over three years. We tested the hypothesis that environmental factors related to human activities may be important in structuring communities in streams. Thus, the objectives of this study were (1) to evaluate the intra and inter-annual variations in Chironomidae communities and (2) to determine whether these temporal variations are associated with environmental factors.

MATERIAL AND METHODS

This study was conducted in the upper portion of the Uruguay River Basin in southern Brazil (27°12'59" and 28°00'47"S, 52°48'12" and 51°49'34"W, Fig. 1). The region is characterized by a subtropical climate (Koppen Cfb) with average annual rainfall of 1912.3 mm and average annual temperature of 17.6°C. The vegetation is a subtropical forest mix. It is mostly composed of species with tropical-subtropical distribution in the Upper Uruguay, and Araucaria Forest with a predominance of Araucaria (Oliveira-Filho et al. 2015). The predominant land use is intensive agricultural practice (~77% of the total area), with soybeans, corn, wheat crops, and large forested areas (Decian et al. 2009). Thus, all 10 selected streams are embedded in a complex agricultural matrix. All streams studied were small-order streams (<3rd order) and had similar limnological characteristics. The average percentage of vegetation in the riparian zone of the streams was 23% (range 11-49%).

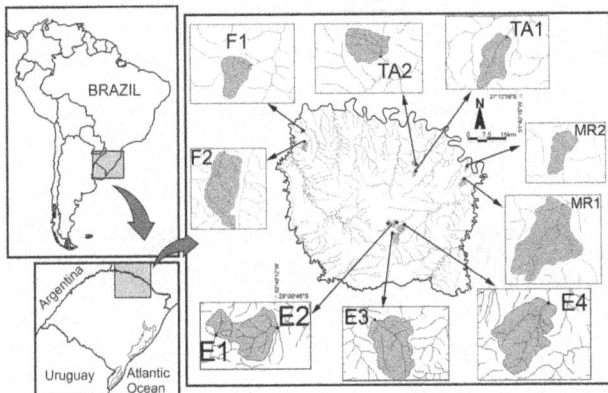

Figure 1. Geographical location of sampling sites at Alto Uruguay region, RS. F: Faxinalzinho, TA: Três Arroios, ERE: Erechim, MR: Marcelino Ramos.

We obtained the following variables from each stream: water temperature, turbidity, conductivity, total dissolved solids, dissolved oxygen and pH, with the aid of a multiparameter analyser HORIBA® U50. A Shimadzu® TOC-VCSH analyzer was used to measure total organic carbon (TOC) and total nitrogen.

Chironomidae larvae were collected in August (winter) and December (summer) of the years 2010, 2011 and 2012. At each stream, three sampling units were obtained with a Surber sampler (mesh 250 μm, area 0.09 m²) on a rock substrate. The material was fixed in the field using 80% alcohol. In the laboratory, Chironomidae larvae were dipped in a of 10% bleach solution of potassium hydroxide for 24 hours. Individuals were then mounted on semi-permanent slides with Hoyer solution and were identified under optical microscope with a magnifi-

cation of 1,000 times. Specimens were identified at the genus level using the identification keys of Trivinho-Strixino & Strixino (1995) and Trivinho-Strixino (2011).

To assess the intra- and inter-annual variations in abiotic variables Multivariate Analysis of Variance (MANOVA) was used. Variations in chironomid abundance and richness between the seasons (intra-annual) and between the years (inter-annual) were evaluated using a repeated measure Analysis of Variance (RM-ANOVA). Non-Metric Multidimensional Scaling (NMDS) (Kruskal 1964) was used to order the chironomid communities. The NMDS was performed with a biological matrix based on the presence or absence of genera in each stream using the Jaccard index. The relationship between environmental and biological data was tested by fitting vectors of environmental variables to the NMDS ordination (function 'envfit' of the vegan package). Analysis of Similarity (ANOSIM) was used to evaluate the level of segregation in community composition between years and within years,. All analyses were performed using R software (R Core Team 2013) with the 'vegan' package (Oksanen et al. 2013).

RESULTS

The studied streams have well-oxygenated (10.85 ± 2.51 mg L⁻¹), slightly acidic water (pH 6.62 ± 0.22) with electrical conductivity of 0.030 ± 0.059 mS cm⁻¹ (mean of three years). The highest average turbidity was recorded in summer (11.75 ± 4.90 NTU). However, the highest average total organic carbon was recorded in winter (218.34 ± 216.16 mg L⁻¹) (Table 1). The total organic carbon was very high in the winter of 2011 (Table 1). The highest average monthly rainfall occurred in 2011 (172.20 ± 85.16 mm) followed by 2010 (122.6 ± 101.7 mm) and 2012 (36.7 ± 60.0 mm; Fig. 2). However, in 2010 and 2012there was as much rainfall in the winter and the summer. In 2011, the difference in rainfall between the winter and summer seasons was ca. 150 mm. Overall, the abiotic variables differed among the years and between the seasons ($F_{(2,56)}$ = 18.22, p = 0.001 and $F_{(1,56)}$ = 25.10, p = 0.001, respectively, Table 2).

We obtained a total of 7,568 chironomid larvae distributed in 49 genera. The highest abundance was recorded in 2012 (3,304 larvae, 43.7% of the total), followed by 2011 (2,430 larvae, 32.1%) and 2010 (1,834 larvae, 24.2%). In two of the three years studied (2010 and 2012, Fig. 3, Table 3) chironomids were more abundant in the winter. The greatest number of chironomid genera (43 genera) was identified in 2011, followed by 2012 (33 genera) and 2010 (25 genera). Thus, abundance varied intra-annually while richness varied intra- and inter-annually (Table 3, Fig. 4).

Among the genera identified, *Pentaneura* Philippi, 1865, *Polypedilum* Kieffer, 1912, and *Rheotanytarsus* Thienemann & Bause in Bause, 1913 were the most frequent in the samples. *Aedokritus* Roback, 1958, *Antillocladius* Saether, 1981, *Denopelopia* Roback & Rutter, 1988, *Djalmabatista* Fittkau, 1968,

Table 1. Mean and standard deviation of limnological variables quantified the drainage areas of the 10 studied streams in the region Alto Uruguay Rio Grande Sul, in the period 2010-2012.

Variables	2010		2011		2012	
	Summer	Winter	Summer	Winter	Summer	Winter
Water temperature (°C)	20.72 ± 0.87	15.05 ± 1.84	21.45 ± 3.03	14.40 ± 1.57	22.78 ± 1.98	16.31±1.79
pH	6.87 ± 0.84	6.45 ± 0.56	6.45 ± 0.59	6.89 ± 0.79	6.71 ± 0.43	6.37±0.48
Electrical Conductivity (mS cm⁻¹)	0.05 ± 0.03	0.05 ± 0.02	0.08 ± 0.04	0.051 ± 0.02	1.52 ± 4.59	0.06±0.03
Turbidity (UNT)	13.37 ± 17.26	3.07 ± 8.99	6.25 ± 2.82	8.46 ± 5.05	15.65 ± 19.87	8.85±5.40
DO (mg L⁻¹)	10.38 ± 1.06	8.36 ± 0.90	9.54 ± 2.57	9.49 ± 0.73	11.92 ± 3.15	15.39±1.67
TDS (mg L⁻¹)	0.04 ± 0.02	0.03 ± 0.01	0.05 ± 0.03	0.03 ± 0.01	0.04 ± 0.02	0.03±0.02
Nitrogen (mg L⁻¹)	15.66 ± 3.37	6.36 ± 2.80	0.44 ± 0.47	1.32 ± 0.68	1.13 ± 0.89	1.05±0.69
TOC (mg L⁻¹)	17.14 ± 5.35	25.05 ± 21.80	83.90 ± 39.06	451.76 ± 100.39	61.28 ± 58.23	178.21±50.15

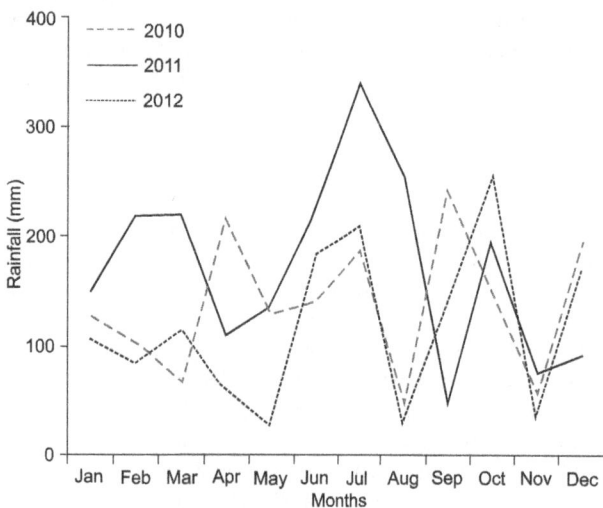

Figure 2. Monthly rainfall in the years 2010, 2011 and 2012 at Alto Uruguay region, RS. The horizontal lines indicate the annual average for the respective years (INMET).

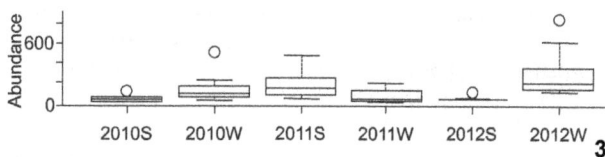

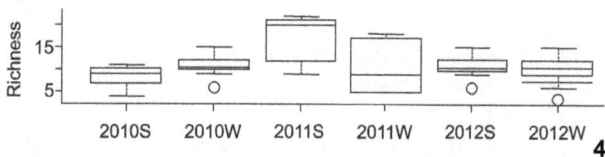

Figures 3-4. Box-plot (median and quartiles) showing the variation of (3) and abundance (4) wealth of inter-annual chironomid larvae (2010, 2011 and 2012) and intra-annual (summer: S, winter: W) in subtropical streams.

Table 2. MANOVA results for limnological intra and inter-annual among the studied streams were considered as factors the years studied (inter-annual) and the seasons (intra-annual).

	DF	SS	MS	F	p
Season	1	1.013	1.013	25.101	0.001
Year	2	1.471	0.735	18.225	0.001
Residuals	56	2.260	0.040	0.476	
Total	59	4.744			

Table 3. Repeated measures ANOVA results for the variation in the abundance (log) and richness of Chironomidae community between seasons and among years at 10 stream sites in Alto Uruguay region, Rio Grande Sul.

	DF	SS	F	p
Abundance (log)				
Year	2	1.01	0.75	0.477
Seasons	1	4.73	7.04	0.010
Year:Seasons	2	32.22	23.97	<0.001
Residuals	54	36.30		
Richness				
Year	2	235.90	8.44	<0.001
Seasons	1	43.35	3.10	0.008
Year:Seasons	2	237.90	8.51	<0.001
Residuals	54	754.50		

Manoa Fittkau, 1963, *Paracladius* Hirvenoja, 1973, *Paramerina* Fittkau & Stur, 1997, and *Pseudochironomus* Riethia Malloch, 1915 occurred only in winter, while *Endotribelos* Grodhaus, 1987, *Microchironomus* Kieffer, 1918, *Parapentaneura* Stur, Fittkau & Serrano, 2006, and *Ubatubaneura* Wiedenbrug & Trivinho-Strixino, 2009 occurred only in summer (Table 4). Community composition was similar between seasons when the three years were analysed together (ANOSIM, R = -0.01, p = 0.596,

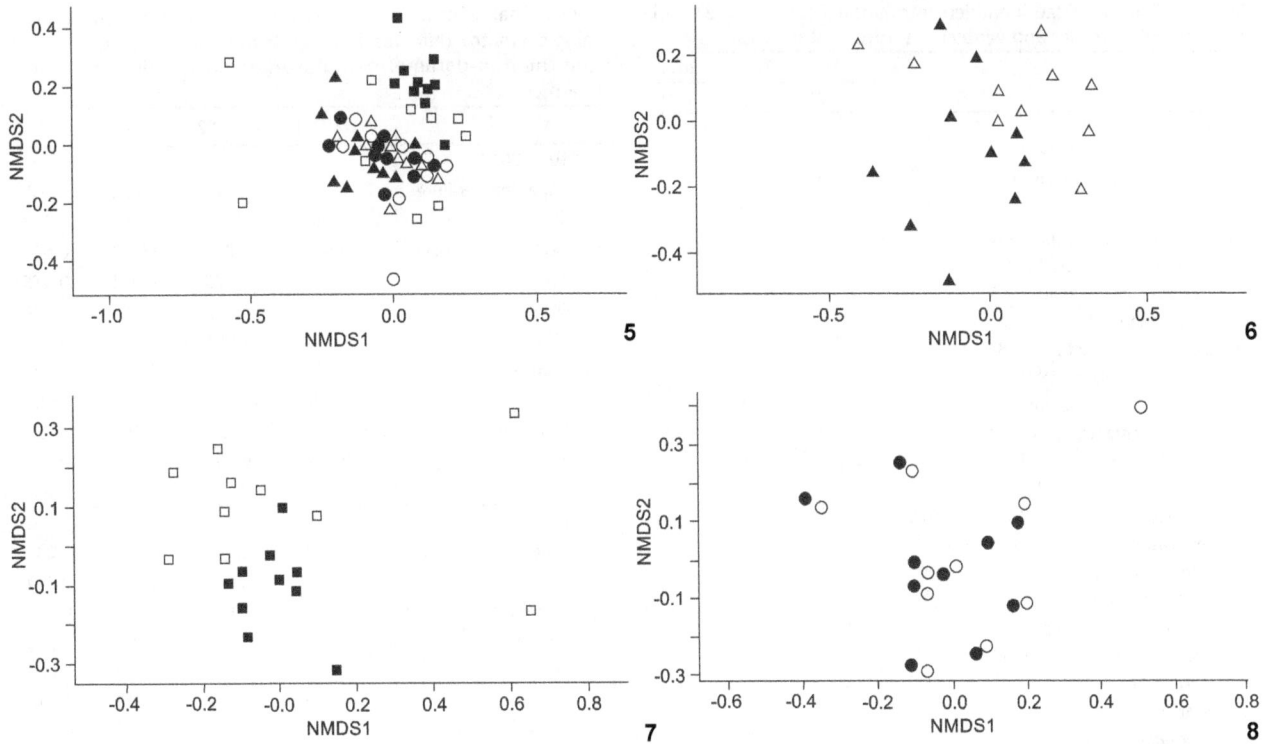

Figures 5-8. Non-Metric Multidimensional Scaling Analysis (NMDS) of the temporal distribution of chironomid community in subtropical streams. (5) inter-annual distribution (2010, 2011, 2012), (6) intra-annual distribution for 2010, (7) intra-annual distribution for 2011, (8) intra-annual distribution for 2012. Open symbols: winner, closed symbols: summer.

Figs. 5-8). However, the generi composition showed segregation between summer and winter in 2011 (ANOSIM, R = 0.32, p = 0.001) and 2010 (ANOSIM, R = 0.12, p = 0.034), but not in 2012 (ANOSIM, R = -0.09, p = 0.976) (Figs. 5-8).

The environmental variables that were correlated with the scores of the first two NMDS dimensions were those that variated seasonally (temperature), human activities (total organic carbon and nitrogen) and rainfall (Table 5). When data from all years were pooled, the community composition correlated with C:N ratio (r = 0.09, p = 0.05) and rainfall (r = 0.11, p = 0.04). Moreover, in 2010, the NMDS scores showed significant correlation with temperature (r = 0.31, p = 0.04) and nitrogen (r = 0.31, p = 0.03). In 2011 there was a relationship between water temperature (r = 0.47, p = 0.004), nitrogen (r = 0.42, p = 0.009), TOC (r = 0.35, p = 0.02) and rainfall (r = 0.30, p = 0.03). In 2012 there was no correlation with any environmental variable (Table 5).

DISCUSSION

Chironomidae abundance and generic richness varied either intra-annually (between seasons) and inter-annually (between years). In our study, the abundance of individuals and number of chironomid genera were associated with limnological characteristics. The effect of these characteristics on the chironomid community may have occurred in association with a significant rainfall event in 2011. Increased precipitation causes changes in aquatic ecosystems, such as changes in the physical and chemical characteristics of the water, as well as changes in the distribution of communities (Smith et al. 2003). When rainfall increases, chemical compounds present in the soil are dragged into the streams and cause changes in the abundance, richness, and composition of chironomid communities (Gresens et al. 2007). The effect of runoff in this study was especially observed in TOC concentrations in the winter of 2011.

In 2010 and 2012 rainfall was normal with respect to historical records. Streams and their chironomid communities were stable as a result of this. Rainfall increases the amount of water and unstable and homogeneous substrates into the streams (Rosa et al. 2013, Salles & Ferreira-Junior 2014). Therefore, the increased rainfall (2011) during the study period was an important contributor to the observed variations in inter-annual richness. Fluctuations in water regimen allow great habitat diversification, i.e. more tolerant species can occupy distinct regions of the drainage area, in addition to changing the limnological characteristics of streams (Shuvart et al. 2005, Aburaya & Callil 2007, Roque et al. 2007). Silva et al. (2014) also observed that Chironomidae diversity was higher during the years when the abiotic charac-

Table 4. Chironomidae identified intra-annual (2010-2012) and inter-annual (summer and winter) in subtropical streams.

	2010 S	2010 W	2011 S	2011 W	2012 S	2012 W
Chironominae						
Aedokritus Roback, 1958				*		
Caladomyia Säwedal, 1981	*	*	*		*	*
Chironomus Meigen, 1803			*	*		
Dicrotendipes Kieffer, Epler, 1988			*	*		
Endotribelos Grodhaus, 1987			*			
Goeldichironomus Fittkau, 1965	*		*			
Manoa Fittkau, 1963			*			
Microchironomus Kieffer, 1918			*			
Parachironomus Lenz, 1921	*		*			
Paratanytarsus Thienemman & Bause, 1951			*	*		
Paratendipes Kieffer, 1911	*	*	*	*	*	*
Phaenopsectra Kieffer, 1921	*		*	*		
Polypedilum Kieffer, 1912	*	*	*	*	*	*
Pseudochironomus, Riethia Malloch, 1915		*	*	*	*	*
Rheotanytarsus Thienemann & Bause, 1913	*	*	*	*	*	*
Saetheria Saether, 1983			*			
Stenochironomus Kieffer, 1919			*			
Tanytarsus Van der Wulp, 1874	*	*	*	*	*	*
Xestochironomus Sublette & Wirth, 1972			*			
Zavreliella Kieffer, 1920			*	*		
Orthocladiinae						
Antillocladius Saether, 1981				*		
Cardiocladius Kieffer, 1912			*	*		
Corynoneura Winnertz, 1846	*	*	*	*	*	*
Cricotopus Van der Wulp, 1874	*	*	*		*	*
Cricotopus, Orthocladius Lopescladius Oliveira, 1967	*	*	*	*	*	*
Gymnometriocnemus Goetghebuer, 1932	*	*	*			
Lopescladius Oliveira, 1967	*	*	*			
Metriocnemus Kieffer 1921			*	*		
Nanocladius Kieffer, 1912	*	*	*	*	*	*
Oriconeura Andersen & Saether, 2005	*	*	*	*	*	*
Orthocladiinae A Kieffer, 1911			*	*		
Orthocladiinae B Kieffer, 1911			*	*		
Paracladius Hirvenoja, 1973			*			
Parakiefferiella Thienneman 1926		*	*	*	*	*
Parametriocnemus Goetghebuer, 1932	*	*	*	*	*	*
Paraphaenocladius Thienemann, 1924			*	*		
Rheocricotopus Thienemann & Harnisch, 2004	*	*	*	*	*	*
Thienemannia Kieffer, 1909			*	*		
Thienemanniella Kieffer, 1911		*	*	*	*	*
Ubatubaneura Wiedenbrug & Trivinho-Strixino, 2009			*			
Tanypodinae						
Denopelopia Roback e Rutter (1988)				*		
Djalmabatista Fittkau (1968)			*			
Hudsonimyia Roback, 1979			*	*		
Labrudinia Fittkau, 1962			*			
Larsia Roback & Coffman (1989)			*	*		
Nilotanypus Kieffer, 1923	*	*	*	*	*	*
Paramerina Stur and Fittkau, 1997				*		
Parapentaneura Stur, Fittkau & Serrano, 2006			*			
Pentaneura Philippi, 1865	*	*	*	*	*	*

Table 5. Analysis of the structure between the abiotic data and the biological matrix (NMDS), inter- and intra-annual variation tested from the non-parametric multivariate analysis in subtropical streams.

	NMDS1	NMDS2	R2	p
2010 to 2012				
Water temperature	0.331	0.943	0.078	0.12
pH	-0.967	-0.251	0.094	0.058
Electrical Conductivity	0.559	-0.828	0.011	0.669
Turbidity	-0.812	-0.583	0.016	0.575
DO	-0.168	-0.985	0.088	0.088
TDS	-0.125	0.992	0.042	0.285
Nitrogen	-0.999	0.043	0.041	0.287
TOC	-0.968	-0.247	0.004	0.907
C:N ratio	-0.288	0.957	0.099	0.050
Water velocity	0.891	-0.452	0.016	0.551
Rainfall	-0.847	-0.531	0.110	0.041
2010				
Water temperature	-0.834	-0.551	0.310	0.040
pH	-0.898	0.439	0.118	0.364
Electrical Conductivity	-0.990	-0.134	0.092	0.457
Turbidity	-0.163	-0.986	0.071	0.535
DO	-0.426	-0.904	0.285	0.064
TDS	-0.880	-0.473	0.111	0.378
Nitrogen	-0.711	-0.702	0.316	0.039
TOC	-0.609	0.792	0.284	0.057
C:N ratio	-0.408	0.912	0.260	0.090
Water velocity	0.749	-0.661	0.062	0.588
Rainfall	-0.695	-0.718	0.051	0.626
2011				
Water temperature	-0.039	-0.999	0.474	0.004
pH	0.590	0.806	0.091	0.441
Electrical Conductivity	0.335	-0.941	0.107	0.378
Turbidity	-0.523	0.851	0.238	0.105
DO	-0.271	0.962	0.065	0.527
TDS	0.342	-0.939	0.110	0.371
Nitrogen	-0.328	0.944	0.426	0.009
TOC	0.113	0.993	0.352	0.021
C:N ratio	0.980	0.194	0.196	0.156
Water velocity	-0.579	0.814	0.065	0.462
Rainfall	0.134	0.990	0.309	0.036
2012				
Water temperature	0.986	-0.165	0.007	0.938
pH	0.753	0.658	0.017	0.854
Electrical Conductivity	0.291	-0.956	0.097	0.338
Turbidity	-0.785	0.618	0.226	0.114
DO	0.947	0.320	0.006	0.949
TDS	0.917	0.397	0.023	0.832
Nitrogen	0.948	0.315	0.012	0.903
TOC	0.142	0.989	0.075	0.543
C:N ratio	-0.671	0.741	0.249	0.104
Water velocity	0.332	0.943	0.039	0.726
Rainfall	0.705	0.708	0.019	0.855

teristics of their studied lake changed. Rainfall is one abiotic variable that can create favorable conditions for certain species, not only as a function of the new habitat conditions (Bispo et al. 2006). Water temperature affects the metabolism of organisms and the availability of food, causing changes in community composition (Hahn & Figi 2007, Gray & Elliott 2009). The highest concentrations of carbon and nitrogen were found in areas with intense human activity (e.g. agricultural practices) (Neill et al. 2001, Silva et al. 2007). Currently, many streams display similar signs of anthropogenic change (mainly as the result of agricultural practices), a phenomenon known as eutrophication of aquatic ecosystems (Galloway et al. 2003, Silveira et al. 2006). The drainage areas of the streams studied were populated with crops and exposed soil (in preparation for cultivation). In these areas, the use of pesticides and fertilizers is high, causing soil contamination particularly when rainfall is intense. These pesticides, plus organic matter and nutrients, are carried into the streams by the rain water.

Organic matter is primarily composed of carbon, but it can be associated with other chemical compounds, such as metals (Ali et al. 2002, Al-Shami et al. 2010a, Sensolo et al. 2012). On the other hand, nitrogen is among the most limiting nutrients to primary productivity and the availability of this nutrient affects the abundance of some aquatic organisms (Galloway et al. 2003). *Cricotopus* species feed on algae, which in turn are dependent on certain concentrations of dissolved nutrients (Sensolo et al. 2012). *Polypedilum* and *Rheotanytarsus* were the most common organisms in all samples. Studies report that these genera are easily sampled in streams and are reported as cosmopolitan/tolerant (Amorim et al. 2004, Marchese et al. 2005, Aburaya & Callil 2007). Furthermore, the high density of individuals of *Rheotanytarsus* is due to their eating habits: they are filter-feeding organisms, consuming exclusively organic matter present in the water (Coffman & Ferrington 1996). *Polypedilum* species stand out for being tolerant to a wide range of environmental conditions, as they may occur both in sites impacted with organic compounds and in non-impacted sites (Heinis & Davids 1993, Rosin & Takeda 2007).

In conclusion, in small temporal scales, local environmental factors have great relative influence on community composition. On the other hand, in larger timescales, climatic factors generate variation (Silva et al. 2014). In this study, we observed that in those three years, with semi-annual collecting, intra-annual variations are most evident in the chironomid communities. In this study, the temporal variation of the community was dependent on climatic variations (rainfall) as well as the changes in the TOC and TN caused by intensive agricultural land use.

ACKNOWLEDGMENTS

DN received financial support from the Program PROSUP/CAPES. RMR receives financial support from CNPq (Proc. 477274/2011-0 and proc. 475251/2009-1). LUH receives financial support from CNPq (Edital Universal, process 471572/2012-8) and FAPERGS (process 12/1354-0). The authors thank Rodrigo Fornel for their help editing images, and Adriano Melo and two anonymous reviewers for their suggestions and criticisms.

LITERATURE CITED

Aburaya FH, Callil CT (2007) Variação temporal de larvas de Chironomidae (Diptera) no Alto Rio Paraguai (Cáceres Mato Grosso, Brasil). **Revista Brasileira de Zoologia 24**(3): 565-572. doi: 10.1590/S0101-81752007000300007

Ali A, Frouz J, Lobinske RJ (2002) Spatio-temporal effects of selected physico-chemical variables of water, algae and sediment chemistry on the larval community of nuisance Chironomidae (Diptera) in a natural and a man-made lake in central Florida. **Hydrobiologia 470:** 181-193.

Al-Shami SA, Rawi CSM, HassanAhmad A, Nor SAM (2010a) Distribution of Chironomidae (Insecta: Diptera) in polluted rivers of the Juru River Basin, Penang, Malaysia. **Journal of Environmental Sciences 22**(11) 1718-1727.

Al-Shami SA, Salmah MRC, Hassan AA, Azizah MNS (2010b) Temporal distribution of larval Chironomidae (Diptera) in experimental rice fields in Penang, Malaysia. **Journal of Asia-Pacific Entomology 13:** 17-22. doi: 10.1016/j.aspen.2009.11.006

Amorim RM, Henriques-Oliveira AL, Nessimian JL (2004) Distribuição espacial e temporal das larvas de Chironomidae (Insecta: Diptera) na seção ritral do rio Cascatinha, Nova Friburgo, Rio de Janeiro, Brasil. **Lundiana 5**(2): 119-127.

Bispo PC, Oliveira LG, Bini LM, Sousa KG (2006) Ephemeroptera, Plecoptera and Trichoptera assemblages from riffles in mountain streams of central Brazil: environmental factors influencing the distribution and abundance of immature. **Brazilian Journal of Biology 66**(2): 611-622. doi: 10.1590/S1519-69842006000400005

Buss DF, Baptista DF, Nessimian JL, Egler M (2004) Substrate specificity, environmental degradation and disturbance structuring macroinvertebrate assemblages inneotropical streams. **Hydrobiologia 518**(1-3): 179-188. doi: 10.1023/B:HYDR.0000025067.66126.1c

Coffman WP, Ferrington Jr LC (1996) Chironomidae, p. 635-754. In: Merritt KW, Cummins RW (Eds.). **An introduction to the aquatic insects of North America.** Dubuque, Kendall, Hunt Publishing.

Decian V, Zanin EM, Henke C, Quadros FR, Ferrari CA (2009) Uso da terra na região Alto Uruguai do Rio Grande do Sul e obtenção de banco de dados relacionado a fragmentos de vegetação arbórea. **Perspectiva 33**(121): 165-176.

Epler J (2001) **Identification manual for the larval Chironomidae (Diptera) of North and South Carolina.** Orlando, Departament of Enviromental and Natural Resources, 526p.

Ferrington LC (2008) Global diversity of non-biting midges (Chironomidae; Insecta-Diptera) in freshwater. **Hydrobiologia 595:** 447-455. doi: 10.1007/s10750-007-9130-1

GALDEAN N, CALLISTO M, BARBOSA FAR, ROCHA LA (2000) Lotic ecosystems of Serra do Cipó, southeast Brazil: water quality and a tentative classification based on the benthic macroinvertebrate community. **Journal Aquatic Ecosystem Health and Management 3:** 545-552. doi: 10.1016/S1463-4988(00)00044-0

GALLOWAY JN, ABER JD, ERSIMAN JW, SEITZINGER SP, HOWARTH RW, COWLING EB, J COSBY (2003) The Nitrogen Cascade. **BioScience 53**(4): 341-356. doi: 10.1641/0006-3568(2003)053[0341:TNC]2.0.CO;2

GRAY JS, ELLIOTT M (2009) **Ecology of Marine Sediments. From Science to Management.** Oxford, Oxford University Press, 2nd ed., 225p.

GRESENS SE, BELT KT, TANG JA, GWINN DC, BANKS PA (2007) Temporal and spatial responses of Chironomidae (Diptera) and other benthic invertebrates to urban stormwater runoff. **Hydrobiologia 575:** 173-190. doi 10.1007/s10750-006-0366-y

HANH NS, FIGI R (2007) Alimentação de peixes em reservatórios brasileiros: alterações e consequências nos estágios iniciais de represamento. **Oecologia Brasiliensis 4**(11): 469-480. doi: 10.4257/oeco.2007.1104.01

HEINIS F, DAVIDS C (1993) Factors Governing the spatial and temporal distribution of chironomid larvae in the Maarseveen Lakes with special emphasis on the tole of oxygen conditions. **Netherlands Journal of Aquatic Ecology 27**(1): 21-34.

HENRIQUES-OLIVEIRA AL, NESSIMIAN JL, DORVILLÉ LFM (2003) Feeding habits of chironomid larvae (Insecta: Diptera) from a stream in the floresta da Tijuca, Rio de Janeiro, Brazil. **Brazilian Journal Biology 63**(2): 269-281. doi: 10.1590/S1519-69842003000200012

HEPP LU, MILESI SV, BIASI C, RESTELLO RM (2010) Effects of agricultural and urban impacts on macroinvertebrates assemblages in streams (Rio Grande do Sul, Brazil). **Revista Brasileira de Zoologia 27**(1): 106-113. doi: 10.1590/S1984-46702010000100016

HEPP LU, LANDEIRO VL, MELO AS (2012) Experimental Assessment of the Effects of Environmental Factors and Longitudinal Position on Alpha and Beta Diversities of Aquatic Insects in a Neotropical Stream. **International Review of Hydrobiology 97**(2): 157-167. doi: 10.1002/iroh.201111405

KRUSKAL JB (1964) Multidimensional caling by optimizing goodness of fit to a nonmetric hypothesis. **Psychometrika 9**(1): 1-27.

MARCHESE MR, WATZEN KM, EZCURRA-DE-DRAGO I (2005) Benthic invertebrate assemblages and species diversity patterns of the upper Paraguay River. **River Research and Applications 21**(5): 485-499. doi: 10.1002/rra.814

NEILL C, DEEGAN LD, CERRI CC, THOMAZ S (2001) Deforestation for pasture alters nitrogen and phosphorus in small Amazonian streams. **Ecological Applications 11**(6): 1817-1828. doi: 10.1890/1051-0761(2001)011[1817:DFPANA]2.0.CO;2

OLIVEIRA-FILHO AT, BUDKE JC, JARENKOW JA, EISENLOHR PV, NEVES DRM (2015) Delving into the variations in tree species composition e richness across South American subtropical Atlantic and Pampean forests. **Journal of Plant Ecology 8**(3): 242-260. doi: 10.1093/jpe/rtt058

OKSANEN J, BLANCHET F, KINDT R, LEGENDRE P, O'HARA RG, SIMPSON GL, SOLYMOS P, STEVENS MHH, WAGNER H (2013) **Vegan: Community Ecology Package.** R package, v. 1.17-0. Available online at: http://CRAN.R-project.org/package = vegan [Accessed: 12/11/2013]

POFF NL, OLDEN JD, VIEIRA NKM, FINN DS, SIMMONS MP, KONDRATIEFF BC (2006) Functional trait niches of North American lotic insects: traits based ecological applications in light of phylogenetic relationships. **Journal of the North American Benthological Society 25**(4): 730-755. doi: 10.1899/0887-3593(2006)025[0730:FTNONA]2.0.CO;2

R CORE TEAM (2013) **A Language and Environment for Statistical Computing. R Foundation for Statistical Computing.** Vienna, ISBN 3-900051-07-0. Available online at: http://www.R-project.org [Accessed: 12/11/2013]

RESTELLO RM, HEPP LU, MENEGATTI C, DECIAN V, HENKE-OLIVEIRA C (2012) Efeito das características da área de drenagem sobre a distribuição de Chironomidae (Diptera) em riachos do Sul do Brasil, p. 324-340. In: SANTOS JE, ZANIN EM, MOSCHINI LE (Orgs.). **Faces da Polissemia da Paisagem – Ecologia, Planejamento e Percepção.** São Carlos, Rima, vol. 4.

ROQUE FO, TRIVINHO-STRIXINO S, MILAN L, LEITE JG (2007) Chironomid species richness in low-order streams in the Brazilian Atlantic Forest: a first approximation through a Bayesian approach. **Journal of North American Benthological Society 26**(2): 221-231.

ROSA BFJV, VASQUES M, ALVES RG (2011) Structure and spatial distribution of the Chironomidae community in mesohabitats in a first order stream at the Poc'o D'Anta Municipal Biological Reserve in Brazil. **Journal of Insect Science 11:** 1-13.

ROSA BFJV, VASQUES M, ALVES RG (2013) Chironomidae (Insecta, Diptera) associated with stones in a first-order Atlantic Forest stream. **Revista Chilena de Historia Natural 86:** 291-300.

ROSIN GC, TAKEDA AM (2007) Larvas de Chironomidae (Diptera) da planície de inundação do alto rio Paraná: distribuição e composição em diferentes ambientes e períodos hidrológicos. **Acta Scientiarum Biological Sciences 29**(1): 57-63. doi: 10.4025/actascibiolsci.v29i1.127

SALLES FF, FERREIRA-JÚNIOR N (2014) Hábitat e Hábitos, p. 39-49. In: HAMADA N, JL NESSIMIAN, QUERINO RB (Eds.). **Insetos aquáticos na Amazônia brasileira: taxonomia, biologia e ecologia.** Manaus, INPA, 724p.

SANSEVERINO AM, NESSIMIAN JL (2001) Haìbitats de larvas de Chironomidae (Insecta, Diptera) em riachos de Mata Atlantica no Estado do Rio de Janeiro. **Acta Limnologica Brasiliensia 13:** 29-38.

SCHEFFER M, VAN NES EH (2007) Shallow lakes theory revisited: various alternative regimes driven by climate, nutrients, depth and lake size. **Hydrobiologia 584:** 455-466. doi: 10.1007/978-1-4020-6399-2_41

SENSOLO D, HEPP LU, DECIAN V, RESTELLO RM (2012) Influence of landscape on assemblages of Chironomidae in Neotropical streams. **Annales de Limnologie – International Journal of Limnology 48**: 391-400. doi: 10.1051/limn/2012031

SHUVARTZ M, OLIVEIRA LG, DINIZ-FILHO JAF, BINI LM (2005) Relações entre distribuição e abundância de larvas de Trichoptera (Insecta), em córregos de Cerrado no entorno do Parque Estadual da Serra de Caldas (Caldas Novas, Estado de Goiás). **Acta Scientiarum Biological Sciences 27**(1): 51-55. doi: 10.4025/actascibiolsci.v27i1.1360

SILVA DML, OMETTO JPHB, LOBO GA, LIMA WP, SCARANELLO MA, MAZI E, ROCH HR (2007) Can Land Use Changes Alter Carbon, Nitrogen and Major Ion Transport in Subtropical Brazilian Streams? **Scientia Agricola 64**(4): 317-324. doi: 10.1590/S0103-90162007000400002

SILVA JS, ALBERTONI EF, SILVA CP (2014) Temporal variation of phytophilous Chironomidae over a 11-year period in a shallow Neotropical lake in southern Brazil. **Hydrobiologia 737**(1): 1-14. doi: 10.1007/s10750-014-1972-8

SILVEIRA MP, BUSS DF, NESSIMIAN JL, BAPTISTA DF (2006) Spatial and temporal distribution of benthic macroinvertebrates in southeastern Brazilian river. **Brazilian Journal of Biology 66**(2): 623-632. doi: 10.1590/S1519-69842006000400006

SIQUEIRA T, ROQUE FO, TRIVINHO-STRIXINO S (2008) Phenological patterns of neotropical lotic chironomids: Is emergence constrained by environmental factors? **Austral Ecology 33**: 902-910. doi: 10.1111/j.1442-9993.2008.01885.x

SMITH H, WOOD PJ, GUNN J (2003) The influence of habitat structure e flow permanence on invertebrate communities in karst spring systems. **Hydrobiologia 510**(1): 53-66. doi: 10.1023/B:HYDR.0000008501.55798.20

SÚAREZ YR (2008) Variação espacial e temporal na diversidade e composição de espécies de peixes em riachos da bacia do Rio Ivinhema, Alto Rio Paraná. **Biota Neotropica 8**(3): 197-204. doi: 10.1590/S1676-06032009000100012

TRIVINHO-STRIXINO S (2011) **Larvas de Chironomidae. Guia de identificação.** São Carlos, Departamento de Hidrobiologia, Laboratório de Entomologia Aquática, UFSCar, 371p.

TRIVINHO-STRIXINO S, STRIXINO G (1995) **Larvas de Chironomidae (Diptera) do estado de São Paulo: guia de identificação e diagnose dos gêneros.** São Carlos, PPG-RRN/UFSCar, 229p.

14

Effect of humic acid on survival, ionoregulation and hematology of the silver catfish, *Rhamdia quelen* (Siluriformes: Heptapteridae), exposed to different pHs

Silvio T. da Costa[1,*], Luciane T. Gressler[2], Fernando J. Sutili[2], Luíza Loebens[1], Rafael Lazzari[1] & Bernardo Baldisserotto[3]

[1]*Departamento de Zootecnia e Ciências Biológicas, Centro de Educação Norte do Rio Grande do Sul, Universidade Federal de Santa Maria. 98300-000 Palmeira das Missões, RS, Brazil.*
[2]*Programa de Pós-graduação em Farmacologia, Universidade Federal de Santa Maria. 97105-900 Santa Maria, RS, Brazil.*
[3]*Departamento de Fisiologia e Farmacologia, Universidade Federal de Santa Maria. 97105-900 Santa Maria, RS, Brazil.*
[*]*Corresponding author. E-mail: silvio.teixeira.da.costa@gmail.com*

ABSTRACT. This study evaluates whether humic acid (HA; Aldrich) protects the silver catfish, *Rhamdia quelen* (Quoy & Gaimard, 1824), against exposure to acidic pH. Survival, levels of Na^+, Cl^- and K^+ plasma, hematocrit, hemoglobin and erythrocyte morphometry were measured. Fish were exposed to 0, 10, 25 and 50 mg L^{-1} HA at four pH levels: 3.8, 4.0, 4.2 and 7.0 up to 96 hours. None of the fish exposed to pH 3.8 survived for 96 hours into the experiment, and survival of fish subjected to pH 4.0 decreased when HA concentration increased. Plasma Na^+ levels decreased when pH was acidic, with no influence of HA, while Cl^- levels declined at low pH with increased HA concentration. The levels of K^+ at pH 4.0 and 4.2 increased without HA. Hematocrit and hemoglobin augmented under the effect of HA. At pH 4.0 and 4.2, erythrocytes of fish not exposed to HA were smaller, an effect that was partially offset by the presence of HA, since the values at pH 7.0 were higher. Although HA showed some positive effects changes in hematological and plasma K^{+a} in silver catfish caused by exposure to acidic pH, the overall findings suggest that HA does not protect this species against acidic pH because it increased mortality and Cl^- loss at pH 4.0.

KEY WORDS. Blood parameters; humic acid; plasma ion levels; survival.

Dissolved organic matter, an integral part of all ecosystems, results from the decay of plant and animal debris (THURMAN 1985). It comprises humic, fulvic, and other organic acids, and is usually quantified as dissolved organic carbon (DOC) (WOOD et al. 2011). DOC is known to positively regulate several biotic/abiotic processes (STEINBERG et al. 2007, WOOD et al. 2011). In blackwaters, such as those found in forest streams in the Amazon, coastal lagoons in southeastern Brazil, Finnish and Swedish lakes, and Canadian wetlands, DOC may range from 10 to 300 mg CL^{-1}, while its average content in freshwater systems elsewhere is 0.5-4.0 mg CL^{-1} (THURMAN 1985, KÜCHLER et al. 2000, FARJALLA et al. 2009). The high levels of DOC account for the acidity of the aquatic environment. In order to thrive in acidic environments, organisms need a certain degree of specialization in their osmoregulatory organs (MATSUO & VAL 2007).

Low pH (pH 4-5) induces ion loss (ZAIONS & BALDISSEROTTO 2000, WOOD et al. 1998, 2002, 2003, GONZALEZ et al. 1998, 2002, BOLNER & BALDISSEROTTO 2007, MATSUO & VAL 2007, DUARTE et al. 2013) and the interference with gill ionoregulatory mechanisms may also trigger hematological disturbances. Ionic dilution,

potentiated by the plasma acidosis prompted by H^+ entry, affects body fluid distribution. This could promote reduction in plasma volume, swelling of erythrocytes or splenic contraction, resulting in elevation of the hematocrit (MILLIGAN & WOOD 1982). Despite being highly responsible for the acidic nature of blackwaters, there is evidence that DOC protects native fish from the deleterious effects of low pH, reducing ion loss (WOOD et al. 1998, 2002, 2003, 2011, GONZALEZ et al. 1998, 2002, MATSUO & VAL 2007). According to some studies, the occurrence of various charged functional groups in the heterogeneous compounds of DOC may change fundamental properties of the gill epithelium, such as the transepithelial potential, thus altering membrane permeability and stimulating ion uptake (WOOD et al. 2011). Humic acid also reduces respiratory stress in fish exposed to slightly acidic water, but increases it at more acidic waters (HOLLAND et al. 2014) and decreases lipid peroxidation and modulates the antioxidant system (RIFFEL et al. 2014). In opposition to the several findings regarding the effects of DOC on ionoregulatory disturbances, respiratory stress and antioxidant system, no evidence has been documented about its influence on the hematology of fish subjected to acidic pH.

This study evaluated whether humic acid (HA), one of the major components of DOC, would offer the silver catfish, *Rhamdia quelen* (Quoy & Gaimard, 1824), protection against the physiological disturbances induced by low pH. This species does not naturally inhabit DOC-enriched, acidic waters, so it is not adapted to such conditions. However, different water quality parameters, including pH and DOC, are present in southern Brazil, where this species is widely cultivated. The outcome of the interaction between such variables should be investigated to improve the rearing conditions of this fish. In laboratory settings, silver catfish juveniles survive for at least 96 hours in the pH 4-9 range (Zaions & Baldisserotto 2000), but exposure to pH 5.0 is enough to reduce growth in this species (Copatti et al. 2005). Therefore, if humic acid has a protective effect on silver catfish exposed to acidic waters, it could reduce the deleterious effect of low pH and improve growth in this species.

MATERIAL AND METHODS

Juvenile silver catfish (n = 240, 73.43 ± 3.5 g, 20.32 ± 1.22 cm, voucher number 19612, Ichthyology Laboratory, Universidade Federal do Rio Grande do Sul) were acquired from a commercial fishery in Santa Maria, southern Brazil, and acclimated in the Laboratório de Fisiologia de Peixes, Universidade Federal de Santa Maria (UFSM) for three weeks. The fish were equally distributed in 8 tanks of 250 L and kept in dechlorinated tap water under constant aeration (22.14 ± 1.5°C, 6.05 ± 0.45 mg L^{-1} dissolved oxygen (DO), pH 7.45 ± 0.13 and hardness 24.7 ± 3.9 mg CaCO$_3$ L^{-1}). The water was totally renewed every second day and siphoning was performed daily two hours after feeding. The fish were fed commercial food for juveniles with 42% crude protein once a day.

Lyophilized HA (CAT: 0.675-2 Aldrich® H1 – HA sodium salt) was the source of DOC used in the tests. It was dissolved in water (the same water used in the acclimation tanks) and agitated for 12 hours in a magnetic stirrer to prepare the stock solution. It was not possible to measure the concentration of DOC in the experimental solutions, but estimation of DOC concentration was made based on the fact that the commercial HA corresponded to ~40% DOC (McGeer et al. 2002). HA was tested at 0 (control), 10, 25 and 50 mg L^{-1} HA, the latter corresponding to the nominal DOC concentration of 20 mg C L^{-1}. These concentrations were chosen because they are within the range observed in the water of the rio Negro Basin (Küchler et al. 2000). At each concentration of HA, four pH ranges were tested, with the following minimum values: 3.8, 4.0, 4.2 and 7.0 (Table 1). The acidic pH tested in the present study were near the most acidic pH (pH 4.0) that allows 100% survival in silver catfish (Zaions & Baldisserotto 2000). A pH meter DMPH-2 (Digimed, São Paulo, Brazil) was used to measure the variable four times a day and adjustments to the minimum values within each range were made with sulfuric acid 1 M when necessary. The water in the experimental aquaria was not renewed during exposure time.

Table 1. Ion levels in water at different pH and humic acid (HA) levels for *Rhamdia quelen*.

HA (mg L^{-1})	pH	Na$^+$ (mg L^{-1})	Cl$^-$ (mg L^{-1})	K$^+$ (mg L^{-1})
0	3.84 ± 0.5	3.1 ± 0.3	5.9 ± 1.0	0.04 ± 0.01
	4.08 ± 0.4	3.2 ± 0.7	6.1 ± 0.9	0.04 ± 0.01
	4.25 ± 0.4	3.4 ± 0.8	6.0 ± 0.9	0.03 ± 0.02
	7.02 ± 0.3	3.3 ± 0.4	6.0 ± 0.8	0.04 ± 0.02
10	3.87 ± 0.5	3.7 ± 0.3	5.6 ± 1.3	0.03 ± 0.01
	4.09 ± 0.6	3.7 ± 0.3	5.7 ± 1.4	0.03 ± 0.01
	4.22 ± 0.3	3.6 ± 0.2	6.8 ± 1.6	0.04 ± 0.02
	7.03 ± 0.4	4.3 ± 0.4	6.0 ± 0.7	0.05 ± 0.02
25	3.83 ± 0.4	3.8 ± 0.6	6.1 ± 0.6	0.04 ± 0.02
	4.05 ± 0.5	4.3 ± 1.1	5.8 ± 0.8	0.04 ± 0.01
	4.27 ± 0.6	3.9 ± 0.5	6.3 ± 1.0	0.05 ± 0.02
	7.05 ± 0.5	3.9 ± 0.2	6.2 ± 1.2	0.03 ± 0.02
50	3.81 ± 0.2	4.9 ± 0.4	6.1 ± 1.6	0.04 ± 0.02
	4.02 ± 0.2	4.7 ± 0.5	6.4 ± 0.8	0.03 ± 0.01
	4.21 ± 0.1	4.8 ± 0.9	6.5 ± 1.3	0.05 ± 0.02
	7.02 ± 0.3	5.5 ± 1.3	6.0 ± 1.6	0.05 ± 0.02

Mean values ± SE (n = 4/group for ions). There was no significant difference between treatments.

Juveniles were fasted for 24 hours prior to being transfered to 40 L aquaria (16 treatments, three replicates of each treatment, five fish per replicate) for the 96-h experiment. Survival was observed four times a day and the dead fish were removed from the aquaria. Fish that survived up to the end of the experimental period were anesthetized with eugenol 50 mg L^{-1} (Cunha et al. 2010) and their blood was rapidly collected from the caudal vein with heparinized syringes. After sampling, fish were killed by sectioning the spinal cord. All procedures were conducted with the approval of the Ethics Committee on Animal Experimentation of the UFSM (registration #128/2010).

DO levels and temperature were measured daily with Orion 810 oxygen meter (Thermo Electron Corporation, Waltham, Al, USA). Water samples were collected every second day to verify total ammonia (Verdouw et al. 1978), un-ionized ammonia, hardness (Eaton et al. 2005), nitrite (Boyd & Tucker 1992), Cl$^-$ (Zall et al. 1956), and Na$^+$ and K$^+$ levels, which were measured in a flame photometer (Micronal B262, São Paulo, Brazil). Details on the composition of the water are provided in Tables 1 and 2. There were no significant differences in water quality parameters between treatments.

To obtain the hematocrit, microcapillary tubes were filled with blood immediately after euthanasia and centrifuged at 10000 Xg for 5 minutes, and the results were obtained using a hematocrit card reader. The concentration of hemoglobin was determined by the cyanmethemoglobin method using a spectrophotometer (Brown 1976). For the morphometric analyses, blood smears were prepared immediately from the whole blood, air-dried, fixed in methanol and stained with May-Grünwald (TAVARES-DIAS et al. 2004). The surface area and the major and minor axes of the erythrocyte as well as of its nucleus were determined (DORAFSHAN et al. 2008). Briefly, ten high-power fields were randomly selected on each blood smear, and morphometry of ten erythrocytes were determined in each of these fields. All analyses were performed using the Zeiss Axio Vision System with Remote Capture 4.7 Rel DC – Cannon Power shot G9.

Blood samples were spun at 3000 Xg for 10 minutes and plasma was stored at -25°C until analyses of Na+, Cl- and K+. The ion levels in the plasma were determined as previously described for the water ion levels.

Homogeneity of variances was assessed via Levene test and the comparison between treatments was carried out by two-way ANOVA and Tukey test. The Kruskal-Wallis test, followed by multiple comparisons of mean ranks, was used for analyses of plasma ion levels (Statistica 7.0 software). Minimum level of significance was 95% (p < 0.05). Data are presented as mean ± standard error (SE).

RESULTS

Survival

None of the fish exposed to pH 3.8 survived the 96 hours of experiment. At pH 4.0 there was a progressive decrease in survival (100, 86, 60 and 40%) with increased HA level (0, 10, 25 and 50 mg L^{-1} HA, respectively). The survival rates at pH 4.2 (93.33%) and 7.0 (100%) were not affected by HA concentration. Survival at pH 4.0 was lower than at pH 4.2 and 7.0 at all treatments with the presence of HA (i.e. 10, 25 and 50 mg L^{-1} of HA) (Fig. 1).

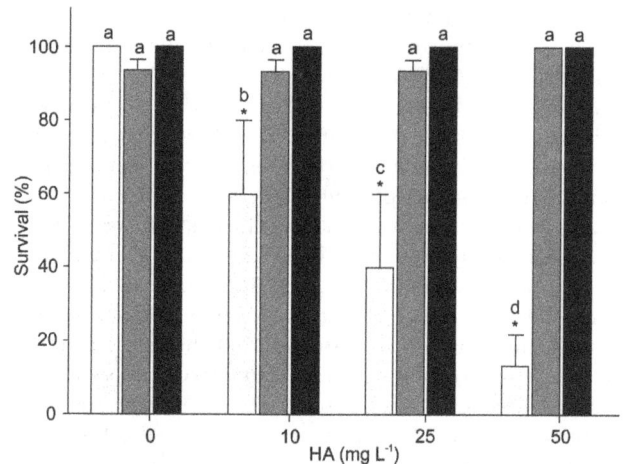

Figure 1. Effect of humic acid (HA) and pH on survival of silver catfish (*Rhamdia quelen*). Different letters indicate significant difference between HA concentrations at the same pH. * indicate significant difference from pH 7.0 at the same HA concentration (p < 0.05). Mean values ± SE (n = 6-15/group). (□) pH 4.0, (▨) pH 4.2, (■) pH 7.0.

Hematocrit and hemoglobin

Overall, HA triggered an increase in the percentages of hematocrit and hemoglobin. The presence of HA promoted an increase in the percentage hematocrit at pH 4.0 and 4.2. Upon exposure to pH 7.0, fish experienced a gradual increase in hematocrit from 0 to 25 mg L^{-1} HA. Hematocrit declined with the

Table 2. Water quality parameters at different pH and humic acid (HA) levels for *Rhamdia quelen*.

HA (mg L^{-1})	pH	Total ammonia (mg L^{-1})	Un-ionized ammonia (mg L^{-1})	Nitrite (mg L^{-1})	Hardness (mg $CaCO_3$ L^{-1})	Temperature (°C)	Dissolved oxygen (mg L^{-1})
0	3.8	0.85 ± 0.0015	0.0342 ± 0.0009	0.3181 ± 0.012	25.8 ± 4.2	21.4 ± 2.1	6.25 ± 0.61
	4.0	0.44 ± 0.0010	0.0251 ± 0.0003	0.3492 ± 0.025	25.7 ± 4.1	21.2 ± 2.2	6.21 ± 0.64
	4.2	0.16 ± 0.0004	0.0270 ± 0.0004	0.3758 ± 0.032	27.3 ± 6.1	21.2 ± 2.6	6.10 ± 0.45
	7.0	0.32 ± 0.0010	0.0279 ± 0.0005	0.3625 ± 0.041	26.1 ± 5.8	21.3 ± 2.6	6.08 ± 0.54
10	3.8	0.76 ± 0.01	0.0027 ± 0.0003	0.4011 ± 0.013	26.8 ± 5.6	20.1 ± 2.0	6.12 ± 0.61
	4.0	0.92 ± 0.021	0.0028 ± 0.0002	0.3442 ± 0.022	25.1 ± 4.8	20.2 ± 2.1	6.14 ± 0.48
	4.2	0.83 ± 0.014	0.0027 ± 0.0004	0.2034 ± 0.024	27.6 ± 5.1	20 ± 2.1	6.18 ± 0.45
	7.0	1.09 ± 0.032	0.1820 ± 0.0019	0.3285 ± 0.045	26.3 ± 4.9	20.4 ± 2.0	6.17 ± 0.48
25	3.8	0.74 ± 0.012	0.015 ± 0.0003	0.4101 ± 0.023	26.1 ± 5.1	21.4 ± 1.9	6.26 ± 0.62
	4.0	0.87 ± 0.017	0.027 ± 0.0003	0.4119 ± 0.025	25.9 ± 5.2	22 ± 1.3	6.11 ± 0.53
	4.2	1.22 ± 0.024*	0.215 ± 0.0150	0.3289 ± 0.034	26.4 ± 5.4	21.5 ± 2.4	6.15 ± 0.60
	7.0	0.81 ± 0.014	0.021 ± 0.0003	0.3032 ± 0.054	26.8 ± 5.1	21.9 ± 2.6	6.14 ± 0.70
50	3.8	0.64 ± 0.011	0.019 ± 0.0006	0.4516 ± 0.024	26.4 ± 5.0	21.1 ± 2.3	6.21 ± 0.65
	4.0	0.66 ± 0.012	0.019 ± 0.0005	0.3442 ± 0.039	27.1 ± 4.9	21.1 ± 2.1	6.24 ± 0.39
	4.2	0.75 ± 0.015	0.021 ± 0.0007	0.2034 ± 0.054	26.1 ± 5.1	21.0 ± 2.0	6.18 ± 0.53
	7.0	0.81 ± 0.014	0.022 ± 0.0009	0.3285 ± 0.061	26.2 ± 5.2	20.9 ± 2.0	6.19 ± 0.55

*Significantly different from pH 4.2 and HA 0 mg L^{-1} (p < 0.05). Mean values ± SE (n = 4/group).

increase in pH at 0 mg L⁻¹ HA. Exposure to 10 mg L⁻¹ HA induced significantly higher hematocrit percentage at pH 4.0 and 4.2, while at 25 mg L⁻¹ HA the pH had negligible influence on hematocrit. Treatment with 50 mg L⁻¹ HA caused a significantly reduction in hematocrit concentration at pH 7.0 (Fig. 2).

There was no difference in the percentage of hemoglobin between HA treatments at pH 4.2. Exposure to pH 4.0 induced a higher hemoglobin level at 25 mg L⁻¹ HA than at 0 mg L⁻¹ HA. When fish were exposed to pH 7.0 the hemoglobin was lower at 50 than at 10 mg L⁻¹ HA, and at 0 mg L⁻¹ HA was also lower than at 25 mg L⁻¹ HA. In the absence of HA the levels of hemoglobin decreased as the pH increased. Exposure to 10 mg L⁻¹ HA did not induce differences in hemoglobin values between the different pH. Hemoglobin levels at 25 mg L⁻¹ HA were higher at pH 4.0 than at pH 4.2 and 7.0. On exposure to 50 mg L⁻¹ HA the hemoglobin levels were higher at pH 4.0 and 4.2 than at pH 7.0 (Fig. 3).

Erythrocyte morphometry

Fish subjected to pH 4.0 showed greater cell area and cell minor and major axes in the presence of HA than in the absence of it. At pH 4.2, cell area was larger at 10 and 50 mg L⁻¹ HA than at 25 mg L⁻¹ HA, and it decreased further when HA was not present. At the same pH (4.2), cell minor axis was bigger at 50 than at 0 and 25 mg L⁻¹ HA; it was also bigger at 10 mg L⁻¹ HA compared to 0 mg L⁻¹ HA. At 0 mg L⁻¹ HA, cell area and its minor and major axes were bigger at pH 7.0 than at pH 4.0. The group exposed to 25 mg L⁻¹ HA presented greater cell area at pH 7.0 than at all the acidic pH and greater cell minor axis at pH 7.0 than at pH 4.2. Fish treated with 50 mg L⁻¹ HA had bigger cell minor axis at pH 4.2 than at pH 4.0, and bigger cell major axis at pH 7.0 comparing with pH 4.0 (Table 3).

Plasma Na⁺, Cl⁻ and K⁺

In Na⁺ levels, no significant differences were observed in fish exposed to the different HA treatments at pH 4.0 and 7.0, but at pH 4.2, exposure to 25 mg L⁻¹ HA increased Na⁺ levels compared to the group non exposed to HA. Silver catfish exposed to pH 4.0 and 4.2 without HA presented significantly lower Na⁺ levels than those at pH 7.0 without HA, but at 10 mg L⁻¹ HA plasma Na⁺ in fish exposed to pH 4.2 were not significantly different from pH 7.0. Fish at 25 mg L⁻¹ HA and pH 4.2 presented significantly higher Na⁺ levels than at pH 7.0 (Fig. 4).

The levels of Cl⁻ at pH 4.0 were significantly greater in fish exposed to 10 mg L⁻¹ HA, while at pH 4.2 and 7.0 HA did not affect significantly plasma Cl⁻. Fish subjected to 10 mg L⁻¹ had higher Cl⁻ levels at pH 7.0 and 4.0 than at pH 4.2. Plasma Cl⁻ levels were significantly lower at 25 mg L⁻¹ HA and pH 4.0 and at 50 mg L⁻¹ and pH 4.0 and 4.2 than at pH 7.0 and the same HA levels (Fig. 5).

K⁺ levels were not affect by HA treatments at pH 7.0. However, significantly higher K⁺ levels were observed at 0 mg L⁻¹ HA than at 10 mg L⁻¹ HA and pH 4.0, and at 25 and 50 mg L⁻¹ HA and pH 4.2. The levels of K⁺ were significantly higher at pH 4.0 and 4.2 than at pH 7.0 in fish kept in water without HA (Fig. 6).

DISCUSSION

All water parameters analyzed were within the limits that permit normal growth and survival of silver catfish (e.g. nitrite and un-ionized ammonia levels below 1.2 mg L⁻¹ and 0.1 mg L⁻¹ respectively) (Lima et al. 2011, Miron et al. 2011).

According to Zaions & Baldisserotto (2000), even though silver catfish presents a marked loss of Na⁺ at pH 4.0, this is the acidic pH threshold for the species survival, at least for 96 h. In the present assessment this assertion was confirmed by the 0% survival of fish exposed to pH 3.8 regardless the HA concentrations. As stated by Wood & McDonald (1982), nonacidophilic species suffocate at pH levels below 4.0 due to gill structural damage, edema and mucification. Moreover, fish mortality in acid waters is largely associated with a failure to

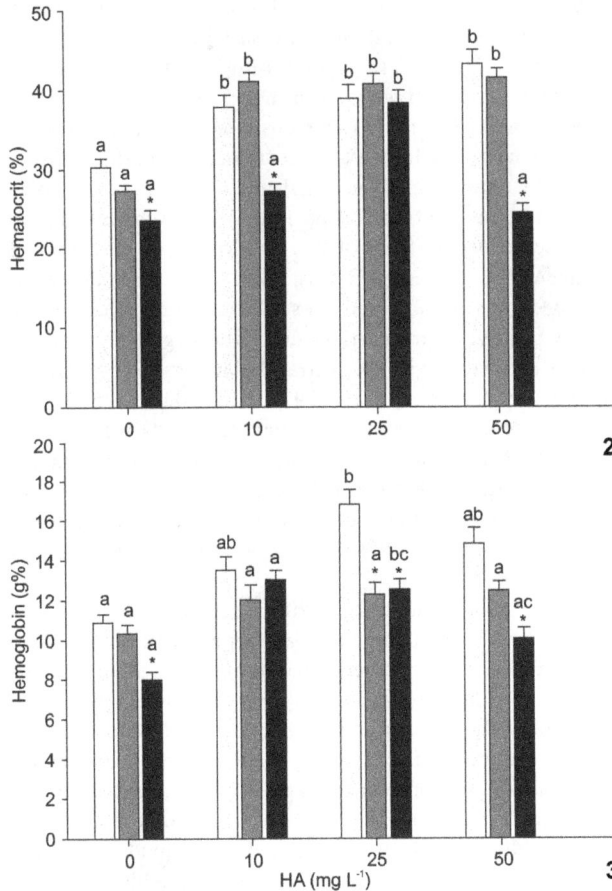

Figures 2-3. Effect of humic acid (HA) and pH on hematocrit (2) and hemoglobin (3) of silver catfish (*Rhamdia quelen*). Different letters indicate significant difference between HA concentrations at the same pH. * indicate significant difference from pH 7.0 at the same HA concentration (p < 0.05). Mean values ± SE (n = 6-15/group). (□) pH 4.0, (▨) pH 4.2, (■) pH 7.0.

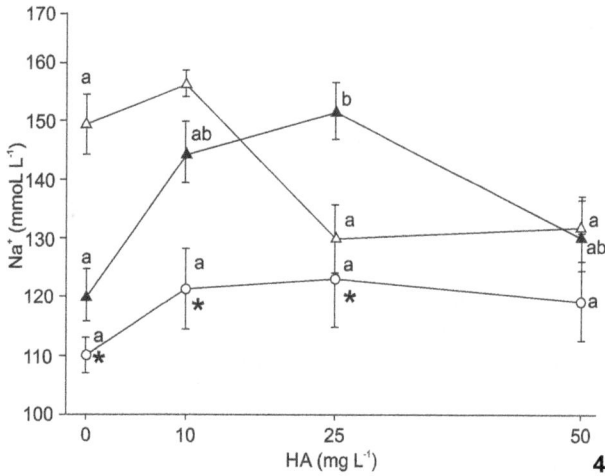

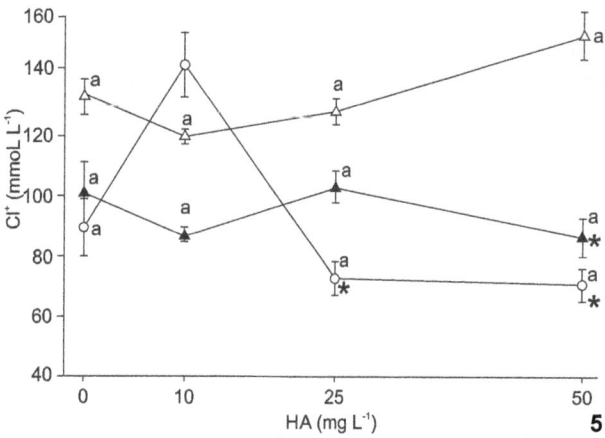

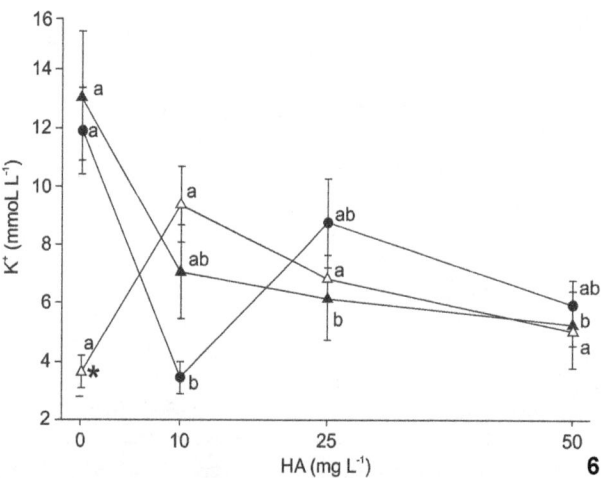

Figures 4-6. Effect of humic acid (HA) and pH on plasma Na+ (4), Cl- (5) and K+ (6) of silver catfish (*Rhamdia quelen*). Different letters indicate significant difference between HA concentrations at the same pH. * indicate significant difference from pH 7.0 at the same HA concentration (p < 0.05). Mean values ± SE (n = 6-15/ group). (○) pH 4.0, (▲) pH 4.2, (△) pH 7.0.

ionoregulate, especially due to stimulation of Na+ efflux (MILLIGAN & WOOD 1982). A study on shiners *Notropis cornutus* (Mitchill, 1817), rainbow trout *Oncorhynchus mykiss* (Walbaum, 1792) and perch *Perca flavescens* (Mitchill, 1814) clearly proved that principle by showing that a great amount of Na+ and Cl- (50-60%) had been lost at death after exposure to pH 4.0 (FREDA & MCDONALD 1988). Similarly, HOLLAND et al. (2014) reported increased morbidity of the eastern rainbow fish *Melanotaenia splendida splendida* (Peters, 1866) as the pH dropped to 3.5-4.0 in the presence of commercial HA, despite having observed a protective effect of the substance at higher acidic levels. The authors suggested that HA may have enhanced the toxicity of low pH by increasing ion loss in the fish.

At pH 4.0, HA displayed a deleterious effect on the physiology of silver catfish; increased concentration of HA was associated with a decline in fish survival. COSTA et al. (pers. comm.) observed that the presence of HA induces proliferation of chloride cells in the lamellae of the gill of silver catfish. Gas transfer in pavement cells might be impaired when chloride cells inundate the lamellae, since thickening of the lamellar epithelium increases blood-to-water diffusion distance (GRECO et al. 1996). BINDON et al. (1994) previously reported a reduction in the lamellar epithelium as a consequence of chloride cell proliferation. Thus, the silver catfish may have been unable to cope with the combination of limited gas exchange, due to increased concentrations of HA, and ionic loss, the result of extreme pH. At higher pH levels, however, the detrimental effect of HA was not observed, most likely because at low pH the excess positive charge titrates away the negatively charged groups associated with the extracellular surface of epithelial membranes (CAMPBELL et al. 1997). This reduces the electrochemical repulsion between the membrane and the negatively charged HA, allowing the two parameters to associate and induce an effect. However, as pH increases the decrease in positive charge means that the HA is now less electrochemically favored to associate with the membrane and thus the effect is diminished.

As stated by ARIDE et al. (2007), acid stress triggers various changes in hematological parameters of freshwater fish. When it takes place, the generated osmotic and ionic gradient favors the entry of water into intracellular space and electrolyte flux in the opposite direction. With that, blood volume decreases, erythrocyte physiology changes, hematocrit, hemoglobin and plasma protein levels rise, and ion loss through the gills is further enhanced (WOOD et al. 1998, ARIDE et al. 2007). Water acidification is also associated with blood acidosis. It affects the oxygenation capacity of the hemoglobin and PO_2 is thus reduced, which in turn triggers an increase in hematocrit and hemoglobin in order to restore proper homeostatic control (MCDONALD & WOOD 1981, DHEER et al. 1987). MILLIGAN & WOOD (1982) found increases in both hematocrit and hemoglobin in rainbow trout during acid exposure. The authors stated that hematocrit elevation probably resulted from a re-

Table 3. Effect of humic acid and pH on erythrocyte morphology of *Rhamdia quelen*.

	Humic acid (mg L⁻¹)			
	0	10	25	50
pH 4.0				
Cell area (µm²)	98.73 ± 15.9ᵃ*	146.82 ± 20.3ᵇ	135.92 ± 13.3ᵇ*	136.30 ± 6.48ᵇ*
Cell minor axis (µm)	9.55 ± 1.00ᵃ*	11.75 ± 0.75ᵇ	11.26 ± 0.70ᵇ	11.32 ± 0.31ᵇ
Cell major axis (µm)	13.32 ± 0.92ᵃ*	16.13 ± 1.11ᵇ	15.67 ± 0.73ᵇ	15.60 ± 0.51ᵇ*
Nucleus area (µm²)	18.34 ± 3.62ᵃ	20.62 ± 4.37ᵃ	21.57 ± 2.09ᵃ	20.41 ± 3.84ᵃ
Nucleus minor axis (µm)	4.16 ± 0.47ᵃ	4.42 ± 0.29ᵃ	4.58 ± 0.19ᵃ	5.77 ± 2.81ᵃ
Nucleus major axis (µm)	5.79 ± 0.51ᵃ	6.12 ± 0.80ᵃ	6.21 ± 0.37ᵃ	7.70 ± 3.66ᵃ
pH 4.2				
Cell area (µm²)	122.70 ± 6.15ᵃ	139.88 ± 7.38ᵇ	132.82 ± 4.57ᶜ*	144.73 ± 6.47ᵇ
Cell minor axis (µm)	10.52 ± 0.36ᵃ	11.45 ± 0.27ᵇᶜ	11.09 ± 0.53ᵃᶜ*	11.94 ± 0.39ᵇ
Cell major axis (µm)	15.11 ± 0.50ᵃ	15.85 ± 0.84ᵃ	15.53 ± 0.52ᵃ	15.74 ± 0.27ᵃ
Nucleus area (µm²)	20.07 ± 1.60ᵃ	20.23 ± 2.61ᵃ	20.87 ± 1.31ᵃ	18.45 ± 1.47ᵃ
Nucleus minor axis (µm)	4.34 ± 0.18ᵃ	4.43 ± 0.32ᵃ	4.48 ± 0.15ᵃ	4.27 ± 0.26ᵃ
Nucleus major axis (µm)	6.03 ± 0.28ᵃ	5.99 ± 0.33ᵃ	6.30 ± 0.39ᵃ	5.73 ± 0.27ᵃ
pH 7.0				
Cell area (µm²)	135.11 ± 19.63ᵃ	151.22 ± 20.83ᵃ	155.43 ± 12.71ᵃ	153.34 ± 14.78ᵃ
Cell minor axis (µm)	11.21 ± 0.95ᵃ	11.88 ± 0.79	12.21 ± 0.57ᵃ	11.81 ± 0.23ᵃ
Cell major axis (µm)	15.52 ± 1.08ᵃ	16.45 ± 1.19ᵃ	16.51 ± 0.66ᵃ	16.85 ± 1.30ᵃ
Nucleus area (µm²)	20.63 ± 3.53ᵃ	23.73 ± 3.63ᵃ	22.43 ± 2.40ᵃ	22.02 ± 1.78ᵃ
Nucleus minor axis (µm)	4.48 ± 0.35ᵃ	4.67 ± 0.38ᵃ	4.67 ± 0.29ᵃ	4.55 ± 0.09ᵃ
Nucleus major axis (µm)	6.04 ± 0.55ᵃ	6.66 ± 0.51ᵃ	6.31 ± 0.32ᵃᵇ	6.38 ± 0.40ᵃ

Different letters indicate significant difference between HA concentrations at the same pH. * Indicate significant difference from pH 7.0 at the same HA concentration ($p < 0.05$). Mean values ± SE (n = 8-15/group).

duction in plasma volume, erythrocyte swelling and release of erythrocytes from the spleen due to increased circulating catecholamines.

In this investigation, both hematocrit and hemoglobin were highly affected by the experimental variables, considering the basal range previously reported for silver catfish, 17.00-34.00 and 4.95-9.09 respectively (TAVARES-DIAS et al. 2002). Some of the groups exposed to higher pH levels increased hematocrit and hemoglobin values in the presence of HA, which may be a result of the before-mentioned limited gas exchange induced by HA. Inefficient gill ventilation triggers mechanisms such as splenic contraction in an attempted to absorb more oxygen, therefore elevating hematocrit and hemoglobin (SAMPAIO et al. 2008). RIFFEL et al. (2014) have similarly reported that the addition of HA to the water, though at low concentrations, induced hematocrit and hemoglobin rises in silver catfish at neutral pH.

Somewhat different results were found at pH 7.0 for the hematocrit in the group subjected to 10 mg L⁻¹ HA and for both the hematocrit and hemoglobin in the group exposed to 50 mg L⁻¹ HA. It seems that the fish in those groups, especially in the latter one, were able to compensate for the decreased

ventilatory drive caused by HA at the neutral pH, which was not observed at 25 mg L⁻¹ HA.

Exposure of tambaqui to an extreme pH of 3.0 had no influence on blood oxygenation or hemoglobin concentration, demonstrating that this fish, which migrates from circumneutral to acidic waters in its natural habitat, does not encounter challenges in oxygen delivery at such pH level (WOOD et al. 1998). Likewise, ARIDE et al. (2007) observed similar hemoglobin levels between tambaqui subjected to either circumneutral or acid pH, though there was elevation in hematocrit during acid exposure.

As already mentioned, MILLIGAN & WOOD (1982) found that acid exposure triggered disturbances in hematological homeostasis and fluid volume distribution in rainbow trout. Elevation in erythrocyte volume in that species was most likely a result of fluid redistribution from extra- to intracellular compartments due to the ionic dilution of the plasma. In contrast, ARIDE et al. (2007) observed no changes in erythrocyte volume in tambaqui subjected to acid exposure. In the present study it was demonstrated that: a) regardless the HA concentration, the size of the erythrocytes and their nuclei remained stable throughout the groups at pH 7.0; b) at pH 4.0 and 4.2, the significant differences indicate smaller values in the absence of HA; and c) within a given concentration of HA, most differences pointed to higher values at pH 7.0 than at acidic pH. The overall response suggests that, unlike the studies cited above, low pH caused a shrinking effect on the erythrocytes of silver catfish. The presence of HA did not fully counteract such outcome, since the differences were significant comparing to pH 7.0 This effect could be due to output of water and hydromineral disturbance, which typically arise from stress in fish (Wendelaar Bonga 1997).

Plasma levels of Na^+ of silver catfish at pH 4.0 and 4.2 were lower than those observed at pH 7.0. The disruptive process in this extreme aquatic environment primarily involves active inhibition of ion uptake and increased ion loss in the gills (Milligan & Wood 1982). Freda & McDonald (1988) observed the complete inhibition of Na^+ influx in shiners and trout exposed to pH 4.0, in addition to an increase in the ion outward flux. Lin & Randall (1993) claimed that inhibition of Na^+ uptake at low pH results from the reduced activity of an apical electrogenic $H^+ATPase$ that energizes an apical Na^+ channel in chloride cells, an effect attributed to the H^+ gradient. Stimulation of Na^+ efflux, which is the primary determinant of low pH tolerance, is usually a consequence of the H^+-induced Ca^{2+} leaching from the paracellular channels in the gills (Gonzalez et al. 1997).

McDonald & Wood (1981) and Wood et al. (1998) stated that disturbance of ionoregulation by high external H^+ is likely to occur in nonacidophilic species when they are subjected to a sudden acid stress. On the other hand, fish that inhabit naturally acidified, diluted waters, such as those found in the Amazon basin or along the eastern coast of the United States, show a greater tolerance to high concentration of water H^+ and have a lower pH threshold at which marked ion losses occur (Gonzalez & Dunson 1989, Gonzalez et al. 1998, Wood et al. 1998, Matsuo & Val 2007). In some species the adaptation to thrive in these waters involves increased branchial affinity for Ca^{2+} at the paracellular junction, thus counteracting low pH-induced displacement (Freda & McDonald 1988, Gonzalez & Dunson 1989). Further, some fish are able to take up ions at high rates when there is high diffusive ion leakage (Gonzalez et al. 1997, 1998), and at least two Amazon species have pH-insensitive Na^+ transporter (Gonzalez & Wilson 2001). For Freda & McDonald (1988), two important abilities may respond for the interspecific differences in acid tolerance: limitation of the ionic leakiness prompted by low pH, and ion transporter recovery from the low pH inhibition.

Besides their own endogenous mechanisms, fish native to DOC-enriched habitats may relay on the great amount of organic substances found there to improve ion homeostasis (Gonzalez et al. 1998, 2002, Wood et al. 2002, 2003). Gonzalez et al. (2002) and Matsuo & Val (2007) observed that the presence of DOC in acidic water reduced both Na^+ influx inhibition and diffusive efflux stimulation in teleosts native to

Amazonia. The role of DOC to bind fish gills at low pH and promote physiological benefits (Campbell et al. 1997) may be comparable with the above-mentioned action of elevated water-borne levels of Ca^{2+}, that is, stabilization of tight junctions and prevention of ion losses. That would override any protective effect otherwise achieved by the divalent ion (Wood et al. 2003, 2011). Besides, it could result from the ability of the organic molecules to bind to ion apical transporters and help concentrate Na^+ and Cl^- ions by complexation, or to help deliver the ions to the uptake sites, which is normally credited to mucus (Gonzalez et al. 2002, Steinberg et al. 2007).

Except when pH was 4.2, at which the presence of HA was associated with a slightly higher Na^+ plasma level in silver catfish, there were no differences in the ion levels between the different HA concentrations at any given pH. This could be explained by the observation that this fish species is not native to waters with high DOC content, so its gill physiology may not be sensitive to the DOC's protective mechanism (Matsuo et al. 2004). Another possibility is that commercial HA is not as useful to silver catfish as natural black water, as observed by Wood et al. (2003) in stingrays (Potamotrygon sp.): HA stimulated Na^+ and Cl^- leakage, probably because its high affinity for cations ends up stripping Ca^{2+} from the gills. Consequently, the authors concluded that this source of DOC may have different binding characteristics than does natural black water DOC.

The inhibitory mechanism of Cl^- uptake under low pH is possibly associated with the already described mechanism of Na^+ uptake inhibition. Besides, it could be due to a reduction in intracellular HCO_3^- at the chloride cells, thus exhausting the apical Cl^-/HCO_3^- exchanger (Wood 2001). Cl^- loss in silver catfish was exacerbated at 25 and 50 mg L^{-1} HA in the fish exposed to pH 4.0, thus demonstrating a greater involvement of HA in Cl^- than in Na^+ flux. Such difference may be linked to their distinct ionoregulatory mechanisms across the gill epithelium, since Cl^- and Na^+ are exchanged for base and acid equivalents, respectively (Goss & Wood 1990).

Comparing with the responses on Na^+ and Cl^- balance, a different effect of HA was observed with regard to the dynamics of K^+ regulation. When there was no HA in the test water, the levels of K^+ in silver catfish exposed to pH 4.0 and 4.2 were higher in comparison to ion levels in fish subjected to pH 7.0. A similar outcome was observed in the pirapitinga, Piaractus brachypomus, in a recent investigation (Garcia et al. 2014). Further, Zaions & Baldisserotto (2000) found lower body levels of K^+ in silver catfish subjected to pH 7.0 than in those subjected to either acidic or alkaline pH. Mathan et al. (2010) also observed higher plasma K^+ levels in the common carp, Cyprinus carpio, after exposure to acidic pH, and suggested that it could be due to a release of K^+ from the muscle cells as H^+ enters them. Another study assessing plasma ion levels in silver catfish exposed to Aldrich HA (0, 2.5 and 5 mg L^{-1}) at pH ~ 7.0 observed a progressive increase in K^+ levels with increased HA concentrations, suggesting that HA could limit gill permeability (RIFFEL et al. 2014). In this study HA did not influence K⁺ levels at pH 7.0, while at the intermediate pH levels (4.0 and 4.2) its presence caused a marked decrease in ion levels. Thus, all concentrations of HA were able to counteract increased K⁺ levels caused by pH 4.0 and 4.2, bringing K⁺ levels back to normal values for the species (BOLNER & BALDISSEROTTO 2007), at pH 7.0.

Low pH exposure induced continuous net branchial losses of Na⁺, Cl⁻ and K⁺, and a progressive decline in plasma Na⁺ and Cl⁻ levels in rainbow trout (MCDONALD & WOOD 1981). In spite of an improved tolerance to acidity reported for fish that are exposed to gradual water acidification in the wild, it is possible that the same fish will undergo ion loss when faced with sudden environmental acidification. For instance, ARIDE et al. (2007) found that plasma levels of Na⁺ and K⁺ in the tambaqui were reduced in acidic water compared to a circumneutral water. WILSON et al. (1999) observed that acid exposure produced different patterns of Na⁺, Cl⁻ and K⁺ fluxes in three Amazon fish, which implies that acid tolerance is not necessarily a typical feature of the fish that inhabit this region. Instead, it is largely related to the occurrence of these fish in the blackwater areas of that ecosystem, which are known to impose higher levels of acidity on the species.

Although HA showed some positive effects on hematological and plasma K⁺ changes provoked in silver catfish by acidic pH exposure, the overall findings suggest that HA does not protect this species against acidic pH burden, since it increased mortality and Cl⁻ loss at pH 4.0.

ACKNOWLEDGMENTS

The authors thank Conselho Nacional de Desenvolvimento Científico e Tecnológico (CNPq) for the research fellowship to B. Baldisserotto and Fundação de Amparo à Pesquisa do Estado do Rio Grande do Sul (FAPERGS) and Coordenação de Aperfeiçoamento de Pessoal de Nível Superior, Brazil (Capes) for the graduate fellowships to L.T. Gressler and F.J. Sutili respectively. This work was funded by CNPq and Fundação de Amparo à Pesquisa do Estado do Amazonas (FAPEAM – INCT ADAPTA).

LITERATURE CITED

ARIDE PHR, ROUBACH R, VAL AL (2007) Tolerance response of tambaqui *Colossoma macropomum* (Cuvier) to water pH. **Aquaculture Research 38**: 588-594. doi: 10.1111/j.1365-2109.2007.01693.x

BINDON SD, GILMOUR KM, FENWICK JC, PERRY SF (1994) The effects of branquial chloride cell proliferation on respiratory function in the rainbow trout *Oncorhyncus mykiss*. **Journal of Experimental Biology 197**: 47-63.

BOLNER KCS, BALDISSEROTTO B (2007) Water pH and urinary excretion in silver catfish *Rhamdia quelen*. **Journal of Fish Biology 70**: 50-64. doi: 10.1111/j.1095-8649.2006.01253.x

BOYD CE, TUCKER CS (1992) **Water quality and pond soil analyses for aquaculture**. Auburn, Alabama Agricultural Experiment Station, Auburn University, 183p.

BROWN BA (1976) **Hematology: Principles and procedures**. Philadelphia, Lea & Febiger, 336p.

CAMPBELL PGC, TWISS MR, WILKINSON KJ (1997) Accumulation of natural organic matter on the surfaces of living cells: implications for the interaction of toxic solutes with aquatic biota. **Canadian Journal of Fisheries and Aquatic Sciences 54**: 2543-2554.

COPATTI CE, CODEBELLA IJ, RADÜNZ NETO J, GARCIA LO, ROCHA MC, BALDISSEROTTO B (2005) Effect of dietary calcium on growth and survival of silver catfish fingerlings, *Rhamdia quelen* (Heptapteridae), exposed to different water pH. **Aquaculture Nutrition 11**: 345-350. doi: 10.1111/j.1365-2095.2005.00355.x

CUNHA MA, ZEPPENFELD CC, GARCIA LO, LORO VL, FONSECA MB, EMANUELLI T, VEECK APD, COPATTI CE, BALDISSEROTTO B (2010) Anesthesia of silver catfish with eugenol: time of induction, cortisol response and sensory analysis of fillet. **Ciência Rural 40**: 2107-2114. doi: 10.1590/S0103-84782010005000154

DHEER JMS, DHEER TR, MAHAJAN CL (1987) Haematological and haematopoetic responses to acid stress in an air-breathing freshwater fish, *Channa punctatus*. **Journal of Fish Biology 30**: 577-588.

DORAFSHAN S, KALBASSI MR, POURKAZEMI M, AMIRI BM, KARIMI SS (2008) Effects of triploidy on the Caspian salmon *Salmo trutta caspius* haematology. **Fish Physiology and Biochemistry 34**: 195-200. doi: 10.1007/s10695-007-9176-z

DUARTE RM, FERREIRA MS, WOOD CM, VAL AL (2013) Effect of low pH exposure on Na⁺ regulation in two cichlid fish species of the amazon. **Comparative Biochemistry and Physiology a-Molecular & Integrative Physiology 166**: 441-448. doi: 10.1016/j.cbpa.2013.07.022

EATON AD, CLESCERI LS, RICE EW, GRENNBERG AE (2005) **Standard methods for the examination of water and wastewater**. Springfield, American Public Health Association, 21st ed., 1600p.

FARJALLA VF, AMADO AM, SUHETT AL, MEIRELLES-PEREIRA F (2009) DOC removal paradigms in highly humic aquatic ecosystems. **Environmental Science and Pollution Research 16**: 531-538. doi: 10.1007/s11356-009-0165-x

FREDA J, MCDONALD DG (1988) Physiological correlates of interspecific variation in acid tolerance in fish. **The Journal of Experimental Biology 136**: 243-258.

GARCIA LO, GUTIÉRRES-ESPINOSA MC, VÁSQUES-TORRES W, BALDISSEROTTO B (2014) Dietary protein levels in *Piaractus brachypomus* submitted to extremely acidic or alkaline pH. **Ciência Rural 44**: 301-306. doi: 10.1590/S0103-84782014000200017

GONZALEZ RJ, DUNSON WA (1989) Acclimation of sodium regulation to low pH and the role of calcium in the acid-tolerant sunfish *Enneacanthus obesus*. **Physiologycal Zoology 62**: 977-992.

GONZALEZ RJ, WILSON RW (2001) Patterns of ion regulation in acidophilic fish native to the ion-poor acidic Rio Negro.

Journal of Fish Biology 58: 1680-1690. doi: 10.1111/j.1095-8649.2001.tb02322.x

GONZALEZ RJ, DALTON VM, PATRICK ML (1997) Ion regulation in ion-poor, acidic water by the blackskirt tetra (Gymnocorymbus ternetzi), a fish native to the Amazon River. Physiological Zoology 70: 428-435.

GONZALEZ RJ, WOOD CM, WILSON RW, PATRICK ML, BERGMAN HL, NARAHARA A, VAL AL (1998) Effects of water pH and calcium concentration on ion balance in fish of the Rio Negro, Amazon. Physiological Zoology 71: 15-22.

GONZALEZ RJ, WILSON RW, WOOD CM, PATRICK ML, VAL AL (2002) Diverse strategies for ion regulation in fish collected from the ion-poor, acidic Rio Negro. Physiological and Biochemical Zoology 75: 37-47. doi: 10.1086/339216

GOSS GG, WOOD CM (1990) Kinetic analysis of the relationships between ion exchange and acid-base regulation at the gills of freshwater fish, p. 119-136. In: TRUCHOT JP, LAHLOU B (Eds.). Animal Nutrition and Transport Processes. 2. Transport, Respiration and Excretion: Comparative and Environmental Aspects. Basel, Karger Publishers.

GRECO AM, FENWICK JC, PERRY SF (1996) The effects of soft-water acclimation on gill structure in the rainbow trout Oncorhynchus mykiss. Cell and Tissue Research 285: 75-82.

HOLLAND A, DUIVENVOORDEN LJ, KINNEAR SHW (2014) The double-edged sword of humic substances: contrasting their effect on respiratory stress in eastern rainbow fish exposed to low pH. Environmental Science and Pollution Research 21: 1701-1707. doi: 10.1007/s11356-013-2031-0

KÜCHLER IL, MIEKELEY N, FORSBERG BR (2000) A contribution to the chemical characterization of rivers in the rio Negro basin, Brazil. Journal of the Brazilian Chemical Society 11: 286-292. doi: 10.1590/S0103-50532000000300015

LIMA RL, BRAUN N, KOCHHANN D, LAZZARI R, RADÜNZ-NETO J, MORAES BS, LORO V, BALDISSEROTTO B (2011) Survival, growth and metabolic parameters of silver catfish, Rhamdia quelen, juveniles exposed to different waterborne nitrite levels. Neotropical Ichthyology 9: 147-152. doi: 10.1590/S1679-62252011005000004

LIN H, DJ RANDALL (1993) Proton ATPase activity in crude homogenates of fish gill tissue: inhibitor sensitivity and environmental and hormonal regulation. Journal of Experimental Biology 180: 163-174.

MATHAN R, KURUNTHACHALAM SK, PRIYA M (2010) Alterations in plasma electrolyte levels of a freshwater fish Cyprinus carpio exposed to acidic pH. Toxicological and Environmental Chemistry 92: 149-157. doi: 10.1080/02772240902810419

MATSUO AYO, VAL AL (2007) Acclimation to humic substances prevents whole body sodium loss and stimulates branchial calcium uptake capacity in cardinal tetras Paracheirodon axelrodi (Schultz) subjected to extremely low pH. Journal of Fish Biology 70: 989-1000. doi: 10.1111/j.1095-8649.2007.01358.x

MATSUO AYO, PLAYLE RC, VAL AL, WOOD CM (2004) Physiological action of dissolved organic matter in rainbow trout in the presence and absence of copper: sodium uptake kinetics and unidirectional flux rates in hard and softwater. Aquatic Toxicology 70: 63-81. doi: 10.1016/j.aquatox.2004.07.005

MCDONALD D, WOOD CM (1981) Branchial and renal acid and ion fluxes in the rainbow trout, Salmo gairdneri, at low environmental pH. Journal of Experimental Biology 93: 101-118.

MCGEER JC, SZEBEDINSZKY C, MCDONALD DG, WOOD CM (2002) The role of dissolved organic carbon in moderating the bioavailability and toxicity of Cu to rainbow trout during chronic waterborne exposure. Comparative Biochemistry and Physiology C 133: 147-160. doi: 10.1016/S1532-0456(02)00084-4

MILLIGAN CL, WOOD CM (1982) Disturbances in haematology, fluid volume distribution and circulatory function associated with low environmental pH in the rainbow trout, Salmo Gairdneri. Journal of Experimental Biology 99: 397-415.

MIRON DS, BECKER AG, LORO VL, BALDISSEROTTO B (2011) Waterborne ammonia and silver catfish, Rhamdia quelen: survival and growth. Ciência Rural 41: 349-353. doi: 10.1590/S0103-84782011000200028

RIFFEL APK, SACCOL EMH, FINAMOR IA, OURIQUE GM, GRESSLER LT, PARODI T, GOULART LOR, LLESUY S, BALDISSEROTTO B, PAVANATO MA (2014) Humic acid and moderate hypoxia alter oxidative and physiological parameters in different tissues of silver catfish (Rhamdia quelen). Journal of Comparative Physiology B 184: 469-482. doi: 10.1007/s00360-014-0808-1

SAMPAIO FG, BOIJINK CL, OBA ET, SANTOS LRB, KALININ AL, RANTIN FT (2008) Antioxidant defenses and biochemical changes in pacu (Piaractus mesopotamicus) in response to single and combined copper and hypoxia exposure. Comparative Biochemistry and Physiology C 147: 43-51. doi: 10.1016/j.cbpc.2012.07.002

STEINBERG CEW, SAUL N, PIETSCH K, MEINELT T, RIENAU S, MENZEL R (2007) Dissolved humic substances facilitate fish life in extreme aquatic environments and have the potential to extend the lifespan of Caenorhabditis elegans. Annals of Environmental Science 1: 81-90.

TAVARES-DIAS M, MELO JFB, MORAES G, MORAES FR (2002) Características hematológicas de teleósteos brasileiros: VI. Variáveis do jundiá Rhamdia quelen (Pimelodidae). Ciência Rural 32: 693-698. doi: 10.1590/S0103-84782002000400024

TAVARES-DIAS M, BOZZO FR, SANDRIN EFS, CAMPOS-FILHO E, MOARES FR (2004) Células sanguíneas, eletrólitos séricos, relação hepato e esplenossomática de carpa comum, Cyprinus carpio (Cyprinidae) na primeira maturação gonadal. Acta Scientiarum Biological Sciences 26: 73-80. doi: 10.4025/actascibiolsci.v26i1.1661

THURMAN EM (1985) Organic geochemistry of natural waters. Dordrecht, Martinus Nijhof, Dr. W. Junk Publishers, 507p.

VERDOUW H, VAN ECHTELD CJA, DEKKERS EMJ (1978) Ammonia determination based on indophenols formation with sodium salicylate. **Water Research 12**: 399-402. doi: 10.1016/0043-1354(78)90107-0

WENDELAAR BONGA SE (1997) The stress response in fish. **Physiology Reviews 77**: 591-625.

WILSON RW, WOOD CM, GONZALEZ RJ, PATRICK ML, BERGMAN HL, NARAHARA A, VAL AL (1999) Ion acid-base balance in three species of Amazonian fish during gradual acidication of extremely soft water. **Physiological and Biochemical Zoology 72**: 277-285.

WOOD CM, MCDONALD DG (1982) Physiological mechanisms of acid toxicity to fish, p. 197-226. In: JOHNSON RE (Ed.). **Acid Rain/Fisheries: Proceedings of an International Symposium on Acid Precipitation and Fishery Impacts in North-Eastern North America**. Ithaca, American Fisheries Society.

WOOD CM, WILSON RW, GONZALEZ RJ, PATRICK ML, BERGAMAN HL, NARAHARA A, VAL AL (1998) Responses of an Amazonian teleost, the tambaqui (*Colossoma macropomum*) to low pH in extremely soft water. **Physiologycal and Biochemical Zoology 71**: 658-670.

WOOD CM (2001) Toxic response of the gill, p. 1-89. In: SCHLENK D, BENSON WH (Eds.). **Target organ toxicity in marine and freshwater teleosts**. London, Taylor & Francis.

WOOD CM, MATSUO AYO, GONZALEZ RJ, WILSON RW, PATRICK ML, VAL AL (2002) Mechanisms of ion transport in *Potamotrygon*, a stenohaline freshwater elasmobranch native to the ion-poor blackwaters of the Rio Negro. **Journal of Experimental Biology 205**: 3039-3054.

WOOD CM, MATSUO AYO, WILSON RW, GONZALEZ RJ, PATRICK ML, PLAYLE RC, VAL AL (2003) Protection by natural blackwater against disturbances in ion fluxes caused by low pH exposure in freshwater stingrays endemic to the Rio Negro. **Physiological and Biochemical Zoology 76**: 12-27. doi: 10.1086/367946

WOOD CM, AL-REASI HA, SCOTT DS (2011) The two faces of DOC. **Aquatic Toxicology 105S**: 3-8. doi: 10.1016/j.aquatox.2011.03.007

ZAIONS MI, BALDISSEROTTO B (2000) Na+ and K+ body levels and survival of fingerlings of *Rhamdia quelen* (Siluriformes, Pimelodidae) exposed to acute changes of water pH. **Ciência Rural 30**: 1041-1045. doi: 10.1590/S0103-84782000000600020

ZALL DM, FISHER M, GARNER MQ (1956) Photometric determination of chlorides in water. **Analytical Chemistry 28**: 1665-1678. doi: 10.1021/ac60119a009

15

Two new species of *Triplectides* (Trichoptera: Leptoceridae) from South America

Ana Lucia Henriques-Oliveira[1,2] & Leandro Lourenço Dumas[1]

[1] Universidade Federal do Rio de Janeiro, Laboratório de Entomologia, Departamento de Zoologia, Instituto de Biologia, Caixa Postal 68044, Cidade Universitária, 21941-971, Rio de Janeiro, RJ, Brazil.
[2] Corresponding author: anahenri@biologia.ufrj.br

ABSTRACT. *Triplectides*, with about 70 extant species, is the most diverse genus within the Triplectidinae. In the Neotropical Region there are 14 species distributed from southern Mexico to Patagonia. Two new species of *Triplectides* from the Neotropics are described and illustrated based on the male genitalia: *Triplectides cipo* **sp. nov.**, from state of Minas Gerais, southeastern Brazil, and *Triplectides qosqo* **sp. nov.**, from province of Cuzco, southern Peru. The news species can be distinguished by the male genitalia: *Triplectides cipo* **sp. nov.** can be recognized by having the inferior appendages with mesal lobes subacute and apical lobes short, and the tergum X robust, with a subtruncate apex and deep mesal notch; *Triplectides qosqo* **sp. nov.** can be recognized by the first article of inferior appendages long and narrow when compared to the others *Triplectides* species and by the tibial spur formula 2,2,4.

KEY WORDS. Brazil; Caddisflies; new species; Peru; Triplectidinae.

Leptoceridae is one of the most diverse families of caddisflies, with almost 2,000 described species (MORSE 2011). The family is divided into two subfamilies: Leptocerinae Leach, with cosmopolitan distribution, and Triplectidinae Ulmer, which is primarily distributed in the Southern Hemisphere (HOLZENTHAL et al. 2007). *Triplectides* Kolenati, 1859 is the most species-rich genus within Triplectidinae, with about 70 species (HOLZENTHAL 1988, MALM & JOHANSON 2008). The genus occurs in Central and South America, Southern-East Asia (India to Japan), and especially in Oceania, where it reaches its highest diversity, with 15 and 25 species recorded from New Caledonia and Australia, respectively (MALM & JOHANSON 2008).

MOSELY (1936) provided the first comprehensive revision of the genus. Later, the Neotropical species were reviewed by HOLZENTHAL (1988). Since then, only one species has been described from the Neotropics (DUMAS &NESSIMIAN 2010). Currently, there are 14 species described from the Neotropical Region, distributed from Southern Mexico to Southern Chile: *Triplectides chilensis* Holzenthal, 1988 (Argentina and Chile), *T. colombicus* Navás, 1916 (Colombia), *T. egleri* Sattler, 1963 (Brazil, Guyana, and Surinam), *T. flintorum* Holzenthal, 1988 (Colombia, Costa Rica, Ecuador, Guatemala, Honduras, Mexico, Nicaragua, Panama, Peru, and Surinam), *T. gracilis* (Burmeister, 1839) (Argentina, Brazil, Paraguay, and Surinam), *T. itatiaia* Dumas & Nessimian, 2010 (Brazil), *T. jaffuelli* Navás, 1918 (Argentina and Chile), *T. misionensis* Holzenthal, 1988 (Argentina and Brazil), *T. neblinus* Holzenthal, 1988 (Venezuela), *T. neotropicus* Holzenthal, 1988 (Brazil and Venezuela), *T. nevadus* Holzenthal, 1988 (Peru and Venezuela), *T. nigripennis* Mosely, 1936 (Argentina and Chile), *T. tepui* Holzenthal, 1988 (Venezuela), and *T. ultimus* Holzenthal, 1988 (Brazil).

In the present work, we describe and illustrate two new species from South America: *Triplectides cipo* **sp. nov.** from the state of Minas Gerais, southeastern Brazil, and *Triplectides qosqo* **sp. nov.** from province of Cuzco, southern Peru.

MATERIAL AND METHODS

Specimens were collected with malaise and light traps and were preserved in 80-96% ethanol. In order to observe the genital structures, the abdomen was removed and cleared using the lactic acid method (BLAHNIK et al. 2007). The abdomens were mounted on temporary slides with glycerin for viewing and drawing, and transferred back to ethanol and permanently stored in micro vials. Pencil illustrations were made under a stereomicroscope or under a compound microscope, both equipped with a camera lucida. Pencil drawings of genital structures were inked with a technical pen, and wing illustrations were made using vector lines in an Adobe Illustrator (v. 16, Adobe Inc.) document. The terminology used in this paper follows that presented by HOLZENTHAL (1988).

Type specimens are deposited in the following collections, as indicated in descriptions: Museo de Historia Natural "Javier Prado", Universidad Nacional Mayor de San Marcos, Lima (MUSM) and Coleção Entomológica Professor José Alfredo Pinheiro Dutra, Departamento de Zoologia, Universidade Federal do Rio de Janeiro, Rio de Janeiro (DZRJ).

TAXONOMY

Triplectides cipo **sp. nov.**
Figs. 1-7

Description. Adult male. General color brown (in alcohol). Antennae, palps and legs golden brown. Head and thorax mostly brown. Forewings with forks I and V present in males; discoidal cell apically large. Hind wings broad, with forks I, III, and V present; fork I with distinct petiole. Length of forewing 10.0-11.0 mm, length of hind wing 8.0-9.0 mm (n = 5) (Fig. 1). Tibial spur formula 2,2,4.

Genitalia. Segment IX, in lateral view, narrow with anterior margin almost straight and enlarged dorsally, posterior margin slightly concave medially (Fig. 3); tergum IX with posterior margin almost rounded, slightly protruding laterally, median process apparently absent (Figs. 2 and 3). Preanal appendages slender, digitate, slightly longer than half length of tergum X, bearing long setae (Fig. 2). Tergum X, in lateral view, wide at base, tapering apically, with apex rounded; in dorsal view, slightly widened apically, apex subtruncate, bearing small setae, with apicomesal excision extending anteriorly at half length of segment (Fig. 3). Inferior appendages, long, slightly surpassing tergum X, bearing long setae; 1st article, as viewed laterally, wide at base, constricted medially, with apical portion narrow; apicodorsal lobe club-like, with long setae; basoventral lobes digitate, bearing long setae; in ventral view, mesal lobes shorter than basoventral lobes, wide at base, tapering apically, with flattened aspect, apex subacute; 2nd article short, wide at base, tapering apically, gradually curved inward, with pointed apex (Fig. 5). Phallic apparatus simple, tubular, with phallotremal sclerite small, rod-like, apically positioned (Fig. 6).

Adult female. General color brown (in alcohol). Antennae, palps, and legs golden brown. Thorax and head brown. Length of forewing 13.5-14.5 mm, length of hind wing 10.5-11.5 mm (n = 3). Tibial spur formula 2,2,4.

Genitalia. Sternum VIII, in ventral view, with a sclerotized plate, dark brown; anterior margin deeply concave, posterior margin truncate with a small mesal cleft, with several short setae (Fig. 7). Sternum IX heavily sclerotized, with small transverse striae apically. Appendages of segment X, in lateral view, short, broad at base, subtriangular and setose (Fig. 6). Sensilla-bearing process absent. Valves ventrolateral, well developed, sclerotized, flap-like, slightly concave with fine setae (Fig. 6). Internal vaginal apparatus long, broad, and sclerotized (Fig. 7).

Holotype male: BRAZIL, *Minas Gerais*: Jaboticatubas (Parque Nacional da Serra do Cipó, Córrego das Pedras, 19°22'16.7"S, 48°36'2.8"W, 766 m), 9-13.xii.2011, Malaise trap, APM Santos, DM Takiya, RR Cavichioli & ML Monné *leg.* (DZRJ). Paratypes: same data as holotype, 1 male, 2 females (DZRJ); *Minas Gerais*: Jaboticatubas (Parque Nacional da Serra do Cipó, Córrego das Pedras, 19°22'16.7"S, 48°36'2.8"W, 766 m), 02-05.iii.2013, Mal-

aise trap, BHL Sampaio, BM Camisão, ALH Oliveira, APM Santos & DM Takiya *leg.* (15 males, 6 females) (DZRJ).

Distribution. Brazil (state of Minas Gerais).

Etymology. The specific epithet, *cipo*, refers to the type locality of the species, Parque Nacional da Serra do Cipó, located in Serra do Espinhaço mountain range.

Remarks. *Triplectides cipo* **sp. nov.** is closely related to *T. flintorum* Holzenthal, 1988 and *T. itatiaia* Dumas & Nessimian, 2010 as evidenced by the wing venation and mesal lobes of inferior appendages. In *T. cipo* the hind wing fork I is petiolate, as in the other two similar species cited above. However, the genital structure of the new species is quite distinct from those of *T. flintorum* and *T. itatiaia*. The mesal lobes of inferior appendages of *T. cipo* are less rounded apically, being subacute. Also, the apical lobes of the inferior appendages of the new species are comparatively shorter than those of *T. flintorum* and *T. itatiaia*. In addition, in *T. cipo* tergum X is robust, with a subtruncate apex and deep mesal notch, whereas in *T. itatiaia* it is rounded apically. *Triplectides flintorum* also has tergum X subtruncate apically but the mesal notch is less deep than in *T. cipo*.

Triplectides qosqo **sp. nov.**
Figs. 8-12

Description. Adult male. General color brown (in alcohol). Antennae and palps brown. Legs light brown. Forewings with forks I and V present in male; discoidal cell slightly narrower at apex. Hind wings broad, with forks I, III, and V present; fork I with distinct petiole. Length of forewing 12.0-13.0 mm, length of hind wing 9.5-10.0 mm (n = 6) (Fig. 8). Tibial spur formula 2,2,4.

Genitalia. Segment IX, in lateral view, annular, narrow with anterior margin almost straight and enlarged dorsally, posterior margin slightly protruded near dorsum (Fig. 10); tergum IX with posterior margin almost rounded, slightly protruding laterally (Figs. 9 and 10). Preanal appendages rounded and slightly flat, more than half length of tergum X, bearing long setae (Fig. 9). Tergum X, in lateral view, wide at base, anterodorsal area less sclerotized, slightly tapering apically, with apex rounded; in dorsal view, apex subtruncate, slightly narrower than base, bearing small setae, with apicomesal excision extending slightly beyond apical third of segment (Figs. 9 and 10). Inferior appendages long, surpassing tergum X, bearing long setae; 1st article, in lateral view, wide at base, constricted before half length of segment, with apical portion narrow; apicodorsal lobe digitate, with long setae; basoventral lobes digitate, tapering to apex, bearing long setae; in ventral view, mesal lobes shorter than basoventral lobes, wide at base, tapering apically, with pointed apex, bent outward, flattened in lateral view; 2nd article short, wide at base, tapering apically, gradually curved inward, with pointed apex (Figs. 10 and 12). Phallic apparatus simple, tubular, with phallotremal sclerite small, rod-like, mesally positioned (Fig. 11).

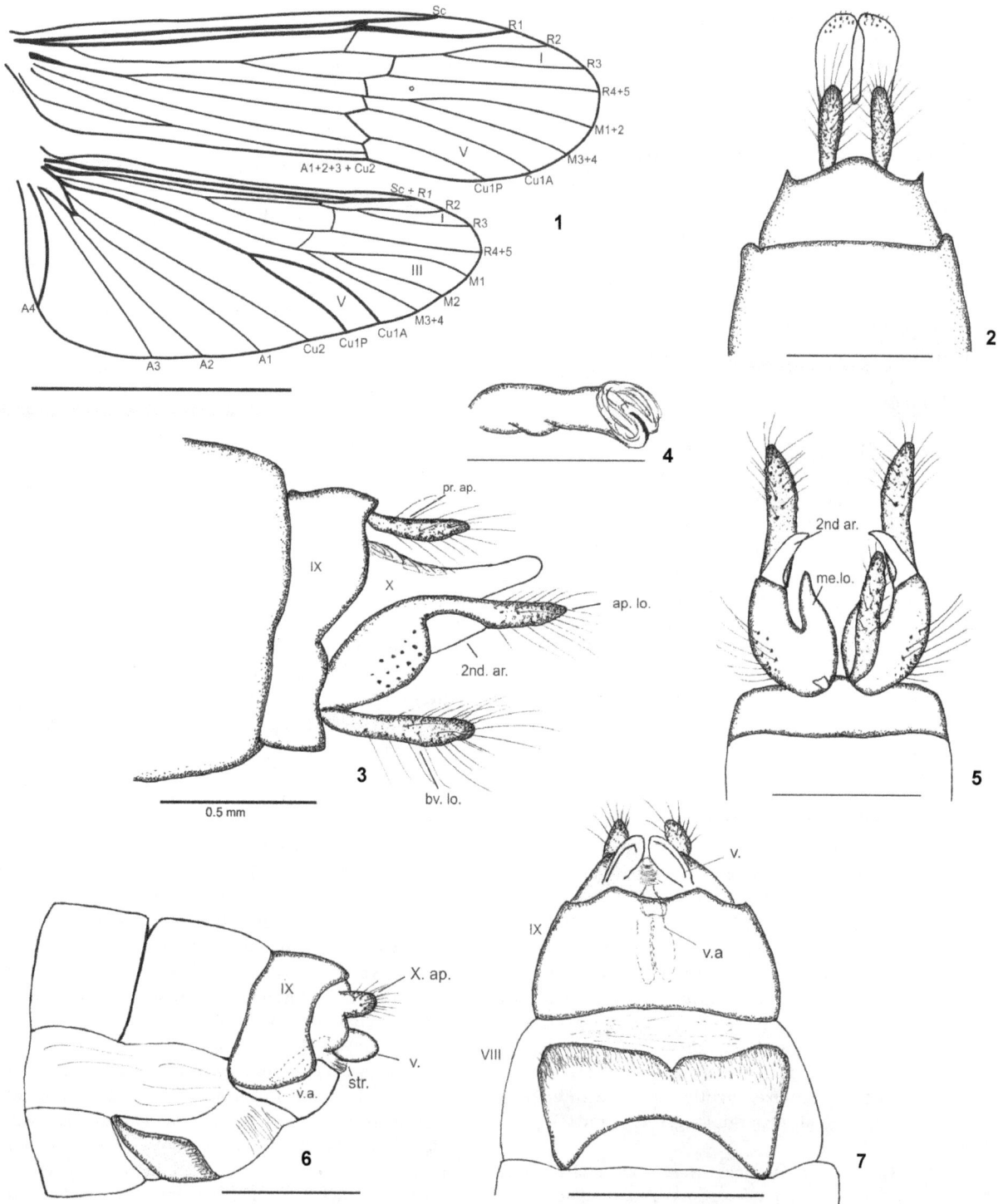

Figures 1-7. *Triplectides cipo* **sp. nov.** (1-5) Male: (1) fore and hind wings; (2) genitalia, dorsal view; (3) genitalia, lateral view; (4) phallic apparatus, lateral view; (5) genitalia, ventral view. (6-7) Female: (6) genitalia, lateral view; (7) genitalia, ventral view. (IX) Tergum IX, (X) Tergum X, (pr. ap.) preanal appendages, (ap. lo.) apicodorsal lobe, (bv. lo.) basoventral lobe, (me. lo.) mesal lobe, (2nd ar.) second article, (X. ap.) appendages of segment X, (v.) valves, (str.) striae, (v.a.) vaginal apparatus. Scale bars: 1 = 5.0 mm, 2-5 = 0.5 mm, 6-7 = 1.0 mm.

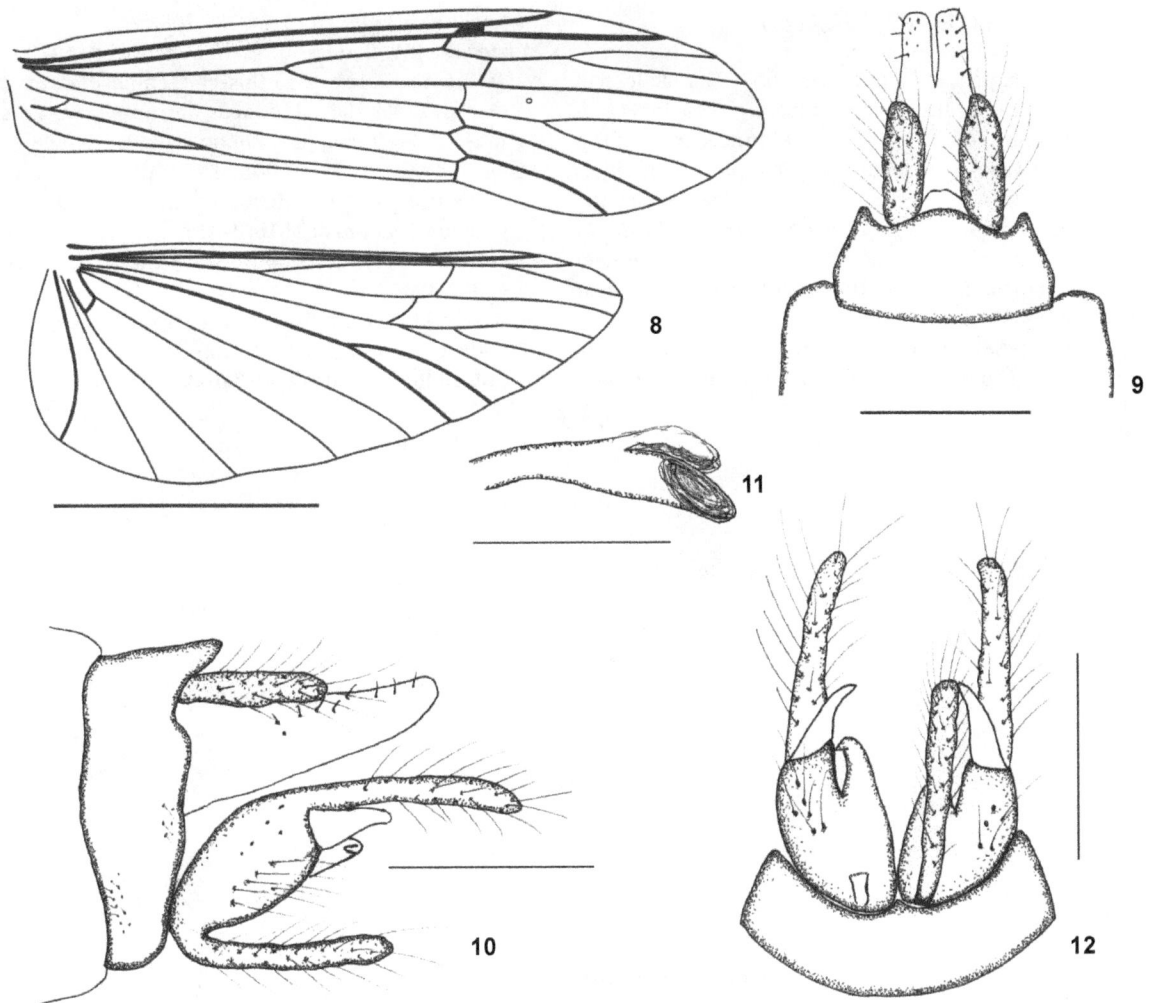

Figures 8-12. *Triplectides qosqo* **sp. nov.**, male: (8) Fore and hind wings; (9) genitalia, dorsal view; (10) genitalia, lateral view; (11) phallic apparatus, lateral view; (12) genitalia, ventral view.

Holotype male: PERU, *Cuzco*: (19rd km W Quincemil, Río Araza tributary, 874 m, 12°20'10"S, 70°50'5"W), Malaise trap, RR Cavichioli, JA Rafael, DM Takiya, APM Santos *leg.* (MUSM). Paratypes: same data as holotype, 1 male (MUSM), 4 males (DZRJ).

Distribution. Peru (province of Cuzco).

Etymology. The specific epithet, *qosqo*, refers to the type locality of the species, province of Cuzco. Cuzco, or Qosqo in Quechua language, is the ancient name for the capital of the great Incan Empire, meaning "navel of the world" in mystical terms.

Remarks. *Triplectides qosqo* **sp. nov.** is similar to *T. nevadus* Holzenthal, 1988, especially in the strongly hooked apices of the mesal lobes of inferior appendages. However, they can be easily separated by the structure of 1st article of inferior appendages, which are much longer and narrower in the new species. Additionally, the tibial spur formula in *T. nevadus* is 0,2,2 or 0,2,4, while in *T. qosqo* it is 2,2,4. The wing venation of the new species is virtually the same of *T. nevadus*.

ACKNOWLEDGEMENTS

We are grateful to Daniela M. Takiya and Allan P. Moreira dos Santos for collecting the specimens described here. The authors are also grateful to two anonymous reviewers for useful suggestions and improvements. Fundação Carlos Chagas Filho de Amparo à Pesquisa do Estado do Rio de Janeiro (FAPERJ) and Coordenação de Aperfeiçoamento de Pessoal de Nível Superior (CAPES) provided financial support. We also thank Instituto Chico Mendes de Conservação da Biodiversidade (ICMBio) for issuing collecting permits.

LITERATURE CITED

BLAHNIK RJ, HOLZENTHAL RW, PRATHER A (2007) The lactic acid method for clearing Trichoptera genitalia, p. 9-14. In: BUENO-SORIA J, BARBA-ALVAREZ R, ARMITAGE B (Eds) **Proceedings of the 12ᵗʰ International Symposium on Trichoptera.** Columbus, The Caddis Press.

DUMAS LL, NESSIMIAN JL (2010) A new long-horned caddisfly in the genus *Triplectides* Kolenati (Trichoptera: Leptoceridae) from the Itatiaia massif, Southeastern Brazil. **Neotropical Entomology 39**(6): 949-951.

HOZENTHAL RW (1988) Sytematics of the Neotropical *Triplectides* (Trichoptera: Leptoceridae). **Annals of the Entomological Society of America 81**(2): 186-208.

HOLZENTHAL RW, BLAHNIK RJ, PRATHER AL, KJER KM (2007) Order Trichoptera Kirby, 1813 (Insecta), Caddisflies. In: ZHANG Z-Q, SHEAR WA (Eds) Linnaeus Tercentenary: Progress in Invertebrate Taxonomy. **Zootaxa 1668**: 639-698.

MALM T, JOHANSON KA (2008) Description of eleven new *Triplectides* species (Trichoptea: Leptoceridae) from New Caledonia. **Zootaxa 1816**: 1-34.

MORSE JC (2011) The Trichoptera world checklist. **Zoosymposia 5**: 372-380.

MOSELY ME (1936) A revision of the Triplectidinae, a subfamily of the Leptoceridae (Trichoptera). **Transactions of the Royal Entomological Society of London 85**: 91-129.

16

Geographic variation in *Caluromys derbianus* and *Caluromys lanatus* (Didelphimorphia: Didelphidae)

Raul Fonseca[1,2] & Diego Astúa[1,3]

[1]*Departamento de Zoologia, Universidade Federal de Pernambuco. Avenida Professor Moraes Rêgo, s/n, Cidade Universitária, 50670-901 Recife, PE, Brazil.*
[2]*Departamento de Zoologia, Universidade do Estado do Rio de Janeiro. 20550-013 Rio de Janeiro, RJ, Brazil.*
[3]*Corresponding author. E-mail: diegoastua@ufpe.br*

ABSTRACT. We analyzed the geographic variations in the shape and size of the cranium and mandible of two woolly opossums, *Caluromys derbianus* and *Caluromys lanatus*. Using geometric morphometrics we analyzed 202 specimens of *C. derbianus* and 123 specimens of *C. lanatus*, grouped in 7 and 9 populations, respectively. We found sexual dimorphism in shape variables only in the dorsal view of the cranium of *Caluromys derbianus*, which is not associated with geographical origin. We detected geographic variation in the size of the mandible in two populations (Nicaragua and Northern Panama), but no geographic variation in shape. The size of the cranium of *C. lanatus* varies significantly, with clinal variation in peri-Amazon populations, with a break between two populations, Bolivia and Paraguay. Shape analyses also revealed some separation between the Paraná population and all other populations. Our results suggest that the available name, *Caluromys derbianus*, should be maintained for all individuals throughout the geographic range of the species. The same is true for *Caluromys lanatus*, which can be separated into two distinct morphologic units, *Caluromys lanatus ochropus*, from the Amazon and Cerrado, and *Caluromys lanatus lanatus*, from the Atlantic forest.

KEY WORDS. Caluromyinae; geometric morphometrics; marsupial; Neotropics; skull; size and shape analysis.

The morphology and/or physiology of organisms usually vary across their distribution range. This is particularly true for species that are distributed over different biomes or biogeographic provinces (THORPE 1987). Such variation in intraspecific characters throughout a species' range is known as geographic variation (MAYR 1977). The study of geographic variation is key for understanding speciation and the role that ecological and geographical features may play in shaping biodiversity (HAFFER 1969, GOULD 1972, EMMONS 1984). Furthermore, geographic variation has been a central theme in evolutionary biology, from the works of Darwin to modern analyses based on molecular approaches (HALLGRÍMSSON & HALL 2005).

Morphological variation across geographical and environmental discontinuities occur in different small mammal groups, such as rodents (e.g., MACÊDO & MARES 1987, LESSA et al. 2005) and marsupials (e.g., LÓPEZ-FUSTER et al. 2000, HIMES et al. 2008). In the latter, variation can be found in external and cranial morphology and morphometric data (LEMOS & CERQUEIRA 2002, LÓPEZ-FUSTER et al. 2002, LÓSS et al. 2011), as well as in genetic characters (COSTA 2003, STEINER & CATZEFLIS 2004, BRAUN et al. 2005).

Woolly opossums of the genus *Caluromys* Allen, 1900 are part of a basal lineage within the living New World Didelphidae opossums (VOSS & JANSA 2009). *Caluromys* currently includes three species, *Caluromys derbianus* (Waterhouse, 1841), *Caluromys lanatus* (Olfers, 1818) and *Caluromys philander*

Linnaeus, 1758 which are widely distributed in forest areas of Central and South America (GARDNER 2008). Variation in external morphological traits has been found in *Caluromys lanatus* (Thomas, 1913) throughout its geographic range. Venezuelan populations of *Caluromys* species (LÓPEZ-FUSTER et al. 2008) also present morphometric variation. This phenotypic diversity lead to the recognition of a number of morphologically distinct groups: eight subspecies of *C. derbianus* (BUCHER & HOFFMANN 1980, GARDNER 2008), four of *C. philander* (CABRERA 1958, GARDNER 2005) and six of *C. lanatus* (CÁCERES & CARMIGNOTTO 2006, GARDNER 2008).

The purpose of this study was to evaluate and to quantify the morphological variation in the size and shape of the cranium and mandible of *Caluromys derbianus* and *Caluromys lanatus* throughout their geographic range. We used geometric morphometric tools to evaluate whether the variation supports the taxonomic status of each species and their currently recognized subspecies.

MATERIAL AND METHODS

We obtained 2D images of the crania in three views (dorsal, ventral and lateral), and lateral images of the mandibles. Only complete adult specimens, i.e., specimens with all three premolars and four molars fully erupted and functional (TRIBE

1990, Astúa & Leiner 2008) were photographed. Specimens analyzed were from the following institutions: Museu Nacional – Universidade Federal do Rio de Janeiro (MN), Museu de Zoologia da Universidade de São Paulo (MZUSP), Museu Paraense Emílio Goeldi (MPEG), Coleção de Mamíferos do Departamento de Zoologia da Universidade Federal de Minas Gerais (UFMG), Museu de História Natural Capão da Imbúia (MHNCI), Museo Argentino de Ciencias Naturales "Bernardino Rivadavia" (MACN), Museo de Historia Natural de la Universidad Nacional Mayor de San Marcos (MUSM), American Museum of Natural History (AMNH), Field Museum of Natural History (FMNH), Louisiana State University, Museum of Natural Science (LSUMZ), Museum of Southwestern Biology (MSB), Museum of Vertebrate Zoology (MVZ), Kansas University, Museum of Natural History (KU) and National Museum of Natural History (USNM).

We digitized a total of 92 landmarks – 28 in dorsal, 28 in ventral, 22 in lateral views of the cranium, and 14 landmarks on the mandible – using TPS Dig (Rohlf 2006) (Fig. 1, Appendix 1). All landmarks were tested for repeatability (Falconer & Mackay 1996), which was set at 85% for inclusion in subsequent analyses.

We applied a a Generalized Procrustes Analysis (GPA) to all landmark configurations (Rohlf & Slice 1990), to remove the effects of isometric size, orientation and position. Conse-

quently, only shape information was retained (Adams et al. 2004, 2013). We obtained two formally independent set of variables, used in the subsequent analyses. One set includes centroid size for all specimens. Centroid size is the univariate size variable resulting from the squared-root of sums of the squared distances between each landmark and the centroid of its configuration. This set was used in the analyses of geographic variation in the size of the studied structures (for more details see Zelditch et al. 2012). GPA also yields the partial warps and uniform components, a set of variables that retain all the information on the shape of the landmark configuration of the studied structures that were used in the analyses of geographic variation in shape. Further detail on the geometric morphometric procedures can be found in Zelditch et al. (2012).

We obtained the geographic coordinates of the collecting localities of each specimen from their skin tags. When coordinates were not in the tags, we used standard ornithological gazetteers (Paynter 1982, 1989, 1992, 1993, 1995, 1997) to recover them. Specimens from different localities were grouped into populations based on the features of the ecoregions (Olson et al. 2001) found in the distribution of both species (specimens from geographically close localities in the same ecoregion were pooled into populations). Next, to increase the sample size of populations resulting from the classification using ecoregions,

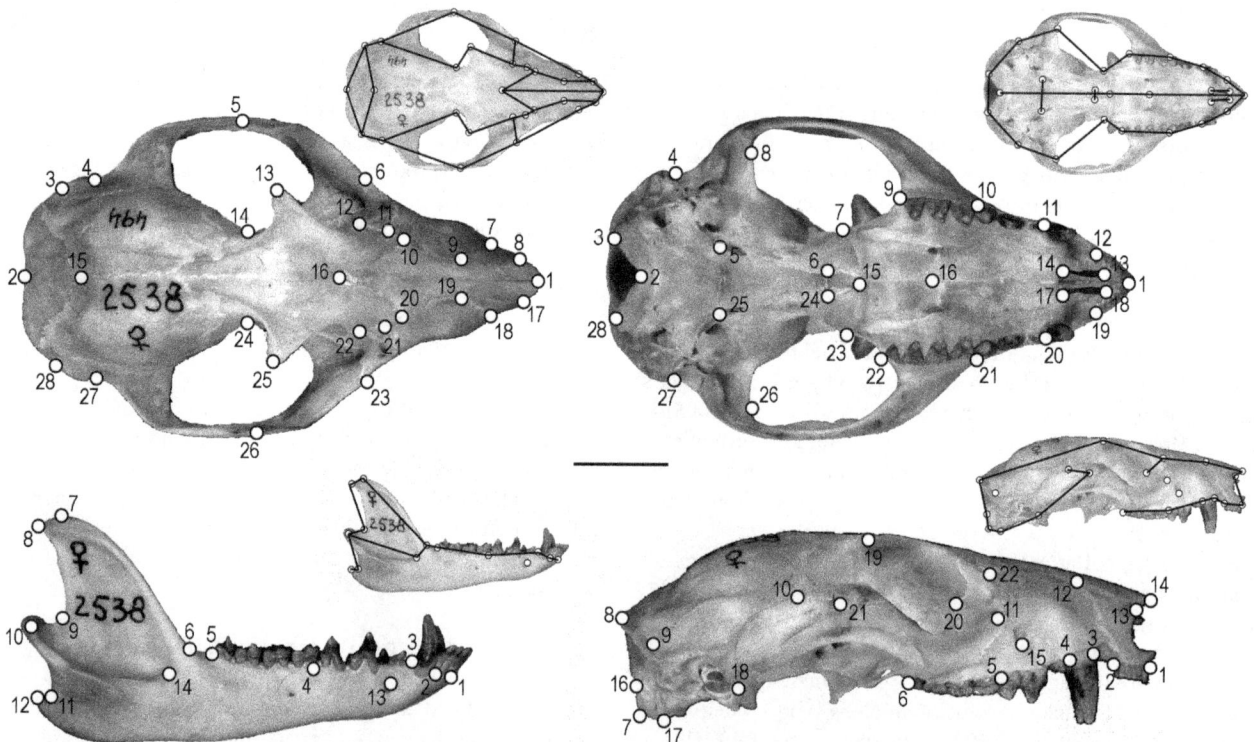

Figure 1. Landmarks used in the cranium and mandible. Smaller versions of each view include landmarks with links, as used deformation grids in subsequent figures. See Appendix 1 for detailed description of landmark locations. Scale bar: 1 cm.

we decided to pool the populations that were geographically closer to each other and which lacked morphometric divergence.

We examined a total of 202 specimens of *Caluromys derbianus* (the number of specimens analyzed in each view may vary because missing structures in one view may preclude the use of a photograph, while the photographs of the same specimen from other views can be used). The specimens were divided into seven populations: Colombian, Ecuadorian and Peruvian individuals (n = 9), Panama-Colombia (n = 16), Southern Panama (n = 22), Northern Panama (n = 51), Nicaragua (n = 54), Honduras (n = 37) and Mexico (n = 15) (Fig. 2). Likewise, we examined a total of 123 specimens of *Caluromys lanatus*, which were divided into 9 populations: Northern Venezuela (n = 11), Southern Venezuela (n = 8), Colombia (n = 17), Northern Peru (n = 22), Iquitos (n = 22), Peru-Bolivia (n = 15), Paraná (n = 8), Trombetas (n = 13), and Cerrado (n = 7) (Fig. 3). The list of all examined specimens with localities is presented in Appendix 2.

Literature information on the absence of sexual dimorphism in both species (Astúa 2010) was obtained from a smaller and geographically restricted dataset. With this in mind we re-evaluated the existence of sexual size dimorphism through a t-test on centroid size, and the existence of sexual shape dimorphism through a Hotteling T² test on shape variables. Since

several populations were represented by only a few specimens, we pooled all males into one group and all females into another regardless of their geographic origin, in order to increase sample size and to avoid a type I error. To evaluate geographic variation in size, we compared populations with ANOVAs on centroid sizes, followed by Tukey *a posteriori* tests. To evaluate geographic variation in shape we compared shape variables between populations using Canonical Variates Analyses (CVA), following Webster & Sheets (2010), given that our total sample size was much larger than $[(2k – 4) + (G – 1)]$, where k is the number of variables and G is the amount of groups analyzed. For each view, this parameter ranged from 30 to 58 for *Caluromys derbianus*, and 32 to 60 for *C. lanatus*, indicating that running a CVA is appropriate. Because all analyses were repeated on four views of both species, we employed Bonferroni correction again, using a significant p-value of 0.0125 (0.05/4).

RESULTS

Sexual dimorphism

Neither species presented sexual dimorphism in size. Sexual dimorphism was observed only in the shape of the dorsal portion of the cranium of *Caluromys derbianus* (Hotteling

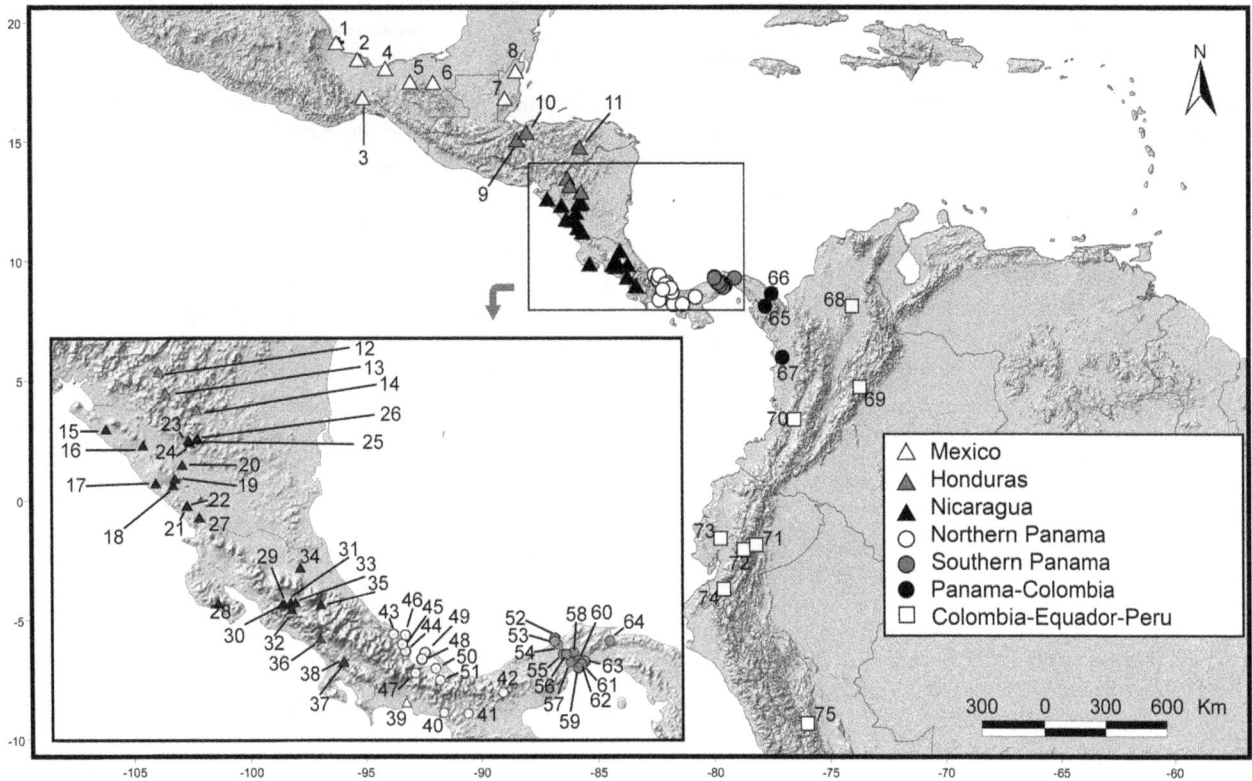

Figure 2. Distribution of the localities of *Caluromys derbianus* with specimens included in this study. Localities were grouped in populations for subsequent analyses, and are labelled accordingly. Numbers indicate localities as listed in Appendix 2.

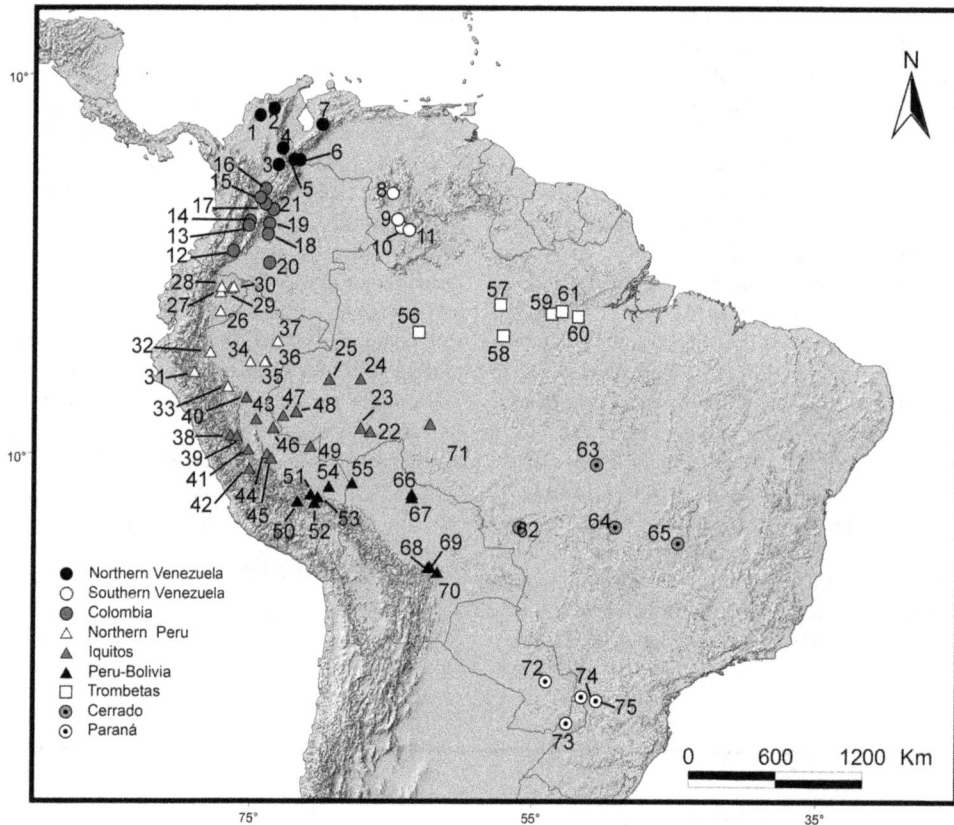

Figure 3. Distribution of the localities of *Caluromys lanatus* with specimens included in this study. Localities were grouped in populations for subsequent analyses, and are labelled accordingly. Numbers indicate localities as listed in Appendix 2.

$T^2 = 0.858$, $F = 1.82$; d.f. = 56, $p < 0.01$, 85 males, 90 females). In view of the absence sexual dimorphism in size and shape variables among individuals in all other views of both species, we decided to pool the sexes together within populations for subsequent analyses. This allowed us to include in the analyses specimens for which the sex was unknown.

Geographic variation in *Caluromys derbianus*

When analyzing size variation, we only found a statistically significant difference in mandible size, between the Nicaragua and Northern Panama populations (ANOVA $F = 2.89$, $p < 0.01031$, $p < 0.002$, *post-hoc* Tukey test). As for shape variation, the CVA scores overlapped considerably, indicating little morphometric divergence in size (Fig. 4). Given that the variation within each population was equal to or larger than the variation between populations, we concluded that the variation is not geographically structured and that the populations cannot be considered morphologically distinct.

Geographic variation in *Caluromys lanatus*

Under all views, size varied geographically (ANOVA, Dorsal: $F = 11.02$, $p < 0.0001$; Lateral: $F = 3.66$, $p < 0.001$; Ventral:

$F = 9.62$, $p < 0.0001$; Mandible: $F = 10.91$, $p < 0.0001$), but no clear grouping was observed among populations. However, a north-south clinal variation in skull size can be inferred, with specimens increasing in size from Colombia (smallest) to Bolivia (largest), with Ecuadorian and Peruvian specimens presenting intermediate sizes. This trend is then interrupted in southern Bolivia, with specimens from southeastern Brazil, Paraguay and Argentina being smaller than their Bolivian counterparts (Fig. 5).

Caluromys lanatus has a conserved skull shape throughout its geographic range. CVA scores show a partial separation of the Paraná population from all others, due to a variation in the morphology of the occipital and posterior roots of the squamosal, which are larger in Paraná specimens than in other individuals. The morphology of the rostrum also varies, with short and narrow nasals and basicranium with short frontals and longitudinal elongation of parietals (visualized through displacement of landmarks at the postorbital constriction) in the dorsal view of the cranium (Fig. 6). Additionally, an increase in occipital width, a more horizontally aligned molar tooth row, and shorter and narrow rostrum are found in Paraná individuals, in lateral view of the cranium (Fig. 7).

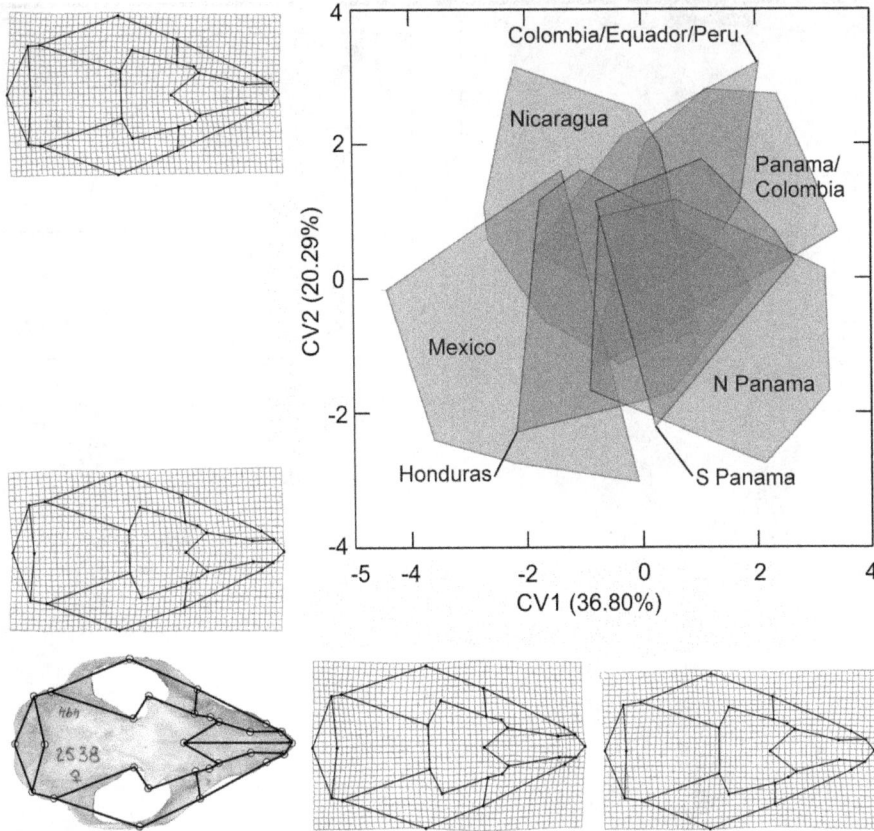

Figure 4. Canonical Variates Analysis on shape variables (partial warps and uniform components) of the skull in dorsal view of *Caluromys derbianus*, using localities as grouping factors, and percentage of variance explained by the first two CVs. Only the convex hulls for each population are shown. Grids indicate deformation associated with the extremes of each CV, from a multivariate regression of shape variables onto CV scores. Overlap for all other views are very similar, therefore only the dorsal view of the cranium is shown.

DISCUSSION

Structured geographic variation in cranial size and shape was not detected in *Caluromys derbianus*. However, it was observed in *Caluromys lanatus* populations. Despite the fact that subspecies have been recognized for *C. derbianus*, its populations belong to a single morphologic unit, which is spread throughout the geographic distribution of the species. Our results also corroborate that *Caluromys lanatus* is one species, but with two distinct morphological groups, one in the Amazon-Cerrado and the other in the Atlantic forest.

The absence of sexual size dimorphism in the skull of these species was already discussed (Astúa 2010), although that analysis, unlike ours, detected significant sexual shape dimorphism in both species.

Geographic variation in *Caluromys derbianus*

We did not find any evidence of structured geographical variation in the size of the skull of *Caluromys derbianus*, despite

its occurrence in the congeneric species *C. lanatus* (this study) and *C. philander* (Olifiers et al. 2004). We were also unable to detect a pattern in the geographic variation of the shape and size of the skull that would match the current taxonomic structures proposed for this species at the subspecific level. Bucher & Hoffmann (1980) and Gardner (2008) divided *Caluromys derbianus* into seven subspecies and one trans-Andean "unspecified" population. These populations were based on morphological differences such as fur color. As we used only cranial quantitative data, it is possible that other characters, particularly in the external morphology, may be the reason for the high number of subspecies. In particular, pelage color, which was not assessed in this study, is well known to vary geographically in this and other marsupial genera (Thomas 1913, Goodwin 1942) and might explain the discrepancy between the existing classification and the one that results from our quantitative results from skull morphology, which failed to support a separation.

The distribution of *Caluromys derbianus* represents a continuum of populations on a N-S stripe, most of which are in

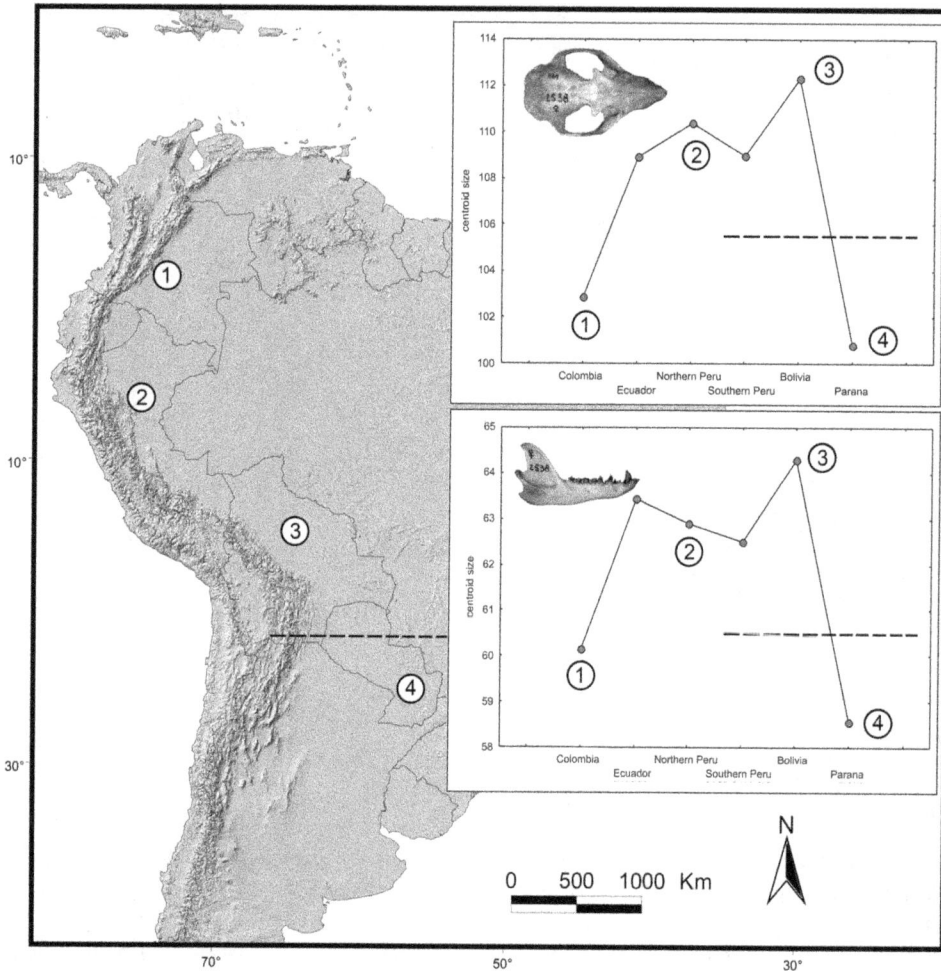

Figure 5. Clinal variation in size of the skull and mandible of populations of *Caluromys lanatus* along the east of the Andes, from Colombia to Bolivia, with a break between Bolivian and Northern Paraguay/Southern Brazil populations, indicated by the dashed line. Numbers in the map refer to the same points in the two graphs.

Central America and the remaining populations in the Andes in South America. The absence of geographic variation in the size of the cranium and mandible shape of this species is noteworthy, since several geographic and ecological discontinuities found throughout its distribution range are believed to cause variation among populations of other taxa (Savage 1987, Pérez-Emán 2005, Castoe et al. 2009).

Geographic variation in *Caluromys lanatus*

Clinal variation occurs throughout the range of many mammals (Storz et al. 2001, Cardini et al. 2007). We found clinal variation in the size of the skull of *Caluromys lanatus* from Andean populations, to the Bolivian-Paraguayan border, coinciding with those populations that overlap less in shape analyses. Even though we have not analyzed molecular data, we believe that the large overlap of CVA scores among all

Amazon populations can be associated with reduced genetic divergence in this species. The latter has been already noted for populations distributed in this area (Patton et al. 2000, Patton & Costa 2003).

The divergence among the Paraná population and the others may correspond to the geographic differences between the Amazon and the Atlantic Rainforest. A similar variation pattern has also been observed to occur in *Didelphis*, *Marmosa*, *Caluromys philander* and *Metachirus nudicaudatus* (Costa 2003, Patton & Costa 2003). Both morphological and genetic divergence were observed in these species. Similar results were also recorded for rodent genera such as *Rhipidomys*, *Oecomys*, *Hylaeamys* and *Euryoryzomys* (Costa 2003). Populations from Paraná are ecologically separated from others by the Chaco – xerophytic plant cover, located in Argentina and Paraguay (Marco & Páezw 2002, Boletta et

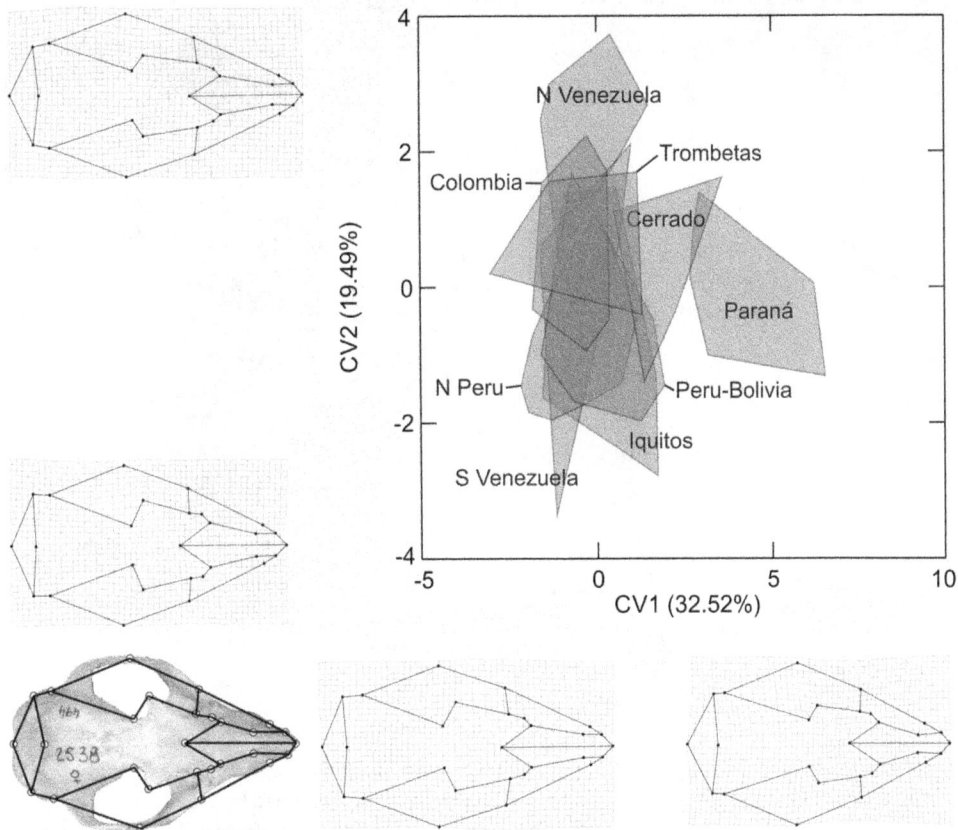

Figure 6. Canonical Variates Analysis on shape variables (partial warps and uniform components) of the cranium in dorsal view of *Caluromys lanatus*, using localities as grouping factors, and percentage of variance explained by the first two CVs. Only the convex hulls for each population are shown. Grids indicate deformation associated with the extremes of each CV, from a multivariate regression of shape variables onto CV scores.

al. 2006), which is characterized by medium and large trees such as Bignoniaceae, Leguminosae and grass fields (PENNINGTON et al. 2000). The increase in the Araucaria cover in the early Holocene (LEDRU 1993, SALGADO-LABOURIAU et al. 1998) over open areas may have served as a bridge between the forested areas of the Atlantic forest and the Amazon (AB'SABER 2000). This plant cover probably allowed the dispersion of *Caluromys lanatus* from the Amazon and Cerrado to the southern Atlantic Rainforest (COSTA 2003, PATTON & COSTA 2003), where these new populations were later isolated by open lands that arose between these areas (LEDRU et al. 1998, VAN DER HAMMEN & HOOGHIEMSTRA 2000, BEHLING 2002). This contact and subsequent isolation hypothesis is particularly likely for *Caluromys lanatus*, since this species is strictly arboreal. Environmental discontinuities that incur in canopy fragmentation may hinder population movements (PIRES et al. 2002), thus providing an effective ecological barrier like the one that has been associated with speciation of the congeneric *Caluromys philander* (LIRA et al. 2007). Morphologi-

cal similarities between populations from Central Brazil and the Amazon may be explained by the fact that Cerrado vegetation may not be uniformly affected by climatic changes (SALGADO-LABOURIAU et al. 1997). At higher altitudes the plant composition was less altered even in the dry periods of the Pleistocene and may have extended to lower areas during cold periods (BUSH et al. 2004). Grassland vegetation may have replaced only low-altitude forests (SALGADO-LABOURIAU et al. 1997, 1998). Due to climatic and pluviometric oscillations, eventual expansions of gallery forests may have created ecological corridors that allowed faunal and floristic population flow among Cerrado, Llanos, Amazonia and even Gran-Sabana (CERQUEIRA 1982, LEDRU 2002, OLIVEIRA-FILHO & RATTER 1995 apud DE OLIVEIRA et al. 2005). Gallery forests house twice as many forest-related species than the entire Cerrado *latu sensu* (JOHNSON et al. 1999). These forested areas may not have been totally affected by climatic changes and may have been used as a corridor that kept Amazonian and Cerrado populations in contact (CARDOSO & BATES 2002).

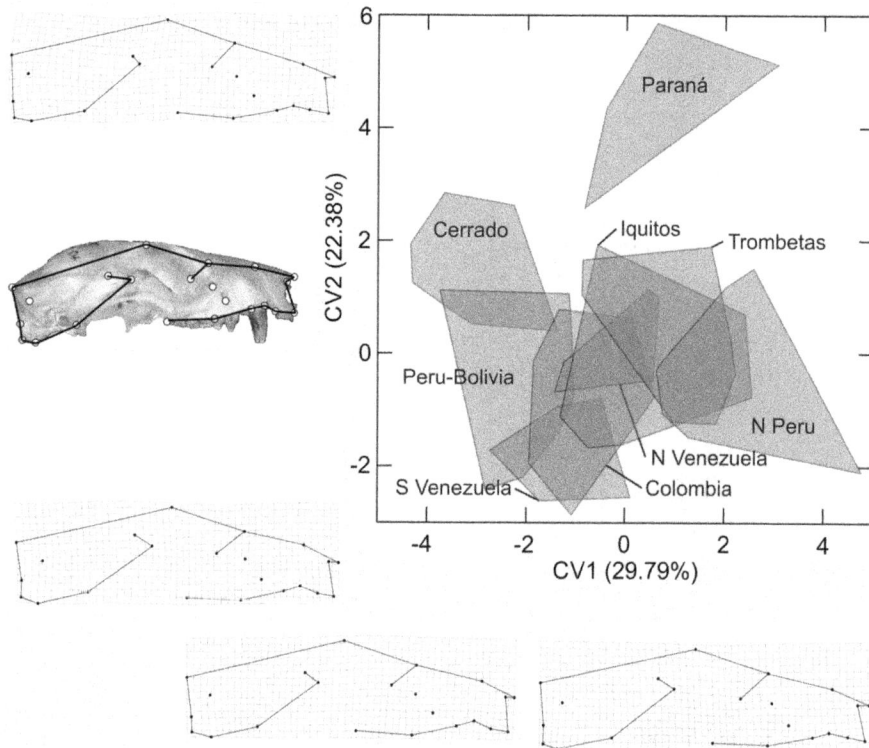

Figure 7. Canonical Variates Analysis on shape variables (partial warps and uniform components) of the cranium in lateral view of *Caluromys lanatus*, using localities as grouping factors, and percentage of variance explained by the first two CVs. Only the convex hulls for each population are shown. Grids indicate deformation associated with the extremes of each CV, from a multivariate regression of shape variables onto CV scores.

Potential implications for the taxonomic classification of *Caluromys derbianus* and *Caluromys lanatus*

The similar skull morphology shared by all populations of *Caluromys derbianus* suggest that the seven subspecies – *C. derbianus aztecus*, *C. d. canutus*, *C. d. centralis*, *C. d. derbianus*, *C. d. fervidus*, *C. d. nauticus* and *C. d. parvidus* – may be considered a unique species on morphometric grounds. Likewise, the lack of geographic variation among the Amazon and Cerrado populations of *Caluromys lanatus* suggest that three of the four subspecies recognized by CABRERA (1958) – *C. lanatus cicur*, *C. lanatus ornatus* and *C. lanatus ochropus* – and four of the six suggested by GARDNER (2008) – *C. lanatus cicur*, *C. lanatus ornatus*, *C. lanatus ochropus* and *C. lanatus vitalinus* can be lumped based on morphometrical data. The geographic variation found in skull morphometric data of individuals from the southern Atlantic Forest also suggest that two subspecies proposed by GARDNER (2008) – *C. lanatus lanatus* and *C. lanatus vitalinus* from southern Brazil can also be lumped.

All these subspecies were described based on external morphological characters, such as body, facial, dorsal, caudal or feet color, characters that usually present geographic variation (THOMAS 1899, 1913, ALLEN 1904, HOLLISTER 1914, GOODWIN

1942). *Caluromys* species were first described based on morphological characters of a single or a few individuals; subspecies were generally described after comparing individual variation with the holotype. For this reason, it cannot ruled out that these subspecies were based on individual variation. In all cases, pending a proper extensive review of coat color or other morphological variation in *Caluromys*, our extensive and quantitative results do not support separation of these taxa.

However, because phenotype is mainly the expression of the underlying genotype, morphological divergence is often interpreted as evidence of specific status. In didelphids, for example, morphological evidence has been used to support splitting of the black-eared and the white-eared opossums (CERQUEIRA & LEMOS 2000, LEMOS & CERQUEIRA 2002) of the genus *Didelphis*, and Bolivian species of *Marmosops* (VOSS et al. 2004). As such, it it possible that the morphologic groups found here may represent distinct species (see, however LÓSS et al. 2011, for a situation where morphologic differentiation does not coincide with species limits). The recognition of southern and southeastern populations of South American didelphids as distinct species appears to be a recurrent pattern that emerges after a deeper analysis of the existing variation, such as in *Phi-*

lander (PATTON & DA SILVA 1997) and *Marmosa* (PATTON & COSTA 2003). Such changes are actually the reflection of our still incomplete knowledge on the taxonomy and systematics of didelphids.

A proper and definite appraisal of the taxonomic status of both woolly opossums would require an integrative approach (including other phenotypical and genetic characters) to unveil their actual status. Especially among *Caluromys lanatus* populations, a molecular approach may be useful to assess if these divergent groups constitute distinct evolutionary lineages that would ultimately validate their status as distinct species. Pending this, we suggest that the available name *Caluromys derbianus* (Waterhouse, 1841) is maintained for all individuals across the geographic distribution of its populations. The name *Caluromys lanatus* (Olfers, 1818) should also be considered valid, with at least two distinct morphometric units, namely *Caluromys lanatus ochropus*, representing Amazon and Cerrado populations, and *Caluromys lanatus lanatus*, encompassing Atlantic forest individuals.

ACKNOWLEDGEMENTS

We are grateful to the following curators and/or collection managers for support and access to the specimens under their care: R. Voss (AMNH), J.L. Patton and C. Conroy (MVZ), J. Salazar-Bravo and W. Gannon (MSB), R. Timm (KU), B. Patterson and M. Schulenberg (FMNH), A. Gardner, L. Gordon and C. Ludwig (USNM), L. Costa, Y. Leite and B. Andrade (UFMG), J.A. Oliveira, L.F. Oliveira, L. Salles and S. Franco (MN), M. de Vivo and J. Barros (MZUSP), S. Aguiar (MPEG), V. Pacheco and E. Vivar-Pinares (MUSM), G. Tebet e T.C. Margarido (MHNCI), and M. Hafner (LSUMZ). RF was supported with a M.Sc. fellowship from CNPq, and funds for travel and equipment to DA were available by FAPESP, FACEPE and an American Society of Mammalogists Grants-in-Aid. DA is currently supported by a CNPq fellowship (306647/2013-3). We are grateful to M. de Vivo, L.P. Costa, M. Montes, I. Bandeira, P. Pilatti, R. Carvalho, A. Cardini, N. Caceres and one anonymous reviewer for their helpful suggestions in the numerous versions of this manuscript.

LITERATURE CITED

ADAMS DC, ROHLF FJ, SLICE DE (2004) Geometric morphometrics: ten years of progress following the 'revolution'. **Italian Journal of Zoology 71**(1): 5-16. doi: 10.1080/11250000409356545

ADAMS DC, ROHLF FJ, SLICE DE (2013) A field comes of age: geometric morphometrics in the 21st century. **Hystrix, the Italian Jornal of Mammalogy 24**(1): 7-14. doi: 10.4404/hystrix-24.1-6283

AB'SABER AN (2000) Spaces occupied by the expansion of dry climates in South America during the Quaternary ice ages. **Revista do Instituto Geológico 21**(1/2): 71-78.

ALLEN JA (1904) Mammals from southern Mexico and Central and South America. **Bulletin American Museum of Natural History 20**(4): 29-80.

ASTÚA D (2010) Cranial sexual dimorphism in New World marsupials and a test of Rensch's rule in Didelphidae. **Journal of Mammalogy 91**(4): 1011-1024. doi: 10.1644/09-MAMM-A-018.1

ASTÚA D, LEINER NO (2008) Tooth eruption sequence and replacement pattern in woolly opossums, genus *Caluromys* (Didelphimorphia: Didelphidae). **Journal of Mammalogy 89**(1): 244-251. doi: 10.1644/06-MAMM-A-434.1

BEHLING H (2002) South and southeast Brazilian grasslands during Late Quaternary times: a synthesis. **Palaeogeography, Palaeoclimatology, Palaeoecology 177**: 19-27. doi: 10.1016/S0031-0182(01)00349-2

BOLETTA PE, RAVELLO AC, PLANCHUELO AM, GRILLI M (2006) Assessing deforestation in the Argentine Chaco. **Forest Ecology and Management 228**: 108-114. doi: 10.1016/j.foreco.2006.02.045

BRAUN JK, VAN DEN BUSSCHE RA, MORTON PK, MARES MA (2005) Phylogenetic and biogeographic relationships of mouse opossums *Thylamys* (Didelphimorphia, Didelphidae) in southern South America. **Journal of Mammalogy 86**(1): 147-159. Available online at: http://jmammal.oxfordjournals.org/content/86/1/147 [Accessed: 20 April 2015]

BUCHER JE, HOFFMANN RS (1980) *Caluromys derbianus*. **Mammalian Species 140**: 1-4.

BUSH MB, DE OLIVEIRA PE, COLINVAUX PA, MILLER MC, MORENO JE (2004) Amazonian paleoecological histories: one hill, three watersheds. **Palaeogeography, Palaeoclimatology, Palaeoecology 214**(4): 359-393. doi: 10.1016/j.palaeo.2004.07.031

CABRERA A (1958) Catalogo de los mamiferos de America del Sur. **Revista del Museo Argentino de Ciencias Naturales "Bernardino Rivadavia", Ciencias Zoológicas 4**(1): 1-46.

CÁCERES NC, CARMIGNOTTO AP (2006) *Caluromys lanatus*. **Mammalian Species 803**: 1-6. doi: 10.1644/803.1

CARDINI A, JANSSON A, ELTON S (2007) A geometric morphometric approach to the study of ecogeographical variation in vervet monkeys. **Journal of Biogeography 34**: 663-678. doi: 10.1111/j.1365-2699.2007.01731.x

CARDOSO JMD, BATES JM (2002) Biogeographic patterns and conservation in the South American Cerrado: A tropical Savanna spot. **BioScience 52**(3): 225-233. doi http://dx.doi.org/10.1641/0006-3568(2002)052[0225:BPACIT]2.0.CO;2

CASTOE TA, DAZA JM, SMITH ER, SASA MM, KUCH U, CAMPBELL JA, CHIPPINDALE PT, PARKINSON CL (2009) Comparative phylogeography of pitvipers suggests a consensus of ancient Middle American highland biogeography. **Journal of Biogeography 36**: 88-103.doi: 10.1111/j.1365-2699.2008.01991.x

CERQUEIRA R (1982) South American landscapes and their mammals, p. 539-539. In: MARES MA, GENOWAYS HH (Eds.). **Mammalian Biology in South American.** Pittsburg, University of Pittsburg, Special Publication Pymatuning Laboratory of Ecology.

CERQUEIRA R, LEMOS B (2000) Morphometric differentiation between Neotropical black-eared opossums, *Didelphis marsupialis* and *D. aurita* (Didelphimorphia, Didelphidae). **Mammalia 64**: 319-327. doi: 10.1515/mamm.2000.64.3.319

COSTA LP (2003) The historical bridge between the Amazon and the Atlantic Forest of Brazil: a study of molecular phylogeography with small mammals. **Journal of Biogeography 30**(1): 71-86. doi: 10.2307/827350

DE OLIVEIRA PE, BEHLING H, LEDRU M-P, BARBERI M, BUSH M, SALGA-DO-LABOURIAU ML, MEDEANIC S, BARTH OM, BARROS MA, SCHELL-YHERT R (2005) Paleovegetação e paleoclimas do Quaternário do Brasil, p. 52-74. In: SOUZA CRG, SUGUIO K, OLIVEIRA AMS, DE OLIVEIRA PE (Ed.). **Quaternário do Brasil**. São Paulo, Holos, Associação Brasileira de Estudos do Quaternário.

EMMONS L (1984) Geographic variation in densities and diversities of non-flying mammals in Amazonia. **Biotropica 16**(3): 210-222. doi: 10.2307/2388054

FALCONER DS, MACKAY TFC (1996) **Introduction to Quantitative Genetics**. Harlow, Longman.

GARDNER AL (2005) Order Didelphimorphia, p. 3-18. In: WILSON DE, REEDER DM (Ed.). **Mammal species of the world: a taxonomic and geographic reference**. Baltimore, Johns Hopkins University Press, 3rd ed.

GARDNER AL (2008) **Mammals of South America**. Chicago, The University of Chicago Press.

GOODWIN GG (1942) Mammals of Honduras. **Bulletin American Museum of Natural History 79**: 107-195.

GOULD SJ, JOHNSTON RF (1972) Geographic variation. **Annual Review of Ecology and Systematics 3**: 457-499. doi: 10.1146/annurev.es.03.110172.002325

HAFFER J (1969) Speciation in Amazonian forest birds. **Science 165**: 131-137. doi: 10.1126/science.165.3889.131

HALLGRÍMSSON B, HALL BK (2005) **Variation. A Central Concept in Biology**. Elsevier Academic Press.

HIMES CMT, GALLARDO MH, KENAGY GJ (2008) Historical biogeography and post-glacial recolonization of South American temperate rain forest by the relictual marsupial *Dromiciops gliroides* **Journal of Biogeography 35**: 1415-1424. doi: 10.1111/j.1365-2699.2008.01895.x

HOLLISTER N (1914) Four new mammals from tropical America. **Proceedings of the Biological Society of Washington 27**: 103-106.

JOHNSON MA, SARAIVA PM, COELHO D (1999) The role of gallery forests in the ditribution of Cerrado mammals. **Revista Brasileira de Biologia 59**(3): 421-427. doi: 10.1590/S0034-71081999000300006

LEDRU M-P (1993) Late Quaternary environment and climatic changes in Central Brazil. **Quaternary Research 39**: 90-98. doi: 10.1006/qres.1993.1011

LEDRU M-P, SALGADO-LABOURIAUB ML, LORSCHEITTERC ML (1998) Vegetation dynamics in southern and central Brazil during the last 10,000yr. B.P. **Review of Palaeobotany and Palynology 99**: 131-142. doi: 10.1016/S0034-6667(97)00049-3

LEDRU MP (2002) Late Quaternary history and evolution of the Cerrados as revealed by palynological records, p. 33-50. In: OLIVEIRA PS, MARQUIS RJ (Eds.). **The Cerrados of Brazil: ecology and natural history of a Neotropical Savanna**. New York, Columbia University Press.

LEMOS B, CERQUEIRA R (2002) Morphological differentiation in the white-eared opossum group (Didelphidae: *Didelphis*). **Journal of Mammalogy 83**(2): 354-369. Available online at: http://jmammal.oxfordjournals.org/content/83/2/354 [Accessed: 20 April 2015]

LESSA G, GONÇALVES PR, PESSÔA LM (2005) Variação geográfica em caracteres cranianos quantitativos de *Kerodon rupestris* (Wied, 1820) (Rodentia, Caviidae). **Arquivos do Museu Nacional 63**(1): 75-88.

LIRA PK, FERNANDEZ FAS, CARLOS HSA, CURZIO PL (2007) Use of fragmented landscape by three species of opossum in southeastern Brazil. **Journal of Tropical Ecology 23**: 427-435. doi: 10.1017/S0266467407004142

LOPEZ-FUSTER MJ, PEREZ-HERNANDEZ R, VENTURA J (2008) Morphometrics of genus *Caluromys* (Didelphimorphia: Didelphidae) in northern South America. **Orsis 23**: 97-114.

LÓPEZ-FUSTER MJ, PÉREZ-HERNANDEZ R, VENTURA J, SALAZAR M (2000) Effect of environment on skull-size variation in *Marmosa robinsoni* in Venezuela. **Journal of Mammalogy 81**(3): 829. doi: 10.1093/jmammal/81.3.829

LÓPEZ-FUSTER MJ, SALAZAR M, PÉREZ-HERNÁNDEZ R, VENTURA J (2002) Craniometrics of the orange mouse opossum *Marmosa xerophila* (Didelphimorphia: Didelphidae) in Venezuela. **Acta Theriologica 47**(2): 201-209. doi: 10.1007/BF03192460

LÓSS S, COSTA LP, LEITE YLR (2011) Geographic variation, phylogeny and systematic status of *Gracilinanus microtarsus* (Mammalia: Didelphimorphia: Didelphidae). **Zootaxa 2761**: 1-33.

MACÊDO RH, MARES MA (1987) Geographic variation in the South American cricetine rodent *Bolomys lasiurus*. **Journal of Mammalogy 68**(3): 578-594. doi: 10.2307/1381594

MARCO D, PÁEZW SA (2002) Phenology and phylogeny of animal-dispersed plants in a Dry Chaco forest (Argentina). **Journal of Arid Environments 52**: 1-16. doi: 10.1006/jare.2002.0976

MAYR E (1977) **Populações, espécies e evolução**. São Paulo, Companhia Editora Nacional, Editora da Universidade de São Paulo.

OLSON DM, DINERSTEIN E, WIKRAMANAYAKE ED, BURGESS ND, POWELL GVN, D'AMICO JA, ITOUA I, STRAND HE, MORRISON JC, LOUCKS CJ, ALLNUTT TF, RICKETTS TH, KURA Y, LAMOREUX JF, WETTENGEL WW, HEDAO P, KASSEM KR (2001) Terrestrial ecoregions of the World: a new map of life on Earth. **BioScience 51**(11): 933-938. doi: http://dx.doi.org/10.1641/0006-3568(2001)051[0933:TEOTWA]2.0.CO;2

OLIFIERS N, VIEIRA MV, GRELLE CEV (2004) Geographic range and body size in Neotropical marsupials. **Global Ecology and Biogeography 13**(5): 439-444. doi: 10.1111/j.1466-822X.2004.00115.x

Patton JL, Costa LP (2003) Molecular phylogeography and species limits in rainforest didelphid marsupials of South America, p. 63-81. In: Jones ME, Dickman CR, Archer M (Eds.). **Predators with pouches: The Biology of Carnivorous Marsupials.** Collingwood, CSIRO Publishing.

Patton JL, Da Silva MNF, Malcom JR (2000) Mammals of the Rio Juruá and the evolutionary and ecological diversification of Amazonia. **Bulletin of the American Museum of Natural History 244:** 1-306.

Patton JL, Da Silva MNF (1997) Definition of species of pouched four-eyed opossums (Didelphidae, *Philander*). **Journal of Mammalogy** 78(1): 90. doi: 10.2307/1382642

Paynter Jr RA (1982) **Ornithological Gazetteer of Venezuela.** Cambridge, Museum of Comparative Biology, Harvard University.

Paynter Jr RA (1989) **Ornithological Gazetteer of Paraguay.** Cambridge, Museum of Comparative Biology, Harvard University.

Paynter Jr RA (1992) **Ornithological Gazetteer of Bolivia.** Cambridge, Museum of Comparative Biology, Harvard University.

Paynter Jr RA (1993) **Ornithological Gazetteer of Ecuador.** Cambridge, Museum of Comparative Biology, Harvard University.

Paynter Jr RA (1995) **Ornithological Gazetteer of Argentina.** Cambridge, Museum of Comparative Biology, Harvard University.

Paynter Jr RA (1997) **Ornithological Gazetteer of Colombia.** Cambridge, Museum of Comparative Biology, Harvard University.

Pennington RT, Prado DE, Pendry CA (2000) Neotropical Seasonally Dry Forests and Quaternary vegetation changes. **Journal of Biogeography** 27(2): 261-273. doi: 10.1046/j.1365-2699.2000.00397.x

Pérez-Emán JL (2005) Molecular phylogenetics and biogeography of the Neotropical redstarts (*Myoborus*; Aves, Parulinae). **Molecular Phylogenetics and Evolution** 37: 511-528. doi: 10.1016/j.ympev.2005.04.013

Pires AS, Lira PK, Fernandez FAS, Schittini GM, Oliveira LC (2002) Frequency of movements of small mammals among Atlantic Coastal Forest fragments in Brazil. **Biological Conservation** 108(2): 229-237. doi: 10.1016/S0006-3207(02)00109-X

Rohlf FJ (2006) **TpsDig.** Stony Brook, Department of Ecology and Evolution, State University of New York.

Rohlf FJ (2011) **TpsRegr.** Stony Brook, Department of Ecology and Evolution, State University of New York.

Rohlf FJ, Slice D (1990) Extensions of the Procrustes method for the optimal superimposition of landmarks. **Systematic Zoology** 39: 40-59. doi: 10.2307/2992207

Salgado-Labouriau ML, Barberi M, Ferraz-Vicentini KR, Parizzi MG (1998) A dry climatic event during the late Quaternary of tropical Brazil. **Review of Palaeobotany and Palynology** 99: 115-129. doi: 10.1016/S0034-6667(97)00045-6

Salgado-Labouriau ML, Casseti V, Ferraz-Vicentini KR, Martin L, Soubiés F, Suguio K, Turcq B (1997) Late Quaternary vegetacional and climatic changes in Cerrado and palm swamp from Central Brazil. **Palaeogeography, Palaeoclimatology, Palaeoecology** 128: 215-226. doi: 10.1016/S0031-0182(96)00018-1

Savage JM (1987) Systematics and distribution of the Mexican and Central American rainfrogs of the *Eleutherodactyllus gollmeri* group (Amphibia: Leptodactylidae). **Fieldiana, Zoology,** 33: 1-57.

Steiner C, Catzeflis FM (2004) Genetic variation and geographical structure of five mouse-sized opossums (Marsupialia, Didelphidae) throughout the Guiana Region. **Journal of Biogeography** 31(6): 959-973. doi: 10.1111/j.1365-2699.2004.01102.x

Storz JF, Balansingh J, Bhat HR, Nathan PT, Doss DPS, Prakash AA, Kunz TH (2001) Clinal variation in body size and sexual dimorphism in an India fruit bat, *Cynopterus sphinx*, (Chiroptera, Pteropodidae). **Biological Journal of Linnean Society** 72: 17-31. doi: 10.1111/j.1095-8312.2001.tb01298.x

Thomas O (1899) Descriptions of new Neotropical mammals. **Annals and Magazine of Natural History** 4(7): 278-288. doi: 10.1080/00222939908678198

Thomas O (1913) The geographical races of the woolly opossum (*Philander laniger*). **Annals and Magazine of Natural History** 12(8): 358-361. doi: 10.1080/00222931308693409

Thorpe RS (1987) Geographic variation: a synthesis of cause, data, pattern and congruence in relation to subspecies, multivariate analysis and phylogenesis. **Bolletino di Zoologia** 54: 3-11. doi: 10.1080/11250008709355549

Tribe CJ (1990) Dental age classes in *Marmosa incana* and other didelphoids. **Journal of Mammalogy** 71(4): 566-569. doi: 10.2307/1381795

Van Der Hammen T, Hooghiemstra H (2000) Neogene and Quaternary history of vegetation, climate, and plant diversity in Amazonia. **Quaternary Science Reviews** 19: 725-742. doi: 10.1016/S0277-3791(99)00024-4

Voss R, Jansa S (2009) Phylogenetic relationships and classification of didelphid marsupials, an extant radiation of New World metatherian mammals. **Bulletin American Museum of Natural History** (322): 1-177. doi: 10.1206/322.1

Voss RS, Tarifa T, Yensen E (2004) An introduction to *Marmosops* (Marsupialia: Didelphidae), with the description of a new species from Bolivia and notes on the taxonomy and distribution of other Bolivian forms. **American Museum Novitates 3466:** 1-40. doi: 10.1206/0003-0082(2004)466<0001:AITMMD>2.0.CO;2

Webster M, Sheets HD (2010) A practical introduction to landmark-based geometric morphometrics. p. 163-188. In: Alroy J, Hunt G (Eds). **Quantitative methods in Paleobiology.** Boulder, The Paleontological Society, Paleontological Society Papers, vol. 16.

Zelditch ML, Swiderski DL, Sheets HD, Fink WL (2012) **Geometric morphometrics for biologists: a primer.** Boston, Elsevier Academic Press.

Appendix 1. Definition of landmarks illustrated in Fig. 1.

Dorsal view of the cranium. 1: Anteriormost point of suture between left and right nasal bones; 2: Posteriormost point of interparietal at the sagital and nuchal crests intersection; 3 and 28: Intersection between interparietal-parietal suture and outline of the brain-case, at the nuchal crest; 4 and 27: Main curve of the squamosal, anteriorly to the post-tympanic process; 5 and 26: Tip of the frontal process of the jugal, on the zygomatic arch; 6 and 23: Lateralmost point of sutures between lacrimal and jugal; 7 and 18: Lateralmost point of sutures between maxilla and premaxilla; 8 and 17: Anteriormost point of suture between nasal and premaxilla; 9 and 19: Point of intersection between sutures of nasal, premaxilla and maxilla; 10 and 20: Intersection between sutures of nasal, frontal and maxilla; 11 and 21: Intersection of sutures between lacrimal, frontal and maxilla; 12 and 22: Posteriormost point of the suture between frontal and lacrimal; 13 and 25: Tip of the orbital process of the frontal; 14 and 24: Postorbital constriction; 15: Intersection between sutures of both parietals and interparietal; 16: Posteriormost point of sutures between both nasals.

Lateral view of the cranium. 1: Anterior base of I1; 2: Posterior base of I5; 3: Anterior base of C, at the junction with maxilla; 4: Posterior base of C, at the junction with maxilla; 5: Anterior base of M1 and posterior base of P3, at the junction with maxilla; 6: Posterior base of M4 at the juction with maxila (posteriormost point of molar series); 7: Posteroventral end of occipital condyle; 8: Posterodorsal end of braincase (posteriormost point of sagital line, junction with nuchal crest); 9: Intersection between sutures of infraparietal, parietal and squamosal; 10: Suture between jugal and squamosal at the dorsal border of the zygomatic arch; 11: Intersection between sutures of jugal, lacrimal and maxilla; 12: Intersection between sutures of nasal, premaxilla and maxilla; 13: Anteriormost point of the sutures of nasal and premaxilla; 14: Anterior tip of nasal; 15: Ventral end oftheinfraorbital fossa; 16: Intersection between exoccipital and occipital condyle; 17: Ventral end of occipital condyle; 18: Tip of postglenoid process; 19: Tip of orbital process of frontal; 20: Intersection of sutures between lacrimal, frontal and palate; 21: Anteriormost point of suture between jugal and squamosal; 22: Intersection of sutures between lacrimal, frontal and maxilla.

Ventral view of the cranium. 1: Point between right and left I1; 2: Anteriormost point of foramem magnum, at the basioccipital; 3 and 28: Posterior end of occipital condyle, at the basioccipital; 4 and 27: Exterior border of braincase, anterior to the posttympanic process; 5 and 25: Sutures between basiocciptal, basephenoid and promontorium; 6 and 24: Posterolateral end of sutures between palateandpterigoyd; 7 and 23: Posterolateral tip of palate; 8 and 26: Anterior base of squamosal process; 9 and 22: Posterolateral base of M4; 10 and 21: Posterolateralbase or M3; 11 and 20: Posterolateral base of C; 12 and 19: Posterolateral base of I5; 13 and 18: Anterior end of incisive foramen; 14 and 17: Posterior end of incisive foramen; 15: Posterior end of suture between palates; 16: Intersection of sutures between maxilla and palate.

Mandible. 1: Anterior base of i1; 2: Anterior base of i4; 3: Anterior base of p1; 4: Anterior base of m1; 5: Posterior base of m4; 6: Intersection between horizontal ramus of the mandible and coronoid process; 7: Uppermost point of coronoid process; 8: Posterior tip of coronoid process; 9: Major curvature between articular process and posterior part of coronoid crest; 10: Labial tip of articular condyle; 11: Posterior base of angular process; 12: Caudal tip of angular process; 13: Upper end part of mental foramen; 14: Anteroventral end of masseteric fossa.

Appendix 2. Specimens examined, by country and locality. Numbers refer to Figs. 2 and 3.

Caluromys derbianus

Belize. 7. Baking Pot (88W55′12″; 16S49′48″) FMNH 106529; **8**. Kate'sLagoon (88W27′36″; 17S58′48″) FMNH 63886.

Colombia. 67. Unguia (77W; 6N) FMNH 69800, 69801, 69802, 69803, 69804; **70**. Cauquita River, South of Cali (76W31′12″; 3S25′12″) AMNH 14189; **68**. Alto Rio Sinú (74W01′12″; 8S09′) FMNH 69327; **69**. Rio Raposo (73W40′48″; 4S46′12″) USNM 334676, 334678.

Costa Rica. 28. Escazu (85W19′48″; 9S55′12″) AMNH 131708, 131710, 131711, 131712, 135329, 137287, 139278; **29**. Piedras Negras (84W19′12″; 9S54′) AMNH 139781, 139783; **30**. Finca La Lola (84W16′48″; 9S54′) LSUMZ 9337; **31**. 2 km NWSanta Ana (84W10′48″; 9S55′48″) LSUMZ 12633; **32**. San Ignacio (84W10′12″; 9S49′12″) USNM 250280; **33**. San Jose (84W06′; 9S55′48″) AMNH 19202, 131709, KU 39247, 60447; **35**. 5 km SE Turrialba (83W40′48″; 9S54′) KU 26927; **37**. Cerro Plano (83W19′48″; 9N) KU 157578, 157579; **38**. Puerto Cortez (83W19′12″; 9S01′12″) AMNH 10057, 139678; **34**. La Selva Biologica Reserve, 35 km S Puerto Viejo, Heredia (83W50′; 10S26′) FMNH 128385; **36**. San Isidro, San Jose (84W17′; 9S54′). **Ecuador. 73**. Vinces (79W43′48″; 1S33′) AMNH 63526; **74**. Zaruma (79W36′; 3S40′48″) AMNH 47194; **72**. Puente delChimbo (78W43′48″; 2S01′12″) AMNH 63525; **71**. Inaza Range (78W10′48″; 1S49′12″) AMNH 10058.

Honduras. 9. Santa Barbara (88W24′; 15S07′12″) AMNH 126134; **10**. Chamelecon (88W; 15S25′12″) USNM 148749; **11**. Olancho (85W45′; 14S48′) AMNH 126980.

Mexico. 1. 20 km ESE San Jesus Carranza (96W07′12″; 19S10′12″) KU 93192; **2**. 3 km SE San AndresTuxtla (95W13′12″; 18S27′) KU 23367, 23368, 23369, 23370, 23371, 23372, 23373;**3**. 16 mi. SMatias Romero, Sarabia, Juchitán (95W01′12″; 16S52′12″) AMNH 185756; **4**. La Venta (94W01′48″; 18S04′48″) USNM 271105, 271106; **5**. 1 mi. E Teapa (92W57′; 17S31′48″) LSUMZ 8105; **6**. Mayan Ruiz (91W58′12″; 17S30′) FMNH 66918.

Nicaragua. 15. Chinandega (87W07'12"; 12S37'12") KU 110661, 105904; **16.** Lake Jiloa (86W31'48"; 12S22'12") AMNH 176710, 176711, 176712, 176714, 176715; **17.** 3 km S4 km WDiriamba (86W19'48"; 11S46'48") KU 110681, 110675, 110679, 114606;**12.** 5 mi. S, Managuá (86W16'48"; 13S30') KU 70160, 70161, 70162, USNM 253050;**13.** 5 km SSabana Grande (86W10'12"; 13S10'12") KU114604, 114605, 96201, 96203, 96208, 97359, 97360, 97361, 97367, 97381, 97382, 97388, 98379, 114603, 116700, 116701, 96209, 96213, 97362, 97365, 97369, 97376, 97377, 97383; **18.** La Calera (86W03'; 11S45') KU 108167,104503,96200, USNM 339889, 339892, 339893; **19.** Chinandega (86W01'12"; 11S51) KU 110661; **20.** Los Cocos, 14 km S Boaco (85W54'; 12S04'12") KU 114597, 114592, 114598, 114599; **21.** Finca Santa Cecilia, 6,5 km NE Guanacaste (85W49'48"; 11S25'48") KU 105906, 105907; **22.** Rivas (85W49'12"; 11S25'48") KU 97389, 105908; **23.** Rio Mico (85W48'; 12S27') KU 105901; **24.** 4 km WTeustepe (85W46'48"; 12S25'12") KU 114591; **25.** Santa Rosa, 17 km S15 km E Boaco (85W40'12"; 12S28'12") KU 110682,110684,110685;**14.** Matagalpa (85W40'12"; 12S55'12") AMNH 28831, 41395, KU 70156, 70157, 114575, 114576, 114580, 114585; **26.** 12 km S13 km E Boaco (85W39'; 12S28'12") KU 114590; **27.** Mecatepe (85W37'48"; 11S15') KU 108165, 108166.

Panama. 43. 7 km SSWChanguinola (82W31'12"; 9S25'48") USNM 315012; **44.** Almirante (82W22'48"; 9S16'48") USNM 315009; **45.** Isla Parida (82W19'48"; 9S10'12") AMNH 18911, 18912; **46.** Bocas del Drago (82W19'48"; 9S25'12") USNM 315011; Divala (82W19'12"; 8S22'48") USNM 243413; **39.** Boquerón (82W19'12"; 8S24') AMNH 18909,18910; **47.** Bocas del Toro (82W10'12"; 8S49'48") USNM 290878, 322943, 322944, 335004, 335005, 335009, 335010, 335011, 335012, 335013, 335014, 335017, 335019, 335020, 449560, 449562, 464247, 578118, 578119, 578934, 578935, 578936, 578939, 578940, 578941, 578942, 578944, 578945, 578946, 578947, 578948, 578950, 578951, 578953, 578954, 578955, 578956, 578957; **48.** Sibube (82W04'12"; 9S03') USNM 335001, 335003; **49.** Cayo Agua (82W01'12"; 9S09') USNM 578116, 335018; **50.** Bisira (81W51'; 8S54') USNM 575393; **51.** Bohio Peninsula, 4,5 km NWFrijoles (81W46'48"; 8S43'12") USNM 503420; **40.** 1/4 mi. WGuabalá (81W43'12"; 8S13'12") USNM 331068;**41.** Isla Cébaco (81W19'48"; 8S12') USNM 360134, 360135, 360136; **42.** La Cascadas (80W46'12"; 8S31'48") USNM 257328; **53.** Fort Sherman, 6 km WCristobal (79W57'; 9S19'48") USNM 456809; **52.** Camp Pina (79W57'; 9S22'12") USNM 306379; **54.** Fort Davis (79W54'; 9S16'48") USNM 297876; **55.**Tabernilla (79W49'12"; 9S07'12") USNM 171033; **56.**Darién (79W46'12"; 9S07'12") USNM 309256, 309257, 309258, 337951, 337952, 337953, 362315, 362316; **57.**Fort Clayton (79W42'; 9N) USNM 302329; **58.**Chagres River station (79W39'; 9S09') AMNH 164491; **59.**Fort Kobbe (79W34'48"; 8S55'12") USNM 301131, 301133, 301134; **60.** Chiva-Chiva (79W34'48";9S01'48") USNM 296344; **61.** Curundu (79W33'; 8S58'48") USNM 296188; **62.** Canal Zone (79W31'48"; 8S58'12") MVZ 183321, 183319, FMNH 30279; **63.** Panama City (79W28'48"; 9S01'12") MVZ 135231, 135233; **64.** France Field (79W04'48"; 9S19'48") USNM 303233; **65.** Jaqué (77W43'48"; 8S07'12") USNM 309256, 309257, 309258, 337951, 337952, 337953, 362315, 362316; **66.** Quebrada Venado (77W28'12"; 8S39') USNM 335021, 335026, 335023, 335024. **Peru. 75.** NE Tingo Maria (75W58'48"; 9S16'48") LSUMZ 17681.

Caluromys lanatus

Argentina. 73. Parque Iguazu (55W; 27S) MACS 21378.

Bolivia. 55. Isla Gargantua (68W34'48"; 12S22'48") MSB 56998; **66.** San Joaquín (64W49'12"; 13S04'12") FMNH 114649; **67.**Estancia Yutiole, 20 km S San Joaquín (64W48'; 13S15') AMNH 215001; **68.**Ichilo (63W46'12"; 17S30') MACS 50181, 50188; **69.**Buena Vista (63W40'12"; 17S27') FMNH 25265; **70.** Santa Cruz de La Sierra (63W10'12"; 17S48') AMNH 133205.

Brazil. 47. Nova Vida, Right bank Juruá river, Acre (72W49'12"; 8S22'12") MVZ 190250, 190251; **48.** Igarapé Porongaba, left bank Juruá river; Acre (72W46'48"; 8S40'12") MVZ 190249; **25.** right bank Juruá river, Amazonas (70W51'; 6S45') MVZ 190247; **24.** Altamira, rightbank, Juruá river, Amazonas (68W54'; 6S34'48") MPEG 28000; **23.** Niteroi, 2o Distrito, Acre (68W24'; 9S02'24") USNM 546177; **22.** Igarapé Grande, Juruá river, Amazonas (67W27'; 9S15') MZUSP 4532; **56.** Estação Ecológica Mamirauá, Japurá river, leftbank, Amazonas (64W25'12"; 3S13'12") MPEG 24566; **71.** UHE Samuel, Rondônia (63W; 11S) MZUSP 27389, 27390; **57.** Balbina, Amazonas (59W16'48"; 1S31'48") MHNCI 1727, 1728; ca. 8 km SLago Sampaio, Wbank Madeira river, Amazonas (59W04'48"; 3S25'12") AMNH 92760; **62.** Jauru river, 2days upper Porto Esperidião, Porto Esperidião, Mato Grosso (57W27'36"; 16S13'12") MS1222; **59.** Villa Bella Imperatriz, Amazonas (56W26'24"; 2S21'36") AMNH 92882, 92883, 92884, 93967; **60.** Rightbank Tapajós river, Pará (54W24'36"; 2S14'24") AMNH 133208; **74.** Flor da Serra, Boa Vista da Aparecida, Paraná (53W24'; 25S25'48") MHNCI 4206, 4207, 4208, 4209; **63.** Baixo Kuluene, Jacaré, Alto Xingu, São Félix do Araguaia, Mato Grosso (53W10'48"; 11S13'48") MS11705; **75.** UH Salto Caxias, Cruzeiro do Iguaçu, Paraná (53W07'48"; 25S37'12") MHNCI 4210, 4211; **64.** Fazenda São Luis, 30 km SBarra do Garças, Mato Grosso (52W09'; 15S31'48") UFMG 2538; **61.** Ilha Boiuçu, Pará (55W27'; 01S55') MZUSP 4531, 4533, 4534, 4883; **65.** Anápolis, Goiás (48W34'48"; 16S12') AMNH 133200, MS20963, 4599, 4782, 4785.

Colombia. 12. Valle de Suaza (76W10'12"; 1S45') FMNH 70994, USNM 541855, 541856; **13.** 5 km SVillavieja, Huila (75W10'12"; 3S19'12") MVZ 114227, 113831, 114223; **14.** Natagaima (75W04'48"; 3S34'48") AMNH 75886, 76768, 76769; **1.** Magdalena (74W30'; 10N) USNM 271317, 280900, 280906; **15.** Cundinamarca (74W25'48"; 4S58'12") USNM 544394, 544395; **16.** Boyaca (74W06'; 5S31'48") FMNH 70995, 70996; **17.** VolcanesTupana, Bogotá (74W04'48"; 4S36') AMNH 143522; **18.** La Macarena, Meta (73W55'12"; 2S45') FMNH 87931; **19.** San Juan de Arama, Meta (73W49'12"; 3S24') FMNH 87927; **2.** Valledupar Distr., Magdalena (73W34'48"; 10S25'12") USNM 280903, 280904, 280907; **21.** Restrepo, Meta (73W34'12"; 4S15') AMNH 136161; **3.** San Gil,

Santander (73W15'; 7N) FMNH 140239; **4.** Sarcula, Norte de Santander (73W; 8N) FMNH 140237; **5.** Toledo (72W15'; 7S18') USNM 544393; **20.** Merida (73W48'; 1N) MZUSP 2529, AMNH 78101.

Ecuador. 26. Pastaza (77W; 1S55'12") FMNH 41444, 43176, 43177; **27.** San Jose, Napo (77W; 0S43'48") AMNH 182938; **28.** Santa Maria, Napo (76W55'12"; 0S25'12"), AMNH 68282, FMNH 58952; **29.** Marián, Napo (76W19'12"; 0S31'12") FMNH 124595; **30.** Limon Cocha, Napo (76W09'; 0S25'12") USNM 528318. **Peru. 31.** Bagua Chica (78W37'48"; 5S37'48") LSUMZ 21880; **32.** La Poza, Rio Santiago (77W37'12"; 4S25'12") MVZ 157608, 157611, 157612; **33.** Tarapoto (76W28'12"; 6S30') MUSM 89, 90, LSUMZ 28420; **38.** Huánuco (76W16'12"; 9S27') FMNH 55409; **39.** Tingo Maria (75W49'48"; 9S30') FMNH 24142, MVZ 140041; **40.** Ucayali (75W15'; 7S10'12") FMNH 55502, 62069, 62070; **41.** Oxapampa (75W04'48"; 10S19'48") USNM 364160; **42.** San Ramon (75W; 11S30') MUSM 1303, FMNH 20787, AMNH 71979, 71984; **34.**Loreto (75W; 5S) AMNH 71979, 71983, 71984, 230001, 273038, 273059; **43.**Yarinacocha (74W36'; 8S30') LSUMZ 14024; **35.** Requena (73W58'48"; 4S58'48") MUSM 11024; **44.** Lagarto, Ucayali (73W52'48"; 10S34'48") AMNH 78951; **45.**Llillapichia river, near "Panguana" Biol. St. (73W37'48"; 10S52'12") MUSM 79; **36.**Nauta (73W33'; 4S31'48") FMNH 87134, 122749; **46.**Santa Rosa (73W30'; 9S) AMNH 75912; **37.** Iquitos (73W15'; 3S46'12") FMNH 87130, 87132, 87133; **50.** Cuzco (72W; 13S30') MUSM 13407; **49.** Balta, Curanja river (71W13'12"; 10S07'48") LSUMZ 14025; **51.** Cosñipata (71W10'48"; 13S04'12") FMNH 84245, 84246; **52.** Marcapata (70W58'12"; 13S34'48") FMNH 68333, 68334; **53.** Quince Mil (70W45'; 13S13'12") FMNH 75087, 75088, 75089; **54.** Albergue, Madre de Diós (70W04'48"; 12S36') MVZ 168852.

Paraguay. 72. Villa Rica, Guairá (56W18'; 24S27') AMNH 66780.

Venezuela. 6. 3 km SNula (71W55'12"; 7S16'48") USNM 416932; **7.** Trujillo (70W30'; 9S25'12") USNM 371280; **8.** San Juan (66W04'12"; 5S15') USNM 406875, 406878; **9.** Amazonas (65W46'12"; 3S39') USNM 388327, 380330; **10.** Esmeralda (65W31'48"; 3S10'12") AMNH 76970; **11.** Boca Mavaca, 68 km SE Esmeralda (65W03'; 3S01'12") USNM 388331, 388332, 388333, 388334.

17

Description of the males of *Lincus singularis* and *Lincus incisus* (Hemiptera: Pentatomidae: Discocephalinae)

Aline S. Maciel[1], Thereza de A. Garbelotto[1], Ingrid C. Winter[1],
Talita Roell[1] & Luiz A. Campos[1,2]

[1]*Departamento de Zoologia, Universidade Federal do Rio Grande do Sul. Avenida Bento Gonçalves 9500, Agronomia, 91501-970 Porto Alegre, RS, Brazil.*
[2]*Corresponding author. E-mail: luiz.campos@ufrgs.br*

ABSTRACT. The Neotropical *Lincus* Stål, 1867 includes 35 species, thirteen of which are known only from females. Several species are vectors of *Phytomonas staheli* McGhee & McGhee, 1979, a trypanosomatid parasitic in palm-trees in South America that causes hart-rot, sudden and slow wilt diseases. The hitherto unknown males of *L. singularis* Rolston, 1983 ("swollen head" species group found in the oil palm *Elaeis guineensis* Jacq.), and *L. incisus* Rolston, 1983 ("hatchet-lobed" species group; found in the coconut tree *Cocos nucifera* L.), are described with emphasis on the morphology of the genitalia, and taxonomic remarks are provided. Males of *L. singularis* can be distinguished from other species included in "swollen head" group by their pronotal lobes with anterior and posterior margins subparallel and projected laterally from the eye margin, while males of *L. incisus* can be distinguished from the species of the "swollen head" group by an obtuse projection with a deepest incision and several additional diagnostic characters of the genitalia.

KEY WORDS. Genitalia; morphology; Ochlerini; stink bugs; taxonomy.

Lincus Stål, 1867 is the richest genus of Ochlerini, comprising 35 species (Campos & Grazia 2006). Even though the genus was described in the 19th century (Stål 1867), most of its 25 species were described in the late 20th century (Rolston 1983, 1989, Dolling 1984), and 13 of them are known only from one sex (Rolston 1983). Species of *Lincus* are found mostly in the Amazon region. There are a few exceptions to this, for instance *Lincus lobuliger* Breddin, 1908, recorded from the Brazilian Atlantic Forest and *Lincus anulatus* Rolston, 1983 and *Lincus discessus* (Distant, 1900) from Central America (Rolston 1983). Several species are sympatric in different countries. Geographic records of *Lincus* are particularly rich in Peru due to the extensive surveys on native palms carried out during the 1980's (Couturier & Kahn 1989, 1992, Llosa et al. 1990). Sixteen species occur in that country, including *Lincus singularis* Rolston, 1983, although it has never been collected on *Elaeis guineensis* Jacq. Suriname comes next in terms of species richness in the Amazon region (Rolston 1983, 1989, Dolling 1984), with six species, including *Lincus incisus* Rolston, 1983.

The association of pentatomids with the transmission of *Phytomonas staheli* McGhee & McGhee, 1979, a trypanosomatid parasitic in plants, has been known for a long time and is well documented (for a review see Camargo 1999 and Mitchell 2004). Several species of *Lincus* play a major role as vectors of hart-rot, and of sudden and slow wilt (also called Marchitez sorpresiva in

Spanish-speaking countries) diseases in palm trees (Arecaceae) in South America, being of economic interest in crops of *E. guineensis* (African oil palm) and *Cocos nucifera* L. (coconut) (Desmier de Chenon 1984, Couturier & Kahn 1989, Perthuis et al. 1985, Panizzi et al. 2000, Di Lucca et al. 2013; for a review see Howard 2001). Although eleven species of *Lincus* have been reported on palm trees (Howard 2001), the genus was not listed as a possible vector of oil palm diseases until the 1980's (Couturier & Kahn 1992). Furthermore, transmission of *Phytomonas* trypanosomatids to palms has been documented in only six species, four of which transmit the parasite to *E. guineensis*: *Lincus lethifer* Dolling, 1984, *L. lobuliger*, *Lincus tumidifrons* Rolston, 1983, and *Lincus spurcus* Rolston, 1983 (Camargo 1999, Di Lucca et al. 2013).

In 2009, the corresponding author received, for identification, specimens of *Lincus* collected from *E. guineensis* palm trees from Palmas del Espino S.A., Peru. These specimens were identified as *L. spurcus* and *L. singularis*, and included the only known males of the latter. Moreover, during the course of this study, we located males of *L. incisus* among specimens of Ochlerini received during the 1990's, two of which from *C. nucifera* crops cultivated by Sococo S.A., Moju, Pará State, Brazil. For the first time, *L. singularis* and *L. incisus* are reported from oil palm and coconut trees, respectively, and their males are described and illustrated for the first time, with emphasis on the morphology of genitalia.

MATERIAL AND METHODS

Five males and one female of *L. singularis* and three males and three females of *L. incisus* were examined in this study. The species were identified based in a revision by Rolston (1983). Observation of specimens, dissection and preservation followed Garbelotto et al. (2013). Measurements are in millimeters (mm) and follow mainly Garbelotto et al. (2013) and Rolston (1983) for: length and width of eye and pronotal lobe, and interocellar distance. The terminology of Baker (1931), Dupuis (1970), Campos & Grazia (2006) and Garbelotto et al. (2013) were adopted for genitalic structures. Photographs were taken using a Nikon AZ100M stereomicroscope and NIS-Elements Advanced Research software. Drawings were made under a stereomicroscope Leica MZ12 coupled with camera lucida and were vectored using Adobe Illustrator. Whenever possible, collection data were georeferenced following Garbelotto et al. (2013); coordinates are in decimal degrees.

Collections' acronyms follow Evenhuis (2014). Voucher specimens are deposited in the entomological collection of the Departamento de Zoologia at Universidade Federal do Rio Grande do Sul (UFRG), Porto Alegre, RS, Brazil.

TAXONOMY

Lincus singularis Rolston, 1983
Figs. 1-9

Lincus singularis Rolston, 1983: 1, 4, 5, 18-20, Figs. 34-35 (female holotype from Chauchamayo, Peru, deposited in USNM 76690, not examined, no paratypes); Couturier & Khan, 1992: 719 (map); Campos & Grazia, 2006: 153 (list).

Description of the male. The color of males is dark brown to fuscous and the general morphology is similar to that described for females by Rolston (1983) (Fig. 1). Genitalia. Pygophore oval, opening of genital cup narrow. Dorsal rim uniformly concave (Fig. 2, dr), bearing 1+1 tufts of setae lateral to segment X. Posterolateral angles rounded, projected distinctly beyond the ventral rim, depressed dorsally (Fig. 2, pa). Basal 1/3 of segment X membranous, lateral margins sinuous tapering to apex (Fig. 2, X). Ventral rim V-shaped, with setae along margin (Fig. 3, vr). Ventral surface tumescent on disc, with 1+1 lateral sulci following ventral rim (Fig. 3, t); ventral surface of posterolateral angles tumescent (Fig. 3). Parameres inconspicuous and covered by segment X, attached to the articulatory apparatus of phallus, subtriangular in lateral view, bearing a dorsal dense tuft of setae on apex (Figs. 4-6). Phallus. Phallotheca globose (Figs. 7-9, ph), strongly sclerotized. Vesica longer than the combined lengths of phallotheca and ductus seminis distalis (Figs. 7-9, v, ds), bearing an dorsal subtriangular process posteriorly directed (Figs. 7-9, dp), and 1+1 lateral processes short and truncate (Figs. 7-9, lp). Free portion of ductus seminis distalis very short, about half the length of the inner portion, projecting ventrad of vesica before the lateral processes (Figs. 7-9, ds).

Male. Measurements (n = 5). Total length 10.75 ± 0.29 (10.37-11.00); width of abdomen 6.62 ± 0.36 (6.12-7.00); head length 1.67 ± 0.08 (1.57-1.76); head width 2.24 ± 0.10 (2.14-2.39); eye length 0.50 ± 0.03 (0.47-0.55); eye width 0.55 ± 0.02 (0.52-0.57); interocellar distance 1.20 ± 0.03 (1.17-1.2); interocular distance 1.21 ± 0.06 (1.13-1.26); pronotum length 2.17 ± 0.14 (1.95-2.27); pronotum width 5.66 ± 0.19 (5.42-5.90); length of pronotal lobe 0.23 ± 0.03 (0.20-0.27); pronotal lobe width 0.17 ± 0.02 (0.15-0.20); scutellum length 4.27 ± 0.25 (3.91-4.60); scutellum width 3.55 ± 0.13 (3.39-3.72); length of antennomers: I 0.77 ± 0.03 (0.75-0.80); II 0.82 ± 0.03 (0.77-0.85); III 1.01 ± 0.04 (0.97-1.07); IV 1.42 ± 0.04 (1.37-1.45); V 1.81 ± 0.11 (1.62-1.92); length of labial segments: I 1.28 ± 0.11 (1.12-1.37); II 2.36 ± 0.09 (2.25-2.37); III 1.87 ± 0.03 (1.82-1.9); IV 1.79 ± 0.06 (1.75-1.90).

Material examined. Peru, *Tocache*: 5 males and 1 female, San Martin (Palmas del Espino S.A., Cultivo Palma Aceitera, parcela A11a [-8.41; -76.41] 500 m a.s.l.), 2009, E. Trindad leg.

Distribution. Peru, Cusco and San Martín regions.

Remarks. Although no phylogenetic hypothesis has been advanced for species of *Lincus*, the genus was recovered in the *Herrichella* Distant, 1911 clade in a cladistic analysis of the Ochlerini (Campos & Grazia 2006). The relationship between *Lincus* and the other members of the clade, however, remained unresolved. More recently, the genus (represented by *L. lobuliger*) was recovered as the sister group of the remaining taxa of the *Herrichella* clade in the phylogenetic analysis of Garbelotto et al. (2013). The monophyly of the genus, however, remains to be tested. Several species of *Lincus* are recognizable by their well-developed pronotal lobes, and all known males have tubular proctiger and reduced parameres (Rolston 1983, 1992). These characters were not used in the phylogenetic studies mentioned above. Regarding the phylogenetic relationships among the species of *Lincus*, Rolston (1983) placed *L. singularis* along with *Lincus parvulus* (Ruckes, 1958) and *L. tumidifrons* in the "swollen head" informal group of species ("species group of convenience" sensu Rolston 1983). This group was characterized by having a tumid vertex. Some features of the pygophore of *L. singularis* are consistent with Rolston's proposal to place the species in it, e.g. the 'V' shape of the ventral rim of the pygophore; subrectangular proctiger with acute apex; and globose phallotheca, the latter also observed in *L. tumidifrons*. *Lincus singularis* can be differentiated from the other species in the "swollen head" group by having the anterior and posterior margins of the pronotal lobes subparallel and each lobe projected laterad of its corresponding eye; the vertex of head not as tumid as in *L. parvulus* and *L. tumidifrons* (Fig. 1; for *L. parvulus* and *L. tumidifrons* see Rolston 1983, Figs. 30 and 36); and the ventral opening of the pygophore is narrower than in those species (Fig. 3, vr; for *L. parvulus* and *L. tumidifrons* see Rolston 1983, Figs. 32 and 41).

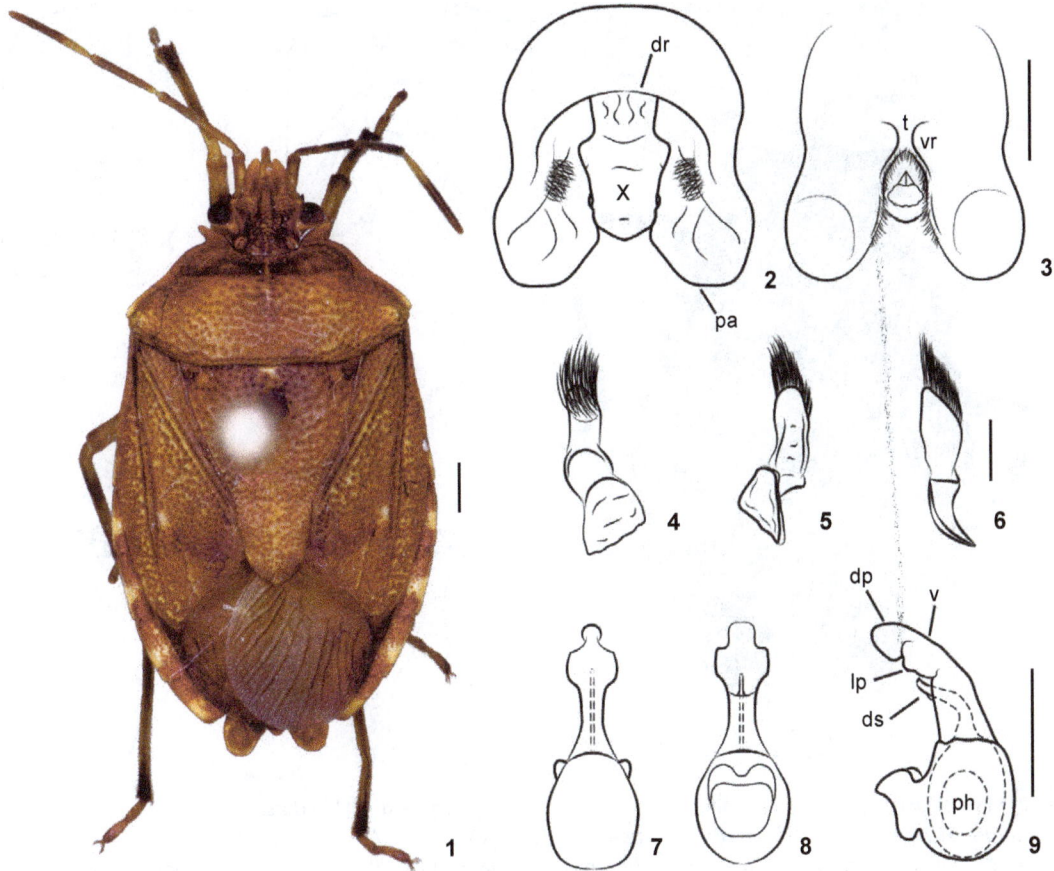

Figures 1-9. Male of *Lincus singularis*: (1) habitus in dorsal view; (2-3) pygophore: (2) dorsal view; (3) ventral view; (4-6) left paramere: (4) dorsal view; (5) ventral view; (6) lateral view; (7-9) phallus: (7) anterior view; (8) posterior view; (9) lateral view. (dp) Dorsal projections, (dr) dorsal rim, (ds) ductus seminis distalis, (lp) lateral projection, (pa) posterolateral angles, (ph) phallotheca, (t) tumescent area, (v) vesica, (vr) ventral rim, (X) segment X. Scale bars: 1-3 = 1 mm, 4-9 = 0.5 mm.

Lincus incisus Rolston, 1983
Figs. 10-18

Lincus incisus Rolston, 1983: 1, 3, 4, 9-10, Figs. 8-9 (female holotype from De Mapane, Suriname, deposited in RMNH, not examined, no paratypes); Campos & Grazia, 2006: 153 (list).

Description of the male. The fuscous general color of male and its general morphology, including the anterolateral margins of pronotum expanded in obtuse angle, posterior to pronotal lobes, are as described for females by ROLSTON (1983) (Fig. 10). Genitalia. Pygophore subrectangular. Surface with short setae. Dorsal rim concave, bearing setae lateral to segment X (Fig. 11, dr). Posterolateral angles obtuse (Fig. 11, pa), depressed, with 1+1 median projections (Fig. 11, mp). Segment X sclerotized, ventrally directed; apex expanded and flattened (Fig. 11, X); anal opening circular, and genital opening in longitudinal slit, both at ventral surface (Fig. 12). Ventral rim concave, with setae along the margin, medially carinated (Fig. 12, vr). Ventral

surface tumescent on disc, with 1+1 lateral sulci following ventral rim (Fig. 12, t). Parameres inconspicuous, attached to the articulatory apparatus of phallus, subrectangular and with an apical tuft of setae (Figs. 13-15). Phallus. Phallotheca globose, strongly sclerotized (Figs. 16-18, ph). Vesica elongated, medially narrowed, longer than the combined lengths of phallotheca and ductus seminis distalis (Figs. 16-18, v); with one globose dorsal projection posteriorly directed (Figs. 16-18, dp); 1+1 lateral globose projections, posteriorly directed (Figs. 16-18, lp); posterior projection truncated, bearing ductus seminis distalis (Figs. 16-18, pp). Ductus seminis distalis antero-dorsally arched toward the projections of vesica (Fig. 18, ds).

Measurements (n = 3). Total length 12.55 ± 0.79 (11.86-13.42); width of abdomen 6.77 ± 0.32 (6.46-7.10); head length 2.18 ± 0.11 (2.06-2.28); head width 2.63 ± 0.06 (2.56-2.68); eye length 0,60 ± 0.03 (0.57-0.63); eye width 0.78 ± 0.03 (0.75-0.82); interocellar distance 0.70 ± 0.01 (0.69-0.71); interocular distance 1.12 ± 0.05 (1.07-1.15); pronotum length 3.47 ± 0.18

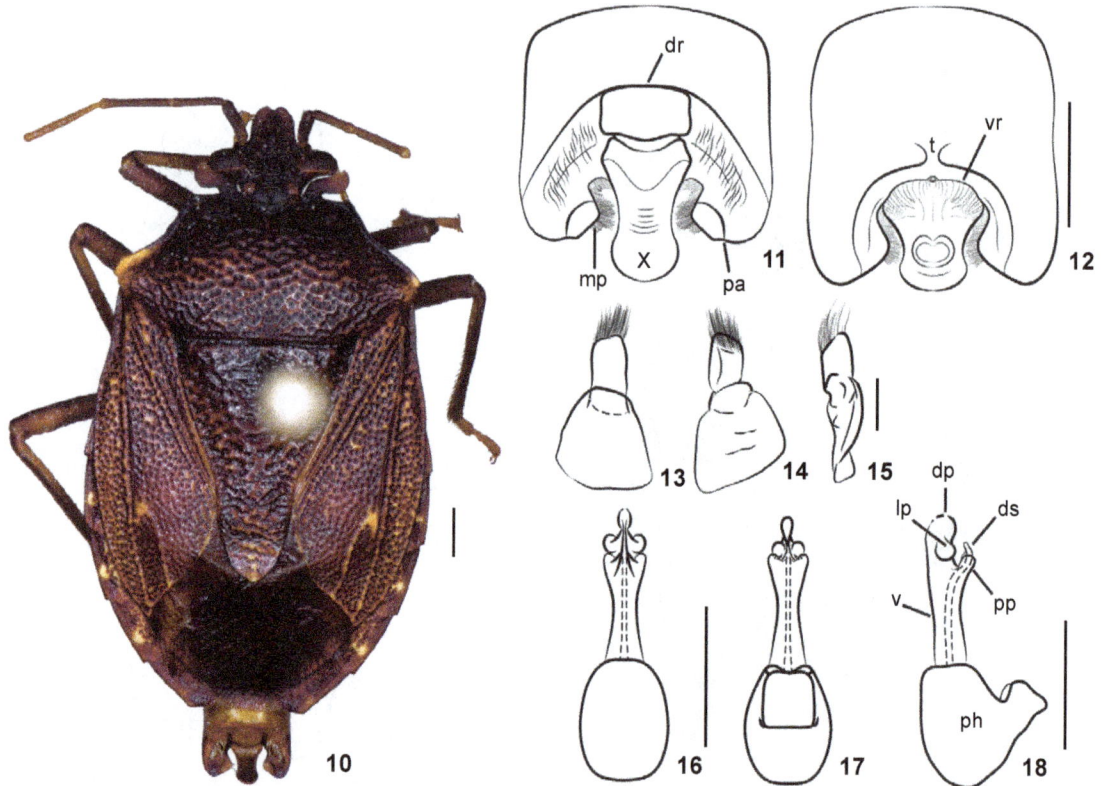

Figures 10-18. Male of *Lincus incisus*: (10) habitus in dorsal view; (11-12) pygophore: (11) dorsal view; (12) ventral view; (13-15) right paramere: (13) dorsal view; (14) ventral view; (15) lateral view; (16-18) phallus: (16) anterior view; (17) posterior view; (18) lateral view. (dp) Dorsal projections, (dr) dorsal rim, (ds) ductus seminis distalis, (lp) lateral projection, (mp) median projection, (pa) postero-lateral angles, (ph) phallotheca, (t) tumescent area, (v) vesica, (pp) posterior projection, (vr) ventral rim, (X) segment X. Scale bars: 10-12 = 1 mm, 13-18 = 0.5 mm.

(3.27-3.59); pronotum width 6.01 ± 0.18 (5.82-6.17); length of pronotal lobe 0.41 ± 0.02 (0.40-0.44); pronotal lobe width 0.85 ± 0.02 (0.80-0.84); scutellum length 4.45 ± 0.30 (4.12-4.70); scutellum width 3.65 ± 0.08 (3.55-3.70); length of antennomers: I 0.97 ± 0.04 (0.92-1.00); II 1.12 ± 0.08 (1.07-1.21); III 1.42 ± 0.07 (1.35-1.50); IV 1.38 ± 0.20 (1.2-1.56); V 1.78 ± 0.00 (1.78-1.78); length of labial segments: I 1.35 ± 0.70 (1.28-1.42); II 2.43 ± 0.14 (2.34-2.60); III 1.97 ± 0.14 (1.85-2.13); IV 1.66 ± 0.05 (1.63-1.72).

Material examined. BRAZIL, *Amazonas*: 3 females, São Miguel da Cachoeira (Cachoeira do Tucano – Pico da Neblina), X.2007, Nogueira & Candiani leg.; *Pará*: 2 males, Moju (Fazenda Sococo) [-2.11; -48.00], 01.XII.1995, P. Lins leg.; 1 male, Tucuruí (Rio Tocantins) [-3.7; -49.7], 20.VII.1984, W. França leg.

Distribution: Suriname, Brazil (Amazonas and Pará States).

Remarks. *Lincus incisus* was placed, along with eight other species (*Lincus convexus* Rolston, 1983, *Lincus croupius* Rolston, 1983, *Lincus fatigus* Rolston, 1983, *Lincus operosus* Rolston, 1983, *Lincus securiger* Breddin, 1904, *Lincus sinuosus* Rolston, 1983, *Lincus spathuliger* Breddin 1908 and *Lincus vandoesburgi* Rolston,

1983), in the "hatchet-lobed" informal group of species (ROLSTON 1983). This placement was justified in view of the anterior pronotal angles resembling a hatchet blade in *Lincus incisus*. Males of *L. incisus* share some genitalic characters with males of *L. convexus*, *L. securiger*, *L. sinuosus* and *L. vandoesburgi*, such as the presence of 1+1 median projections at posterolateral angles, ventrally directed; and an elongated phallus bearing apical projections. The ductus seminis distalis bent toward the projections of the vesica is also observed in *L. vandoesburgi*. Within the hatchet-lobed group, *L. incisus* and *L. vandoesburgi* share the pronotal margin posterior to the lobes expanded on each side into an obtuse projection. The incision between each lobe and the anterolateral margin of the pronotum is deepest in *L. incisus*, reaching half the width of an eye (Fig. 10; ROLSTON 1983, Figs. 1, 8). *Lincus incisus* can also be distinguished from *L. vandoesburgi* by the more convex apical margin of the posterolateral angles of the pygophore, and by a more developed median projection (Figs. 11-12, mp; for *L. vandoesbugi* see ROLSTON 1983, Fig. 2). Among the other species in the hatchet-lobed group with known males, *L. incisus* differs from *L. sinuosus* by the median projection below the apical margin of the posterolateral angles of the pygophore (Figs. 11-12, mp; for *L. sinuosus* see Rolston 1983, Fig. 17); from *L. convexus* and *L. securiger* by the more developed median projection and the ventral opening of the pygophore broader and shallower (Figs. 11-12, mp, vr; for *L. convexus* and *L. securiger* see Rolston 1983, Figs. 23 and 26). Notwithstanding the placement of *L. incisus* within the "hatchet-lobed" group, it is noticeable that the shape of segment X, with an expanded and flattened apex, is also a feature of some species of the "big-eyed" group (Rolston 1983), such as *Lincus lethifer* Dolling, 1984, *Lincus substyliger* Rolston, 1983 and *Lincus subuliger* Breddin, 1908. *Lincus incisus* is recorded for the first time in Brazil.

ACKNOWLEDGMENTS

We thank the curators of the scientific collections for the loan of specimens. We also thank Conselho Nacional de Desenvolvimento Científico e Tecnológico (CNPq) for the scholarships granted to T.A. Garbelotto (process 142448/2011-7), Coordenação de Aperfeiçoamento de Pessoal de Nível Superior (CAPES) to T. Roell, and Universidade Federal do Rio Grande do Sul (UFRGS) to I.C. Winter and the funding from CNPq (process 305367/2012-9) as fellowship grant to L.A. Campos.

LITERATURED CITED

Baker AD (1931) A study of the male genitalia of Canadian species of Pentatomidae. **Canadian Journal of Research** 4(3): 148-220. doi: 10.1139/cjr31-013

Camargo EP (1999) *Phytomonas* and other trypanosomatid parasites of plants and fruit. **Advances in Parasitology 42**: 29-112. doi: 10.1016/S0065-308X(08)60148-7

Campos LA, Grazia J (2006) Análise cladística e biogeografia de Ochlerini (Heteroptera, Pentatomidae, Discocephalinae). **Iheringia, Série Zoologia 96**(2): 147-163. doi: 10.1590/S0073-47212006000200004

Couturier G, Kahn F (1989) Bugs of *Lincus* spp. vectors of Marchitez and Hartrot (oil palm and coconut diseases) on *Astrocaryum* spp., Amazonian native palms. **Principes 33**(1): 19-20.

Couturier G, Kahn F (1992) Notes on the insect fauna on two species of *Astrocaryum* (Palmae, Cocoeae, Bactridinae) in Peruvian Amazonia, with emphasis on potential pests of cultivated palms. **Bulletin de l'Institut Français d'Études Andines 21**(2): 715-725.

Desmier de Chenon R (1984) Recherches sur le genre *Lincus* Stål, Hemiptera Pentatomidae Discocephalinae, et son role éventuel dans la transmission de la Machitez du palmier à huile et du Hart-Rot du cocotier. **Oléagineux 39**(1): 1-6.

Di Lucca AGT, Chipana EFT, Albújar MJT, Peralta WT, Piedra YCM, Zelada JLA (2013) Slow wilt: another form of Marchitez in

oil palm associated with trypanosomatids in Peru. **Tropical Plant Pathology 38**(6): 522-533. doi: 10.1590/S1982-56762013000600008.

Dolling WR (1984) Pentatomid bugs (Hemiptera) that transmit a flagellate disease of cultivated palms in South America. **Bulletin of Entomological Research 74**(3): 473-476. doi: 10.1017/S000748530001573X

Dupuis C (1970) Heteroptera, p. 190-208. In: Tuxen SL (Ed.). **Taxonomist's glossary of genitalia of insects.** Copenhagen, Munksgaard, 359p.

Evenhuis NL (2014) **Abbreviations for insect and spider collections of the world.** Available at: http://hbs.bishopmuseum.org/codens/codens-inst.html [Accessed: 12 November 2014]

Garbelotto TA, Campos LA, Grazia J (2013) Cladistics and revision of *Alitocoris* with considerations on the phylogeny of the *Herrichella* clade (Hemiptera, Pentatomidae, Discocephalinae, Ochlerini). **Zoological Journal of the Linnean Society** 168(3): 452-472. doi: 10.1111/zoj.12032

Howard FW (2001) Sap-feeders on palms, p. 109-232. In: Howard FW, Moore D, Giblin-Davis RM, Abad RG (Eds). **Insects on palms.** New York, CABI Publishing, XIII+403p. doi: 10.1079/9780851993263.0109

Llosa JF, Couturier G, Kahn F (1990) Notes on the ecology of *Lincus spurcus* and *L. malevolus* (Heteroptera: Pentatomidae: Discocephalinae) on Palmae in forests of Peruvian Amazonia. **Annales de laSociété Entomologique de France (Nouvelle Série) 26**(2): 249-254.

Mitchell PL (2004) Heteroptera as vectors of plant pathogens. **Neotropical Entomology 33**(5): 519-545. doi: 10.1590/S1519-566X2004000500001

Panizzi AR, McPherson JE, Javahery JM, McPherson RM (2000) Stink Bugs (Pentatomidae), p. 421-474. In: Schaefer CW, Panizzi AR (Eds). **Heteroptera of economic importance.** Boca Raton, CRC Press, 856p.

Perthuis B, Desmier de Chenon R, Merland E (1985) Mise en évidence du vecteur de la Marchitez sorpresiva du palmier à huile, la punaise *Lincus lethifer* Dolling (Hemiptera Pentatomidae Discocephalinae). **Oléagineux 40**(10): 473-475.

Rolston LH (1983) A revision of the genus *Lincus* Stål (Hemiptera: Pentatomidae: Discocephalinae: Ochlerini). **Journal of the New York Entomological Society 91**(1): 1-47.

Rolston LH (1989) Three new species of *Lincus* (Hemiptera: Pentatomidae) from palms. **Journal of the New York Entomological Society 97**(3): 271-276.

Rolston LH (1992) Key and dignoses for the genera of Ochlerini (Hemiptera: Pentatomidae: Discocephalinae). **Journal of the New York Entomological Society 100**(1): 1-41.

Stål C (1867) Bildrag till Hemipterernas systematik. Conspectus generum Pentatomidum Americae. **Öfversigt af Kongliga Vetenskaps-Akademiens Förhandlingar 24**(7): 522-532.

18

New morphological data on *Solariella obscura* (Trochoidea: Solariellidae) from New Jersey, USA

Ana Paula S. Dornellas[1,2] & Luiz R.L. Simone[1]

[1]*Museu de Zoologia, Universidade de São Paulo. Caixa Postal 42494, 04218-970 São Paulo, SP, Brazil.*
[2]*Corresponding author. E-mail: dornellas.anapaula@usp.br*

ABSTRACT. Anatomical data on *Solariella obscura* (Couthouy, 1838) are presented and analyzed. The main features of this species, when compared with other known trochoids, are: ctenidium with thick lamellae; enlarged ureter (that may indicate sexual dimorphism) instead of a modified urogenital papilla; odontophore very different from other trochoids such as *Calliostoma*, *Agathistoma*, *Monodonta*, and *Gaza*, with short m6, large mj and m4 pairs and absent m8 pair and posterior cartilages; esophageal valve surrounding the odontophore ventrally; anterior and mid-esophagus composed of several thin folds and a very wide cerebral ganglion. *Solariella obscura* differs from *Solariella varicosa* (Mighels & Adams, 1842) by having lower spire, spiral cords weaker on the base and axial rib oblique. There are no differences between *S. obscura* and *S. varicosa* in the external morphology and radula. These internal anatomical data are described for the first time for a solariellid and might improve our understanding of the relationships within this taxon.

KEY WORDS. Anatomy; comparative data; North Atlantic; redescription.

Living Solariellidae is distributed in offshore waters worldwide and the fossil record is primarily in low latitudes during the Paleogene to higher latitudes in the Neogene (HICKMAN & McLEAN 1990). It comprises *Solariella* Wood, 1842, *Archiminolia* Iredale, 1929, *Bathymophila* Dall, 1881, *Hazuregyra* Shikama, 1962, *Ilanga* Herbert, 1987, *Microgaza* Dall, 1881, *Minolia* Adams, 1860, *Minolops* Iredale, 1929, *Spectamen* Iredale, 1924, *Zetela* Finlay, 1927 (HICKMAN & McLEAN 1990, WILLIAMS 2012).

Species belonging to Solariellidae are characterized by small, nacreous shells, short and straight radula, absent cephalic lappets, oral surface of snout with elongated papillae, anterior end of foot with well-developed lateral horns, and neck lobes bearing one or two tentacles (HERBERT 1987, HICKMAN & McLEAN 1990, WILLIAMS et al. 2008). However, some shell characters in this family are convergent in both trochid subfamilies, Umboniinae and Margaritinae. Therefore, it has been inferred that shell characters in Solariellidae are misleading, and cannot be distinguished without knowledge of the external anatomy or radula (HICKMAN & MCLEAN 1990).

The generic name *Solariella* was used in the past in reference to a wide variety of trochids with a round aperture and broad umbilicus. The type species, *Solariella maculata* Wood, 1842, is a fossil from the Pliocene, which makes it impossible to ascertain the state of its radular and soft parts. Nevertheless, recent species that resemble it and which are similar to the North Atlantic *Solariella amabilis* (Jeffreys, 1865) have been placed in *Solariella* by HERBERT (1987), HICKMAN & MCLEAN (1990), and WARÉN (1993). Thus, *Solariella* sensu stricto has three pairs of epipodial tentacles, no prominent epipodial lobes and a radula with nascent lateromarginal plates (HERBERT 1987, MARSHALL 1999).

Herein we describe and analyze the external morphology and certain anatomical structures of the North Atlantic *Solariella obscura* (Couthouy, 1838). In order to achieve new insights that might prove useful in future taxonomic studies.

MATERIAL AND METHODS

Specimens preserved in 70% ethanol were extracted from their shells and subsequently dissected and photographed under the stereomicroscope. The terminology of the odontophore muscles follow DORNELLAS & SIMONE (2013). All drawings were made with the aid of a camera lucida. Samples examined with the SEM (radulae and protoconchs) were mounted on stubs and coated with gold-palladium alloy. The specimens were analyzed and photographed under a stereomicroscope.

Anatomical abbreviations: af, afferent vein; ai, intestinal loop; an, anus; ac, anterior cartilage of odontophore; ax, axis; bc, subesophageal connective; cb, cerebrobuccal connective; cc, cerebral commissure; cd, cerebropedal connective; cg, cerebral ganglion; cm: collumelar muscle; cp, cerebropleural connective; ct, ctenidium; cv, ctenidial vein; df, dorsal fold; dg, digestive gland; ef, esophageal fold; ep, epipodium; es, esophagus; et, epipodial tentacle; ev, esophageal valve; ft, foot; go, gonad; hg, hypobranchial gland; ho, horn; jw, jaws; la, left auricle; lg, labial ganglion; mb, mantle border; mj, jugal muscles; mt, mantle; nl, neck lobe; om, ommatophore; oa, opercular pad; os, osphadia; pc, pericardium; pe, pedal ganglion; pl, pleural ganglion; ps, papillary sac; pt, postoptic tentacle; ra, radula; ru, right auricle; rk, right kidney; ro, rod; sc, spiral caecum; sg, salivary gland; sk, skeletal rod; sm, stomach; sn, snout; st, statocysts; te, cephalic tentacle; ur, ureter; ve, ventricle.

Institutional acronyms. (USNM/NMNH) National Museum of Natural History, Smithsonian Institution, Washington, DC.

Material analyzed. Types: United States, between Marblehead and Nahant, Massachusetts Bay, 7 specimens, MCZ 154825. United States, off New Jersey, North Atlantic Ocean, 39°02′54″N 73°47′06″W, 6 specimens, USNM 828340; 3919′18″N 73°10′06″W, 2 specimens, USNM 828343.

TAXONOMY

Solariella obscura (Couthouy, 1838)
Figs. 1-34

Turbo obscurus Couthouy, 1838: 100 (pl. 3, fig. 12).
Solariella obscura: Tryon, 1889: 308 (pl. 57, figs. 44, 45); Locard, 1903: 43; Cushman, 1906: 16; Odhner, 1912: 70; Johnson, 1915: 89; Smith, 1951: 79 (pl. 31, fig. 19; pl. 71, fig. 16); Lopes & Cardoso, 1958: 62; MacGinite, 1959: 80; Talmadge, 1967: 236; Abbott, 1974: 40 (fig. 271); Procter, 1993: 172.
Margarites albula Gould, 1861: 36.
Margarita obscura: Gould, 1870: 283 (fig. 545).
Margarita bella Verkrüzen, 1875: 236.
Margarita obscura var *cinereaeformis* Leche, 1878: 45 (pl. 2, fig. 24).
Margarita obscura var *intermedia* Leche, 1878: 45 (pl. 2, fig. 25).
Solariella laevis Friele, 1886: 14.
Solariella obscura var *bella* (Verkrüzen): Tryon, 1889: 310 (pl. 64, figs. 57, 58); Blaney, 1906: 111; Johnson, 1915: 89; Lopes & Cardoso, 1958: 62.
Machaeroplax obscura var *planula* Verrill, 1882: 532; Johnson, 1915: 89 (pl. 17, fig. 6); Lopes & Cardoso, 1958: 63.
Machaeroplax obscura var *carinata* Verrill, 1882: 532; Johnson, 1915: 90; Lopes & Cardoso, 1958: 62.
Solariella obscura var *multilirata* Odhner, 1912: 79; Lopes & Cardoso, 1958: 62.

Type. Cotypes, 7 shells, MCZ 154825.

Type locality. Between Marblehead and Nahant, Massachusetts Bay, USA. In fish stomach.

Distribution. Arctic circumpolar; south of New England; eastern and western Greenland; Iceland; Canada, Labrador; USA, Maine, Massachusetts, Connecticut (WARÉN 1993, ROSENBERG 2009).

Description. Shell (Figs. 1-10, 22, 23): up to 5½ whorls, 8 mm in height and 9 mm in diameter; deeply umbilicate; shape trochoid, whorls rounded or angular, suture impressed. Color grayish to pinkish tan, peristome thin and nearly complete, often worn, revealing iridescent color. Protoconch (Figs. 22, 23) of 1 whorl lighter-colored, smooth; about 250 μm diameter. Spire sculptured by weakly beaded spiral cords; about 12 cords on last

whorl. Some specimens with strong cord on middle whorl (Figs. 3, 6), forming carinate shoulder. Base weakly convex, with about 20 cords, thinner than those on spire; smooth. Strong beaded cord surrounding umbilical area; umbilical area with 7-10 cords. Umbilicus wide and deep, about 20% maximum shell width, funnel-shaped. Aperture rounded, inner surface iridescent; ~75% of shell length, ~55% of shell width.

Head-Foot (Figs. 24, 25, 34). Head bulging approximately in middle region of head-foot. Snout wide, cylindrical; distal end wider than base; distal surface papillated; papillae long, thin, cylindrical, with rounded tip. Outer lips mid-ventrally incomplete, mouth located in middle portion of ventral surface of snout. Pair of cephalic lappets absent. Cephalic tentacles (Fig. 24: te) ~20-30% longer than snout, sometimes asymmetrical in relation to one another, covered by small papillae, dorso-ventrally flattened, grooved, narrowing gradually up to lightly pointed tip. Ommatophore on outer base of cephalic tentacles, length 1/5 of tentacle length. Eyes dark, rounded, occupying anterior edge of ommatophore.

Foot thick, occupying half of total head-foot length; whitish, non papillated; anterior end truncated and drawn out laterally into two long processes (horns). Epipodium (Fig. 25: ep) surrounding latero-dorsal region of mesopodium, equidistant from sole and base of ommatophores. Right neck lobe with two tentacles, anterior tentacle postoptic (Fig. 25: pt); left neck lobe with one tentacle. Three pairs of epipodial tentacles (Figs. 24, 25) symmetrical on both sides; epipodial sense organs at base of epipodial tentacles. Opercular pad (Figs. 25, 34: oa) rounded, located in median dorsal region, with free edge in posterior area; posterior end with several chevron furrows, apex pointed posteriorly and two pairs of longitudinal furrows on median line. Furrow of pedal glands along entire anterior edge.

Operculum (Figs. 11, 12). Up to ~2.5 mm in diameter and ~7 whorls, closing entire aperture, corneous, thin, with central nucleus. Inner side convex, outer side concave. Color yellowish gold.

Mantle organs (Figs. 13, 26-28). Pallial cavity of 3/4 whorl. Mantle border (Fig. 26: mb) thick, white; anterior end papillated, occupying 1/3 of mantle border. Gill located on left side of pallial cavity; less than half of width and height of pallial cavity; projecting anteriorly, sustained by gill rod, lacking suspensory membrane (Fig. 26: sk). Anterior end of gill narrow, with acuminate tip. Ventral lamella larger than dorsal lamella (Figs. 27). Afferent gill vessel ~3/4 of gill's length, originating in transverse pallial vessel, running along distal region of central axis of gill. Transverse pallial vessel ~1/5 of afferent vessel length, originating in left nephrostome and discharging in afferent gill vessel. Ctenidial vein (efferent gill vessel) length more than twice afferent vessel length, running along basal region of central axis of gill; posterior end of vein (half) free from gill filaments, lying parallel to afferent vessel up to pericardium. Osphradium rounded, whitish, located at base of gill rod. Hypobranchial glands (Figs. 26: hg) on both sides of rectum; more developed on left side. Rectum occupying 1/4 of pallial cavity

Figures 1-17. *Solariella obscura*. (1-12) Shell and operculum: (1-4) Cotypes MCZ 154825, apertural and umbical views: (1-2) 7.2 mm height x 8.5 mm width, (3-4) 7.2 mm x 8.6 mm. (5-12) USNM 828340, shell and operculum; (5-10) shell, apertural, lateral and umbilical views: (5-7) 5.8 mm x 5.5 mm, (8-10) 8.9 mm x 9.1 mm; (11-12) operculum, external and internal views, 3 mm diameter. (13-17) Anatomy: (13) pallial cavity, ventral view; (14) jaws, ventral view; (15) foregut, anterior and mid-esophagus opened longitudinally, ventral inner view, odontophore removed; (16) pedal and pleural ganglia *in situ*, dorsal view; (17) anterior odontophore cartilages, muscles removed, ventral view. Scale bars: 14, 16 = 0.2 mm; 13, 15, 17 = 0.5 mm.

width, sigmoid; posterior region under kidneys. Anus siphoned, preceded by pleated and short free end, located on posterior right side of pallial cavity. Kidneys length more than half of rectum length, located on posterior region of pallial cavity.

Visceral mass (Fig. 34). Pericardium and posterior portion of right kidney exposed on pallial cavity roof. Stomach and spiral caecum (sc) located 1/3 whorl posterior to pallial cavity. Digestive gland (dg) located on left side and gonad (go) on right side, both posterior to right kidney.

Circulatory and excretory systems (Figs. 26, 28). Pericardium located between pallial cavity and visceral mass (Fig. 26: pc), close to median line and immediately posterior to kidneys. Left side of pericardium receiving ctenidial vein and right side receiving right pallial vein. Ventricle volume 1/3 of pericardium volume; surrounding rectum and flanked anteriorly by left auricle and posteriorly by right auricle; left auricle ventral, triangular, occupying about half of pericardium volume; right auricle weak, smaller than left one (Fig. 28). Papillary sac (or left kidney) base oval, wide, gradually narrowing towards

anterior portion, ending at left nephrostome; inner wall covered by numerous thin, long papillae. Right kidney (Fig. 28: ur, rk) divided into two regions; anterior region as hollow tube (ureter), right nephrostome located in anterior region; ureter might be as large as papillary sac (probably males) (Fig. 26) or twice papillary sac width (probably females) (Fig. 28); no mucus observed in females ureter. Posterior region spreading around visceral mass immediately beneath mantle, encircling inner surface of columellar muscle. Kidney expanding ventrally, covering half of right surface of adjacent visceral hump.

Digestive system (Figs. 14, 15, 17, 18-21, 29-33). Oral tube length ~½ odontophore length; walls with circular muscles (Figs. 29, 30); basal region with thick oblique fibers (Fig. 31: mj), originating gradually from dorsal surface, close to median line; fibers running posteriorly towards both sides, inserting in ventral surface of odontophore. Jaw plates (Fig. 14) thin, light brown, rounded, occupying half of odontophore length. Pair of dorsal folds starting posteriorly to jaws (Fig. 15: df), with each fold bending and forming two partially overlapping

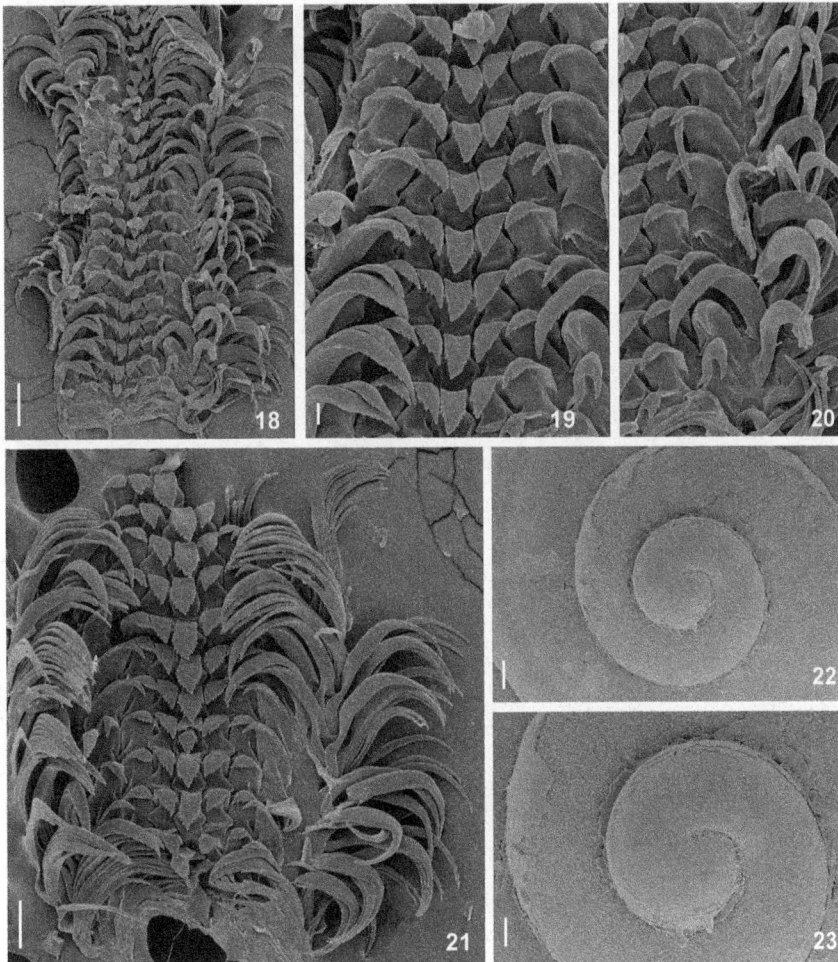

Figures 18-23. *Solariella obscura*, structures under SEM. (18-21) Radular ribbon: (18) middle region, whole view; (19) detail of central area, rachidian and lateral teeth; (20) detail of lateral and marginal teeth; (21) middle region, whole view; (22-23) Protoconch, apical views. Scale bars: 18 = 100 μm, 19 = 20 μm, 20 = 40 μm, 21-23 = 60 μm.

slits. Series of transverse muscles separating outer surface of esophagus from odontophore. Odontophore about 1/3 longer than snout. Odontophore muscles (Figs. 29-33): m1 (Fig. 29): series of small jugal muscles connecting buccal mass with adjacent inner surface of snout and haemocoel; m4 broad and long pair of dorsal tensor muscles of radula and subradular membrane, originating on ventral and lateral surface of anterior cartilages, at some distance from median line, inserting along dorsal region of subradular membrane (exposed inside buccal cavity), with portion in radular ribbon preceding buccal cavity; m5: pair of large accessory dorsal tensor muscles of radula, originating on posterior surface of anterior cartilages, running firstly towards dorsal and median regions and subsequently anteriorly, with insertion in radular ribbon region; m6: horizontal muscle, uniting over half of ventral edges of both anterior cartilages; m7: very small, thin pair of muscles, origi-

nating in middle region of inner ventral surface of radular sac, running anteriorly (insertion not observed); m10: broad pair of ventral protractor muscles of odontophore, originating in ventral region of mouth and buccal sphincter, running posteriorly, inserting in posterior region of anterior cartilage; m11: pair of thin ventral tensor muscles of radula, originating in middle region of ventral surface of anterior cartilage, running anteriorly and covering m6 and ventral surface of anterior cartilage, inserting on distal edge of subradular membrane; m11a; very long and thin pair of oblique ventral tensor muscles of radula, originating on anterior haemocoelic surface near pleural ganglia, running dorsally between anterior edge of anterior cartilages, inserting on distal edge of subradular membrane. Non-muscular structures of odontophore: ac: pair of anterior cartilages (Figs. 17, 33), antero-posteriorly elongated, flat, with anterior and posterior ends rounded, same length as

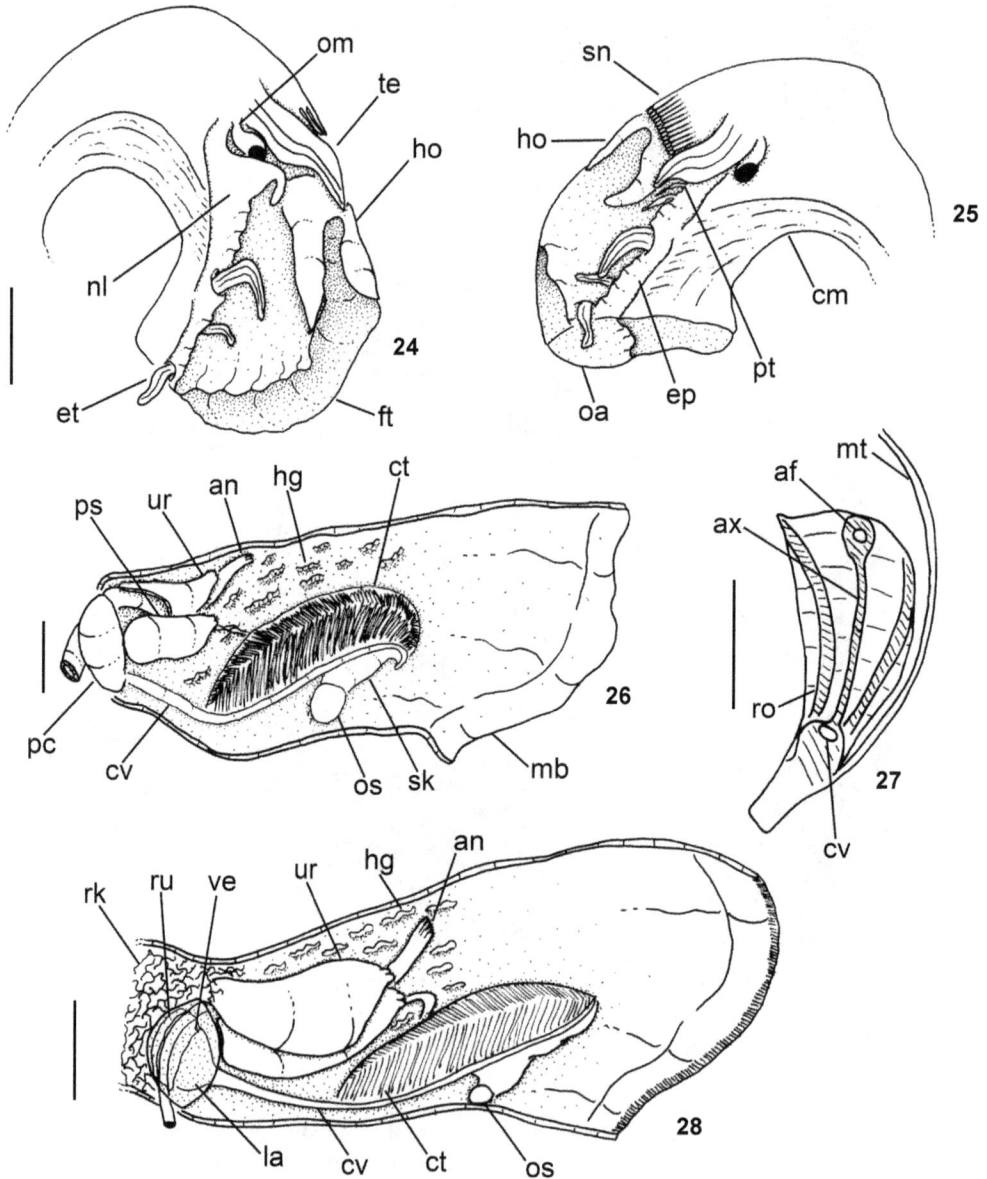

Figures 24-28. Anatomy of *Solariella obscura*: (24-25) head-foot, right and left views; (26) pallial cavity roof, male, ventral view; (27) transverse section in middle region of ctenidium; (28) pallial cavity roof, female, ventral view. Scale bars: 24-26, 28 = 1 mm; 27 = 0.5 mm.

odontophore. Pair of posterior odontophore cartilages absent. Odontophore cartilages whitish, rough; br: subradular membrane covering most of the exposed surface of odontophore in buccal cavity, where most of intrinsic odontophore muscles insert; sc: subradular cartilage maintaining radular ribbon.

Radular sac short (Fig. 31: ra), as long as odontophore. Radular nucleus (Fig. 32) located on ventral side of odontophore. Central complex (rachidian and laterals) well-developed, with interlocking process and correspondent sockets (Fig. 18); shafts expanding laterally, hood-shaped. Rachidian large (Fig.

19), triangular, cutting edge with projection turned posteriorly (almost 90°) and covering posterior end of preceding tooth; tip narrowly tapered, serrated; base and cusp with within-column interaction. Four lateral teeth (Figs. 20, 21); cusps oriented toward midline of radula, most strongly serrate along their outer margins; three inner lateral teeth similar to rachidian in shape; outermost lateral teeth broad, large, length twice of inner teeth length. Lateromarginal plate not observed. Marginal teeth (Fig. 18) as long as outermost lateral teeth, slender, serrate, ~10-12 teeth pairs. Anterior esophagus with esophageal

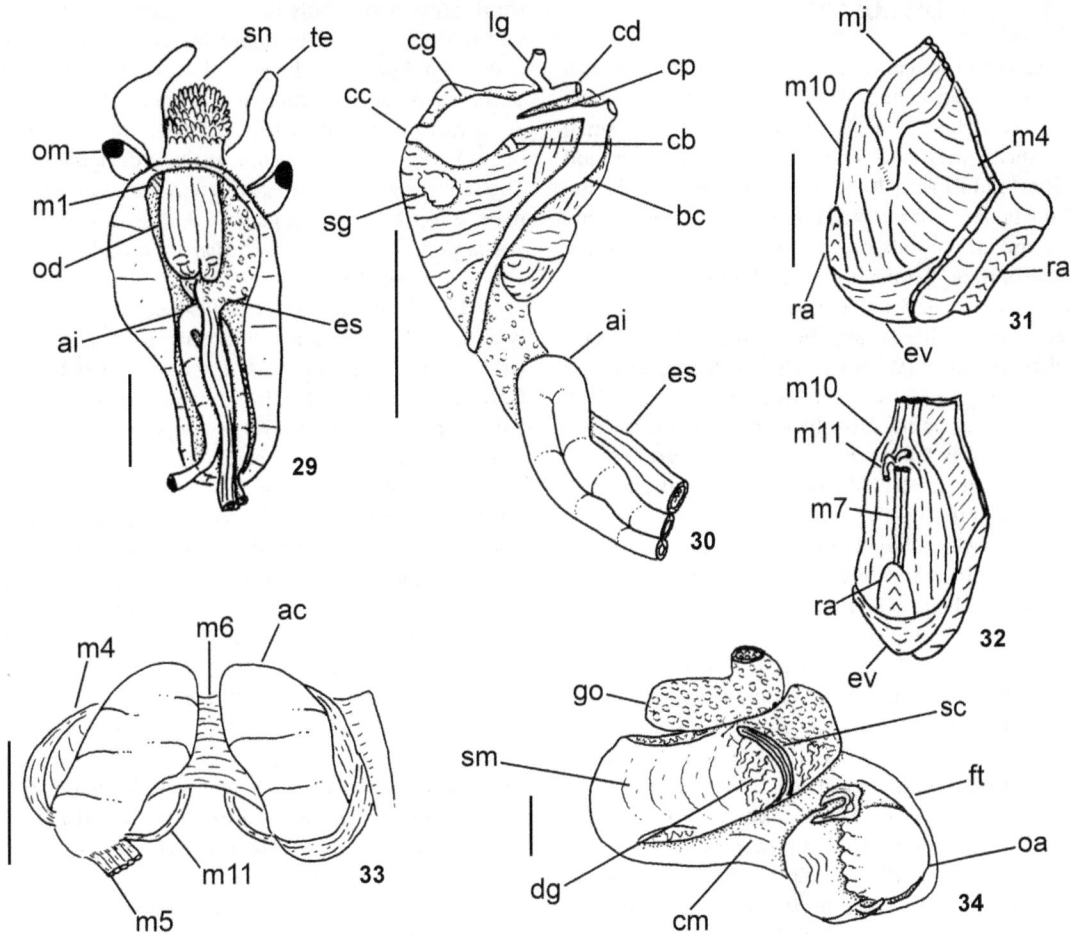

Figures 29-34. Anatomy of *Solariella obscura*: (29) head-foot haemocoel, ventral view, foot and columellar muscle removed; (30) buccal mass and central nervous system, right view; (31-32) odontophore, left, and ventral views; (33) odontophore, dorsal view, radular ribbon removed, left m5 extracted and reflected, m10 extracted; (34) head-foot and visceral mass, whole apertural view. Scale bars: 29-30, 34 = 1 mm; 31-33 = 0.5 mm.

valve covering ventral surface of odontophore. Anterior and mid esophagus (Fig. 15) with folds forming shallow chambers; epithelium entirely covered by villous papillae. Posterior esophagus narrow, with some thin longitudinal folds on inner surface. Stomach not observed. Spiral caecum with ½ counter clockwise (in dorsal view) whorls. Intestine (Figs. 26, 28-30) very wide, running anteriorly forward inside to cephalic haemocoel, bending abruptly, forming wide loop (Fig. 30: ai); anterior region of visceral mass with small loop surrounding kidney and pericardium, exiting in right-posterior corner of pallial cavity. Rectum and anus described above (pallial organs).

Central nervous system (Figs. 16, 30). Nerve ring surrounding anterior half of buccal mass. Cerebral ganglia rounded, located in lateral region of buccal mass (Fig. 30: cg), size ~1/3 of odontophore size; commissure thick, long, dorso-ventrally flattened; cerebropleural and cerebropedal (Fig. 30: cp, cd)

connectives long, thin, originating in anterior region of cerebral ganglia and running ventrally and back to pedal and pleural ganglia. Labial ganglia (Figs. 30: lg) 1/6 of cerebral ganglia, located in ventro-lateral region of buccal mass, anteriorly to cerebral ganglia; connected to cerebral ganglia by short cerebrolabial connective. Buccal ganglia posterior to cerebral ganglia; connected to cerebral ganglia by a buccolabial connective. Pleural and pedal ganglia (Fig. 16: pl, pe) close to each other, located inside pedal musculature immediately below ventral surface of haemocoel; both of about half size of cerebral ganglion. Pedal commissure thick, very short. Pedal nerve running forward from each pedal ganglion, surrounding medial pedal blood sinus. Supra-esophageal connective emerging from right pleural ganglia. Subesophageal connective (Fig. 30) emerging from left pleural ganglia. Statocysts (Fig. 16: st) rounded, bright, located very close to posterior side of pedal ganglia.

DISCUSSION

The organs and systems of *S. obscura* are congruent with the features of solariellid mentioned by previous authors (HERBERT 1987, HICKMAN & McLEAN 1990, WILLIAMS 2012), such as: small and nacreous shells; presence of a ring of digitate papillae around the snout; short radula with 20-30 transverse teeth rows; anterior end of foot bilobed, forming the horn; long and thick cephalic tentacles, and an eye-stalk much shorter than the cephalic tentacle. Some of these features are not exclusive to solariellids, however, especially when compared with other trochoids. The presence of digitate papillae around the snout can also be found in *Gaza* Watson, 1879 (SIMONE & CUNHA 2006), and the Umboniinae also have a bilobed foot (HICKMAN & McLEAN 1990, pers. obs.). The radula, on the other hand, is the main structure for characterizing this family and its genera (HERBERT 1987, MARSHALL 1999), being short (20-30 rows of teeth), with reduced number of marginal teeth (~10 per half row).

The Artic species *Solariella varicosa* (Mighels & Adams, 1842) is similar in shape to *S. obscura*, but differs by having a taller spire, strong oblique axial rounded ribs and stronger spiral cords on the base. The distributions of both species overlaps in the Artic, south of Labrador and northern Canada (WARÉN 1993). The neck lobe shows a variety of shapes among trochoids and might be used to diagnose subfamilies, genera and even species (HICKMAN & McLEAN 1990, DORNELLAS & SIMONE 2013). They are usually digitate, fringed or smooth, as well as symmetrical or asymmetrical to each other according to the taxa. The neck lobes of solariellids show some inter-generic variation in shape (HERBERT 1987, MARSHALL 1999) and are characterized by the presence of one or two short tentacles, the right neck lobe bearing the postoptic tentacle located at its anterior edge (Fig. 25: pt). The neck lobe of *S. obscura* was reported as being virtually identical to that of *S. varicosa* (WARÉN 1993: 161, figs. 4a, b).

Regarding the pallial cavity, the ctenidium of *S. obscura* has a thick lamella (Fig. 13), as is the case in *Solariella carvalhoi* Lopes & Cardoso, 1958 (pers. obs.), when compared with other trochoids such as *Calliostoma* Swainson, 1840, *Monodonta* Lamarck, 1789, *Lithopoma* Gray, 1850, *Agathistoma* Olsson & Harbison, 1953, *Gaza* (FRETTER & GRAHAM 1962, RIGHI 1965, MONTEIRO & COELHO 2002, SIMONE & CUNHA 2006, DORNELLAS & SIMONE 2013). The enlarged ureter in some specimens may indicate sexual dimorphism in *Solariella* (Fig. 28), differently from other vetigastropods in which the females have modified urogenital papillae (WOODWARD 1901, FRETTER & GRAHAM 1962, MONTEIRO & COELHO 2002, DORNELLAS 2012, DORNELLAS & SIMONE 2013). However, this structure differs in shape among vetigastropods (see DORNELLAS & SIMONE 2013).

The odontophore of *S. obscura* is different from that of other trochoids such as *Calliostoma*, *Agathistoma*, *Monodonta* and *Gaza* (FRETTER & GRAHAM 1962, RIGHI 1965, SIMONE & CUNHA 2006, DORNELLAS 2012, DORNELLAS & SIMONE 2013): the m6 is

shorter (occupying only half of the cartilages' length); the mj and m4 pairs are larger (more than twice) than the current size; the m8 pair and the posterior cartilages are lacking. The buccal cavity differs from that of other trochoids by the esophageal valve surrounding the odontophore ventrally (Figs. 31, 32: ev), also observed in *S. carvalhoi* (personal observation). The salivary gland, located in latero-dorsal area of the buccal mass, is rounded and concentrated, similar to that observed in *Tegula viridula* (Gmelin, 1791), *Monodonta labio* (Linnaeus, 1758) and *Lithopoma olfersii* (Philippi, 1846) (RIGHI 1965, pers. obs.).

Usually, the anterior and mid-esophagus of vetigastropods is composed of four folds that compartmentalize it (FRETTER & GRAHAM 1962, FRETTER 1964, HASZPRUNAR 1988, SASAKI 1998, DORNELLAS & SIMONE 2013). In *S. obscura*, on the other hand, the anterior and mid-esophagus are composed of several thin folds (Fig. 15). The presence of papillate glands covering the inner wall of that region, which is also present in *S. obscura*, is a morphological synapomorphy of the clade Vetigastropoda (HASZPRUNAR 1988, SALVINI-PLAWEN 1988, SASAKI 1998).

The radula of *S. obscura* is a typical solariellid radula, comprising a straight and short radular ribbon (as long as the odontophore), triangular rachidian, with the outermost lateral tooth being larger than the innermost teeth, and reduced number of marginal teeth (~10 teeth along the same row). In *Solariella*, the radula is characterized by the presence of well-developed, elongate, cuspless lateromarginal plates (HERBERT 1987). *Solariella obscura* and *S. varicosa* lack lateromarginal plates (WARÉN 1993, FRETTER & GRAHAM 1977), whereas all southern *Solariella* bear lateromarginal plates (MARSHALL 1999).

Despite the gastric spiral caecum being a variable structure, a large spiral caecum is considered derived within Vetigastropoda. This structure opens ventrally in the posterior end of the stomach, more or less as continuous extensions of the typhlosoles (STRONG 2003). The 0.5 whorl long caecum observed in *S. obscura* (Fig. 31: sc) can also be found in *Calliostoma depictum* Dall, 1927, but the number of spiral caecum whorls seems to be an inter-specific feature rather than a generic one, because it may vary among congeners such as in *Calliostoma* and *Lithopoma* (MONTEIRO & COELHO 2002, DORNELLAS & SIMONE 2013).

The central nervous system of *S. obscura* demonstrates a trochoid pattern (SASAKI 1998) but the cerebral ganglion is proportionally wider (Fig. 30) when compared to those described in other vetigastropods (FRETTER & GRAHAM 1962, SASAKI 1998, SIMONE & CUNHA 2006, SIMONE 2008, DORNELLAS 2012, DORNELLAS & SIMONE 2013).

Solariellidae has been recently recognized as a family (BOUCHET et al. 2005), with molecular studies supporting this rank (WILLIAMS et al. 2008, WILLIAMS 2012). As discussed above, several features seem to be exclusive of Solariellidae such as the above-mentioned radular pattern and external morphology. These features have also been further used to trace patterns between solariellid genera (HERBERT 1987). On the other hand, there is no described data about the internal anatomy for any solariellid, and the new data described herein for *S. obscura* might improve our understanding about the relationships within this taxon, at generic and even suprageneric levels.

ACKNOWLEDGEMENTS

We thank Jerry Harasewych (NMNH) for lending us specimens; FAPESP (Fundação de Amparo à Pesquisa do Estado de São Paulo) process numbers 2010/18864-3; 2012/25173-2 for supporting our research; Lara Guimarães (MZSP) for help with the SEM; Diogo Couto for taking photos of the type, and Daniel Cavallari for helping with grammar revision.

LITERATURE CITED

ABBOTT RT (1974) **American Seashells.** New York, Van Nostrand Rheinhold, 2nd ed., 663p.

BLANEY D (1906) Shell-bearing Mollusca of Frenchman's Bay, Maine. **The Nautilus 19**(10): 110-111.

BOUCHET P, ROCROI J, FRÝDA J, HAUSDORF B, PONDER W, VALDÉS A, WARÉN A (2005) Classification and nomenclator of gastropod families. **Malacologia 47**(1-2): 1-397.

COUTHOUY JP (1838) Descriptions of new species of Mollusca and shells, and remarks on several polypi found in Massachusetts Bay. **Boston Journal of Natural History 2:** 53-111.

CUSHMAN JA (1906) The Pleistocene deposits of Sankoty Head, Nantucket, and their fossils. **Publications of the Nantucket Maria Mitchell Association 1**(1): 1-21.

DORNELLAS APS (2012) Description of a new species of *Calliostoma* (Gastropoda, Calliostomatidae) from southeastern Brazil. **ZooKeys 224:** 89-106. doi: 10.3897/zookeys.224.3684

DORNELLAS APS, SIMONE LRL (2013) Comparative morphology and redescription of three species of *Calliostoma* (Gastropoda, Trochoidea) from Brazilian coast. **Malacologia 56**(1-2): 267-293.

FRIELE H (1886) Mollusca II. **The Norwegian North-Atlantic Expedition, 1876-1878 3:** 1-35.

FRETTER V (1964) Observations on the anatomy of *Mikadotrochus amabilis* Bayer. **Bulletin of Marine Science of the Gulf and Caribbean 14**(1): 172-184.

FRETTER V, GRAHAM A (1962) **British prosobranch molluscs. Their functional anatomy and ecology.** Ray Society Publications, London, XVI+755p.

FRETTER V, GRAHAM A (1977) The prosobranch molluscs of Britain and Denmark. Part 2. Trochacea. **Journal of Molluscan Studies 3**(Suppl.): 39-100.

GOULD AA (1861) Descriptions of shells collected by the North Pacific Exploring Expedition. **Proceedings of the Boston Society of Natural History 8:** 14-40.

GOULD AA (1870) **Report on the Invertebrata of Massachusetts.** Boston, 2nd ed., 524p.

HASZPRUNAR, G (1988) On the origin and evolution of major gastropod groups, with special reference to the *Streptoneura*. **Journal of Molluscan Studies 54:** 367-441.

HERBERT DG (1987) Revision of the Solariellinae (Mollusca: Prosobranchia: Trochidae) in southern Africa. **Annals of the Natal Museum 28**(2): 283-382.

HICKMAN CS, MCLEAN JH (1990) Systematic revision and suprageneric classification of trochacean gastropods. **Science Series, Natural History Museum of Los Angeles Country 35:** 1-169.

JOHNSON CW (1915) Fauna of New England 13. List of Mollusca. **Occasional Papers of the Boston Society of Natural History 7:** 1-231.

LECHE W (1878) Öfversigt öfver de af Svenska Expeditionerna till Novaja Semlja och Jenissej 1875 och 1876 Insamlade: Hafs-Mollusker. **Kongliga Svenska Vetenskaps-Akademiens Handlingar 16**(2): 1-86.

LOCARD A (1903) **Conquilles des Mers D'Europe. Turbinidae.** Lyon, Société d'Agriculture, Sciences et Industrie de Lyon, p. 1-66.

LOPES HS, CARDOSO PS (1958) Sobre um novo gastrópodo brasileiro do gênero *Solariella* Wood, 1842 (Trochidae). **Revista Brasileira de Biologia 18**(1): 59-64.

MACGINITE B (1959) Marine Mollusca of Point Barrow, Alaska. **Proceedings of the United States National Museum 10**(3412): 59-208.

MARSHALL BA (1999) A Revision of the Recent Solariellinae (Gastropoda: Trochoidea) of the New Zealand Region. **The Nautilus 113**(1): 4-42.

MONTEIRO JC, COELHO ACS (2002) Comparative morphology of *Astraea latispina* (Philippi, 1844) and *Astraea olfersii* (Philippi, 1846) (Mollusca, Gastropoda, Turbinidae). **Brazilian Journal of Biology 62**(1): 135-150. doi: 10.1590/S1519-69842002000100016

ODHNER NH (1912) Northern and Artic invertebrates in the collection of the Swedish state Museum. **Kungliga Svenska Vetenskapsakademiens Handlingar 48**(1): 1-93.

PROCTER W (1993) **Biological survey of the Mount Desert Region, part V – Marine Fauna.** Philadelphia, Wistar Institute of Anatomy and Biology, 402p.

RIGHI G (1965) Sobre *Tegula viridula* (Gmelin, 1971). **Boletim da Faculdade de Filosofia, Ciências e Letras da Universidade de São Paulo (Zoologia) 25:** 325-390.

ROSEMBERG G (2009) **Malacolog 4.1.1: A Database of Western Atlantic Marine Mollusca.** Avalaible online at: http://www.malacolog.org [Accessed: 22 December 2014]

SALVINI-PLAWEN L VON (1988) The structure and function of molluscan digestive systems, p. 301-379. In: Trueman ER, CLARKE MR (Eds) **The Mollusca: Form and Function.** San Diego, Academic Press, vol. 11, 504p.

SASAKI T (1998). Comparative anatomy and phylogeny of the recent Archaeogastropoda (Mollusca: Gastropoda). **The University of Tokyo Bulletin 38:** 1-224.

SIMONE LRL (2008) A new species of *Fissurella* from São Pedro e São Paulo Archipelago, Brazil (Vetigastropoda, Fissurellidae). **The Veliger 50**(4): 292-304.

SIMONE LRL, CUNHA CM (2006) Revision of genera *Gaza* and *Callogaza* (Vetigastropoda, Trochidae), with description of a new Brazilian species. **Zootaxa 1318**: 1-40.

SMITH M (1951) **East coast marine shells.** Ann Arbor, Edwards Brothers, 4th ed., 314p.

STRONG EE (2003) Refining molluscan characters: morphology, character coding and a phylogeny of the Caenogastropoda. **Zoological Journal of the Linnean Society 137**: 447-554.

TALMADGE RR (1967) Notes on the Mollusca of Prince William sound, Alaska. Part II. **The Veliger 9**(2): 235-238.

TRYON GW (1889) **Manual of Conchology, with illustrations of the species. Trochidae, Stomatiidae, Pleutotomariidae, Haliotidae.** Philadelphia, Published by the Author, vol. 1, #11, 519p.

VERKRÜZEN TA (1875) Bericht über einen Schabe – Ausflug in Sommer 1874. **Jahrbücher der Deutschen Malakozoologischen Gesellschaft 2**: 229-240.

VERRILL AE (1822) Catalogue of marine Mollusca added to the fauna of the New England region, during the past ten years. **Transactions of the Connecticut Academy of Arts and Sciences 5**: 451-587.

WARÉN A (1993) New and little known Mollusca from Iceland and Scandinava. Part 2. **Sarsia 78**: 159-201.

WILLIAMS ST (2012) Advances in molecular systematics of the vetigastropod superfamily Trochoidea. **Zoologica Scripta 41**(6): 571-595. doi: 10.1111/j.1463-6409.2012.00552.x

WILLIAMS ST, KARUBE S, OZAWA T (2008) Molecular systematics of Vetigastropoda: Trochidae, Turbinidae and Trochoidea redefined. **Zoologica Scripta 37**(5): 483-506. doi: 10.1111/j.1463-6409.2008.00341.x

WOODWARD MF (1901) The anatomy of *Pleurotomaria beyrichii*, Hilg. **Bulletin of the Museum of Comparative Zoology 8**: 215-268.

19

The activity time of the lesser bamboo bat, *Tylonycteris pachypus* (Chiroptera: Vespertilionidae)

Li-Biao Zhang[1,*], Fu-Min Wang[2], Qi Liu[1] & Li Wei[3]

[1]*Guangdong Public Laboratory of Wild Animal Conservation and Utilization & Guangdong Key Laboratory of Integrated Pest Management in Agriculture, Guangdong Entomological Institute, Guangzhou 510260, China.*
[2]*Guangdong Provincial Wildlife Rescue Center, Guangzhou 510520, China.*
[3]*College of Ecology, Lishui University, Lishui 323000, China.*
Corresponding author. E-mail: zhanglb@gdei.gd.cn

ABSTRACT. The activity time of the lesser bamboo bat, *Tylonycteris pachypus* (Temminck, 1840), was investigated at two observation locations in southern China: Longzhou and Guiping. Two bouts of activity (post dusk and predawn), with an intervening period of night roosting at diurnal roosts, were identified. The period of activity within each bout was usually less than 30 minutes. The activity periods of individuals belonging to the Longzhou population right after dusk and just before dawn lasted longer than those of the the Guiping population. We also found that the nocturnal emergence time of *T. pachypus* from the Longzhou population happened earlier than in the Guiping population. These findings indicate that the activity time of *T. pachypus* was quite short at night, and that different locations may affect the nocturnal activity rhythm of this species.

KEY WORDS. Activity period; emergence; return; *Tylonycteris pachypus*.

Most bats are nocturnal, foraging at night and resting in roosts during the day. Between activity bouts, they also spend time in night roosts (ANTHONY & KUNZ 1997). The patterns of nocturnal activity vary dramatically among different species. O'SHEA & VAUGHAN (1977) have reported that the pallid bat *Antrozous pallidus* (Le Conte, 1856) utilizes two foraging periods with an intervention period of night roosting. Some other species, such as *Euderma maculatum* (J.A. Allen, 1891), spend the entire night flying and foraging (WAI-PING & FENTON 1988). Likewise, the duration of bouts has also been found to be different. For instance, *Eptesicus fuscus* (Beauvois, 1796) spends only 2 hours flying each night (BRIGHAM 1991), while *Nacteris grandis* spends even less time in this activity (FENTON et al. 1990).

The timing and pattern of bat nocturnal activity may be influenced by environmental factors such as light levels (LEE & MCCRACKEN 2001), prey abundance (ERKERT 1982), temperature (CATTO et al. 1995), cloud (KUNZ & ANTHONY 1996) and rain (MCANEY & FAIRLEY 1988). Moreover, intrinsic biological factors such as predation risk (MCWILLIAM 1989, SPEAKMAN 1991), colony size, age, sex, the reproductive status of individuals (AVERY 1986, RYDELL 1989, KORINE et al. 1994, CLARK et al. 2002, O'DONNELL 2002), and interspecific competition (SWIFT & RACEY 1983, BONACCORSO et al. 2006) may also impact the nocturnal activity of bats. The intensity of competition between or among sympatric related species is expected to be greater because they are morphologically similar, which is assumed to reflect niche similarity (FINDLEY & BLACK 1983, ALDRIDGE & RAUTENBACH 1987, ARITA

1997). When common resources are limited, the niche theory predicts that resource partitioning (such as spatial and temporal niche) is necessary for species to coexist within a guild. They may forage in different habitats (ARLETTAZ 1999) and then feed on different diet items (ARLETTAZ et al. 1997), or forage during different times (BONACCORSO et al. 2006). In contrast, when their shared resources are not limited, species may forage concomitantly and for longer periods.

The lesser bamboo bat, *Tylonycteris pachypus* (Temminck, 1840) (Chiroptera: Vespertilionidae), and its sibling species, *Tylonycteris robustula* Thomas, 1915, are genetically closely related. The two species have similar morphological features (MEDWAY & MARSHALL 1978). According to our field observations, which have been published in another contribution, the two species rarely roost together in the same internode, although they overlap in distribution. Also, they can alternate their use of the same internodes at different times or seasons (ZHANG et al. 2004) and forage on similar categories of insects that occur sympatrically (ZHANG et al. 2005a).

The nocturnal activity rhythm of *T. pachypus* and *T. robustula* is still poorly known. In the present study we investigated the activity time (emergence time and foraging duration) of the lesser bamboo bat, *T. pachypus*, in two populations inhabiting Longzhou and Guiping Counties, Guangxi, south China, approximately 250 km apart from each other. *T. pachypus* is sympatric with *T. robustula* in Longzhou County, but occurs alone in Guiping County.

MATERIAL AND METHODS

The study was carried out from March to November, in 2008 in Guangxi Province, south China. Two locations (Longzhou County, 21°10′N, 106°50′E, 116 m in elevation, and Guiping County, 23°09′N, 110°10′E, 68 m in elevation) were selected. ZHANG et al. (2005a) had previously described the climate and vegetation of Longzhou County as having average annual temperature of 22.8°C and average annual precipitation of 1,180 mm (FANG 1995). Guiping County has a similar climate but the habitat is hilly rather than the typical karst of Longzhou County, with average annual temperature of 21.4°C and average annual precipitation of 1,727 mm (FANG 1995). Within both study areas, the bamboo, *Bambusa spinosa* Roxb, is abundant. This plant has enough internodes to provide enough suitable roost sites for both bat species. During data collection, *T. pachypus* and *T. robustula* were never found in the same roost at the same time, except on one occasion, when a single *T. pachypus* male and a single *T. robustula* male roosted in the same internode. In Longzhou County, bamboo is found in and around villages, while in Guiping County it is distributed along streams in areas that are somewhat far from villages.

Nocturnal observations were conducted from dusk to the next morning. Observers were split into two groups (two persons per group) and each group observed the activity time of *T. pachypus* in the two locations, at the same time. Normally, bamboo bats are faithful to their bamboo internodes for a short period (ZHANG et al. 2004), which allowed us to continuously observe groups in the same internode from dusk to dawn. We selected a fixed bamboo forest in each location to conduct our observations. The bamboo bat colony in each forest had more than two hundred individuals. We swapped among different bamboo internodes on different days during the same month, and the size of the group in each of these internodes was normally over eight bats. In each location, the time of emergence and returning were recorded, respectively. Emergence time was defined as the time when the first bat individual flied out of its bamboo internode; the returning time was defined as the time when the first bat individual flied back into the bamboo roost, or attempted to do so. If no bat emerged from a bamboo internode after 20 minutes (confirmed by using a wire to check bat lairs), we terminated the observation on emergence time. Likewise, if no bat returned back to its bamboo internode we terminated the observation on returning time. The duration of activity was defined as the time period between emergence and returning, since it is difficult to observe the entire behavior of the bat after it flies out. Additionally, *T. pachypus* forages around or over their roost bamboo forest; for example, their foraging sites are nearby their roosts (ZHANG et al. 2007). We also recorded air temperature at sunset and sunrise, and position of the roost site (via GPS, eTrex, Garmin Corp., Taiwan). The time of sunset and sunrise were read from a GPS. During the observation period, nocturnal observations were conducted for at least one week per month, normally in the middle of the month. Since *T. pachypus* is sympatric with *T. robustula* in Longzhou, we identified and confirmed the identify of *T. pachypus* using characteristics of its echolocation calls using a sound detector (D-980, Petterson Electronic AB, Uppsala) when individuals flied out or returned. This study was conducted according to the protocols approved by the Guangdong Entomological Institute Administrative Panel on Laboratory Animal Care.

A total of 75 night observations were conducted both in Longzhou and Guiping. When it was raining during the bat's normal activity period, the data of the corresponding night were omitted from both sites in the analysis. As a result, data on twelve night observations were discharged, and 63 night observations, one week per month, were analyzed. All data were tested for normality and homogeneity of variances using Kolmogorov-Smirnov test and Bartlett test. We conducted Independent-Samples t-test for comparisons of emergence time, returning time and duration of activity between the Longzhou and Guiping populations, respectively. Then for comparisons of monthly variations in activity duration, we used One-Way ANOVA. Regression was used to analyze activity time and local air temperature. All statistical analyses were performed in SPSS 17.0 for Windows. Descriptive data were expressed as Mean ± SD, and the significant difference at 95% confidence level was calculated by Central Limit Theorem.

RESULTS

Based on our observations, the lesser bamboo bat *T. pachypus* normally feeds in and around bamboos in the forests in which they roost, and fly to foraging sites directly from their roost. The activity time of *T. pachypus* individuals in the two populations was characterized by two distinct bouts, one immediately after dusk and the second just before dawn. The duration of the activity within each bout was usually less than 30 minutes. The dusk (23.7 ± 8.39 min) and predawn (27.9 ± 14.79 min) activity periods of *T. pachypus* individuals were significantly longer in Longzhou than in Guiping (dusk: 21.2 ± 8.09 min; predawn: 25.4 ± 14.43 min) (dusk: t = 2.374, p < 0.05; predawn: t = 2.857, p < 0.05) (Fig. 1). A strong positive correlation was also observed between activity duration and air temperature at dusk and predawn in the two populations, respectively (Longzhou: r_{dusk} = 0.891, $r_{predawn}$ = 0.904, respectively, both p < 0.001; Guiping: r_{dusk} = 0.881, $r_{predawn}$ = 0.837, both p < 0.001) (Figs. 2-3). In addition, we found that the durations of activity of *T. pachypus* varied significantly in both populations from one month to another (Longzhou: F = 4.069, p < 0.001; Guiping: F = 3.619, p < 0.001) (Figs. 2-3).

The time of emergence obtained for the Longzhou population at post dusk (12.7 ± 0.62 min after sunset) was significantly earlier than for the Guiping population (14.7 ± 0.48 min) (t = 2.37, p < 0.05), but the returning time at predawn did not significantly differ between them (Longzhou: 17.3 ± 0.94

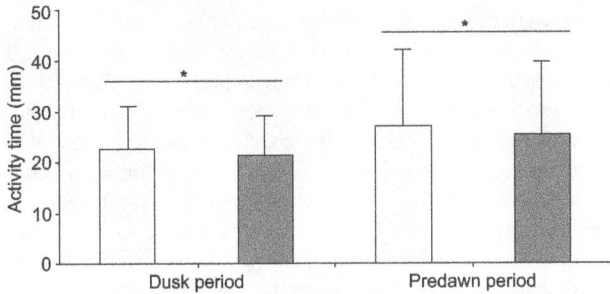

Figure 1. The foraging time of dusk (n = 63) and predawn (n = 63) periods of *Tylonycteris pachypus* in Longzhou County (empty column) and Guiping County (grey column), Guangxi, south China. All data are expressed as Mean ± SD. *p < 0.05.

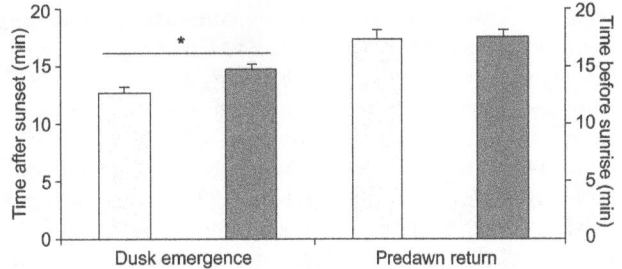

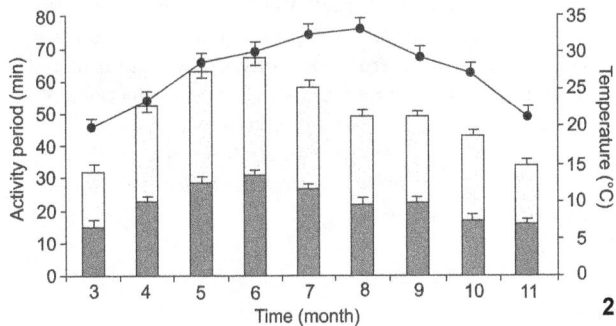

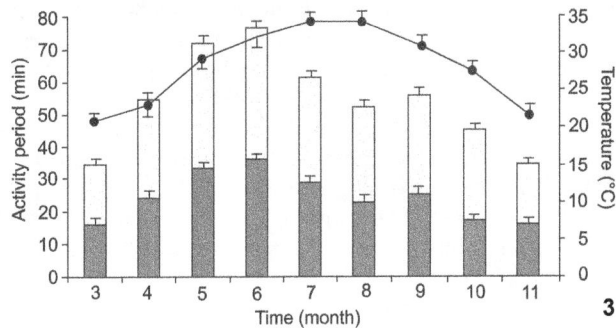

Figure 4. Time of dusk emergence (n = 63) and predawn returning (n = 63) of *Tylonycteris pachypus* in Longzhou County (empty column) and Guiping County (grey column), Guangxi, south China. All data are expressed as Mean ± SD. *p < 0.05.

Figure 2-3. The total foraging time over one night (average value of seven nights for each month) of *Tylonycteris pachypus* in Longzhou County (2) and Guiping County (3). Solid portions indicate dusk foraging periods while hollow portions indicate dawn foraging periods. The dots indicate the average temperature.

min before sunrise, Guiping: 17.5 ± 0.62 min) (t = 1.38, p > 0.05) (Fig. 4). We also found that there is a significant correlation between the time of evening emergence and the average air temperature at sunset, and between time of predawn return and average air temperature at sunrise for both populations, respectively (Longzhou: $r_{emergence}$ = 0.882, r_{return} = -0.915, respectively, both p < 0.001; Guiping: $r_{emergence}$ = 0.735, r_{return} = -0.826, both p < 0.001).

DISCUSSION

Mammals that are characterized by energetically expensive modes of locomotion and which encounter limited (temporally) food supply tend to regulate their foraging behavior. Bat species usually emerge to feed after sunset and return before sunrise (Bateman & Vaughan 1974, O'Shea & Vaughan 1977, Duvergé et al. 2000). The timing of these activity bouts correspond to the time at dusk and predawn when insects are most abundant (Kunz 1974, Racey 1982, Racey & Swift 1985, Rydell 1993, Swift 1997). Based on previous studies, most bat species spend more than one hour each night in predation activities, and their nocturnal activity rhythm can be highly variable among different species. For instance, O'Shea & Vaughan (1977) have reported that the activity of pallid bats was characterized by two foraging periods with an intervening period of night roosting. Anthony et al. (1981) have also reported that the little brown bat typically has two foraging periods at night, which are divided by a short break in night roost. In the present study, we found that the night foraging time patterns of *T. pachypus* were characterized by two bouts: dusk and predawn, respectively. Within each bout, activity duration (time period from emergence to returning) was remarkably short, less than 30 minutes. This may be correlated with the fact that the lesser bamboo bat is one of the smallest bat species, and therefore individuals can spend relatively short periods of activity time to balance energy intake and the costs to maintain their high metabolic rate during flight (Speakman 2005). On the other hand, the roosting behavior of *T. pachypus* may also contribute to their short activity time. *T. pachypus* roosts within bamboo internodes, and these restricted spaces may limit the activity of bats and subsequently reduce their energy consumption. Moreover, the short foraging distance away from their internode roosts decrease flight time as well as energy consumption (Zhang et al. 2007).

Our results indicate that in Longzhou County *T. pachypus* emerges earlier and forages for longer periods of time than that in Guiping County. On average, the activity time of individuals in Longzhou each night is approximately five minutes, which is longer than that in Guiping. Although this increase is small in magnitude, it represents a 10% increase in average activity time. The different activity behavior in different locations may have resulted from many factors such as the effect of interspecific competition (Swift & Racey 1983, Bonaccorso et al. 2006), variations in prey abundance (Erkert 1982), predation risk (McWilliam 1989, Speakman 1991), and colony size (Avery 1986, Rydell 1989, Korine et al. 1994, Clark et al. 2002, O'Donnell 2002). Both *T. pachypus* and *T. robustula* eat similar categories of insects when they occur sympatrically, although the prey of the later is somewhat larger than that of the former (Zhang et al. 2005a). In conclusion, in Longzhou where both sibling species are sympatric, the activity time of *T. pachypus* may be affected by *T. robustula*.

Many authors have documented seasonal fluctuations in emergence and returning time of bats (Bateman & Vaughan 1974, O'Shea & Vaughan 1977, Duvergé et al. 2000). Pallid bats emerge earlier in the summer sunset when compared with spring and autumn sunsets (O'Shea & Vaughan 1977). Lee & McCracken (2001) have also reported that the timing of evening emergence and returning of the Mexican free-tailed bats, *Tadarida brasiliensis* (I. Geoffroy, 1824), are correlated with the time of sunset and sunrise, and that bats are more likely to emerge earlier in relation to earlier sunset time during late summer, when compared with spring to early summer. They also return progressively later at dawn, which is associated with sunrise in the entire season. We found that the dawn returning times of *T. pachypus* were correlated with sunrise from March to November. In this study, *T. pachypus* emerged only 5-10 minutes after sunset during summer, but 10-20 minutes during spring and autumn. Pallid bats emerged 20-40 minutes after sunset (O'Shea & Vaughan 1977) and greater horseshoe bats emerged 45-53 minutes after sunset (Duvergé et al. 2000). Kunz et al. (1995) have pointed out that increased energetic demands during pregnancy and lactation could cause females, especially those lactating, to spend twice or even three times as much time foraging than bats that are neither pregnant nor lactating. As a result, emerging earlier in the evening and returning later at dawn will give these bats more time to feed, but at the cost of a greater danger of predation (Fenton 1995, Speakman et al. 1995, Rydell et al. 1996). Our results indicate that the activity time of flat-headed bats is longer in summer than in spring and in autumn. This may be influenced by prey activity, reproductive status and energy demands (Richards 1989). Female flat-headed bats usually become pregnant in May, and give birth and lactate in June (Zhang et al. 2005b). During this time, reproductive females would require more energy to maintain the increased physiological requirements of pregnancy and lactation, which is reflected in their increased activity time.

Seasonal fluctuations in activity patterns can be influenced by many factors, such as ambient temperature and/or prey availability (Anthony et al. 1981, Rydell 1989, Maier 1992). In the present study, the bamboo bats emerged earlier and returned latter when the ambient temperature was higher, which resulted in a positive association between high ambient temperature and increased activity. In March and November, when temperatures were relatively low, the activity periods of bamboo bats were obviously short than from April to October. Reduced activity in lower temperatures has also been demonstrated for many other insectivorous bat species (e.g., Anthony et al. 1981, Kronwitter 1988, Rydell 1989, Maier 1992, Catto et al. 1995). This negative influence may be a reflection of decreased food availability, because insect activity throughout the night is positively correlated with temperatures (Taylor 1963, Lewis & Taylor 1964).

In conclusion, our findings suggest that the lesser bamboo bat spends relatively short time being active, even less than half hour each bout. This behavior may be correlated with the high-energy demand for this tiny mammal to be able to fly. Secondly, our findings suggest that the activity behavior of the lesser bamboo bat, including emergence time and activity duration, varies in different locations. The variation in activity may result from many factors such as the effect of sibling sympactric species, and variations in prey abundance, predation risk, and colony size in different locations. Further studies should be conducted to confirm the factors that influence the variations in bat activity.

ACKNOWLEDGEMENTS

We thank Hui Qin, Shen-Ming Huang (Guangxi Normal University) for their assistance in the field, and Stuart Parsons (University of Auckland) and Yi Chen for assistance in writing this manuscript. This study was financed by the Special Planning of Major Scientific and Technological Production (Cultivation Project), Guangdong Academy of Sciences (ZDCCYD201307), Special Foundation for Innovative Scientists of Guangdong Entomological Institute (GDEI-cxrc201303), and Science & Technology Planning Project of Guangdong (2013B050800024). Li-Biao Zhang and Fu-Min Wang contributed equally to this paper.

LITERATURE CITED

Aldridge HDJN & Rautenbach IL (1987) Morphology, echolocation and resource partitioning in insectivorous bats. **Journal of Animal Ecology 56**: 763-778. doi: 10.2307/4947.

Anthony ELP, Kunz TH (1997) Feeding strategies of the little brown bat, *Myotis lucifugus*, in southern New Hampshire. **Ecology 58**: 775-786. doi: 10.2307/1936213

Anthony ELP, Stack MH, TH Kunz (1981) Night roosting and the nocturnal time budget of the little brown bat, *Myotis lucifugus*: effects of reproductive status, prey density, and environmental conditions. **Oecologia 51**: 151-156. doi: 10.1007/BF00540593

Arita E (1997) Species composition and morphological structure of the bat fauna of Yucatan, Mexico. **Journal of Animal Ecology 66**: 83-97. doi.org/10.2307/5967

ARLETTAZ R (1999) Habitat selection as a major resource partitioning mechanism between the two sympatric sibling bat species *Myotis myotis* and *Myotis blythii*. **Journal of Animal Ecology** 68: 460-471. doi: 10.1046/j.1365-2656.1999.00293.x

ARLETTAZ R, PERRIN N, HAUSSER J (1997) Trophic resource partitioning and competition between the two sibling bat species *Myotis myotis* and *Myotis blythii*. **Journal of Animal Ecology** 66: 897-991. doi: 10.2307/6005

AVERY MI (1986) Factors affecting the emergence times of *Pipistrelle* bats. **Journal of Zoology (London)** 209: 293-296. doi: 10.1111/j.1469-7998.1986.tb03589.x

BATEMAN GC, VAUGHAN TA (1974) Nightly activities of mormoopid bats. **Journal of Mammalogy 55**: 45-65. doi: 10.2307/1379256

BONACCORSO FJ, WINKELMANN JR, SHIN D, AGRAWAL CI, ASLAMI N, BONNEY C, HSU A, JEKIELEK PE, KNOX AK, KOPACH SJ, JENNINGS TD, LASKY JR, MENESALE SA, RICHARDS JH, RUTLAND JA, SESSA AK, ZHAUROVA L, KUNZ TH (2006) Evidence for exploitative competition: comparative foraging behavior and roosting ecology of short-tailed fruit bats (Phyllostomidae). **Biotropica** 39: 249-256. doi: 10.1111/j.1744-7429.2006.00251.x

BRIGHAM RM (1991) Flexibility in foraging and roosting behavior by the big brown bat (*Eptesicus fuscus*). **Canada Journal of Zoology 69**: 117-121. doi: 10.1139/z91-017

CATTO CMC, RACEY PA, STEPHENSON PJ (1995) Activity patterns of the serotine bat (*Eptesicus serotinus*) at a roost in southern England. **Journal of Zoology (London)** 235: 635-644. doi: 10.1111/j.1469-7998.1995.tb01774.x

CLARK BS, CLARK BK, LESIE JR DM (2002) Seasonal variation in activity patterns of the endangered Ozark big-eared bat (*Corynorhinus townsendii ingens*). **Journal of Mammalogy 83**: 590-598. doi: 10.1644/1545-1542

DUVERGÉ PL, JONES G, RYDELL J, RANSOME RD (2000) Functional significance of emergence timing in bats. **Ecography 23**: 32-40. doi: 10.1034/j.1600-0587.2000.230104.x

ERKERT HG (1982) Ecological aspects of bat activity rhythms, p. 201-242. In: KUNZ TH (Ed.). **Ecology of bats.** New York, Plenum Press. doi: 10.1007/978-1-4613-3421-7-5

FANG WZ (1995) **Natural resource of Guangxi, China.** Beijing, China Environment Science Press, 369p.

FENTON MB (1995) Constraint and flexibility-bats as predators, bats as prey. **Symposium of the Zoological Society of London 67**: 277-290.

FENTON MB, SWANEPOEL CM, BRIGHAM RM, CEBEK E, HICKEY MB (1990) Foraging behavior and prey selection by large slit-faced bats (*Nycteris grandis*; Chiroptera: Nycteridae). **Biotropica 22**: 2-8. doi: 10.2307/2388713

FINDLEY JS, BLACK H (1983) Morphological and dietary structuring of a Zambian insectivorous bat community. **Ecology 64**: 625-630. doi: 10.2307/1937180

KORINE C, IZHAKI I, MAKIIN D (1994) Population structure and emergence order in the fruit-bat (*Rousettus aegyptiacus*: Mammalia, Chiroptera). **Journal of Zoology (London) 232**: 163-174. doi: 10.1111/j.1469-7998.1994.tb01566.x

KRONWITTER F (1988) Population structure, habitat use, and activity patterns of the noctule bat, *Nyctalus noctula* Schreib. 1774 (Chiroptera: Vespertilionidae) revealed by radio-tracking. **Myotis 26**: 23-85.

KUNZ TH (1974) Feeding ecology of a temperate insectivorous bat (*Myotis velifer*). **Ecology 55**: 693-711.

KUNZ TH, ANTHONY ELP (1996) Variation in the timing of nightly emergence behavior in the little brown bat, *Myotis lucifugus* (Chiroptera: Vespertilionidae), p. 225-235. In: GENOWAYS HH, BAKER RJ (Eds.). **Contribution in Mammalogy: a memorial volume honoring Dr. J. Knox.** Lubbock, The Museum of Texas Tech University, 315p.

KUNZ TH, WHITAKER JR JO, WADANOLI MD (1995) Dietary energetics of the insectivorous Mexican free-tailed bats (*Tadarida brasiliensis*) during pregnancy and lactation. **Oecologica 101**: 407-415. doi: 10.1007/BF00329419

LEE Y-F, MCCRACKEN GF (2001) Timing and variation in the emergence and return of Mecican free-tailed bat, *Tadarida brasiliensis mexicana*. **Zoological Study 40**: 309-316.

LEWIS T, TAYLOR LR (1964) Diurnal periodicity of flight by insects. **Transactions of the Entomological Society of London 116**: 396-435. doi: 10.1111/j.1365-2311.1965.tb02304.x

MAIER C (1992) Activity patterns of pipistrelle bats (*Pipistrellus pipistrellus*) in Oxfordshire. **Journal of Zoology (London) 116**: 396-435. doi: 10.1111/j.1469-7998.1992.tb04433.x

MCANEY CM, FAIRLEY JS (1988) Activity patterns of the lesser horseshoe bat *Rhinolophus hipposideros* at summer roosts. **Journal of Zoology (London) 216**: 325-338. doi: 10.1111/j.1469-7998.1988.tb02433.x

MCWILLIAM AM (1989) Emergence behavior of the bat *Tadarida (Chaerephon) pumila* (Chiroptera: Molossidae) in Ghana, West Africa. **Journal of Zoology (London) 219**: 698-701. doi: 10.1111/j.1469-7998.1989.tb02615.x

MEDWAY L, MARSHALL AG (1978) Roost-site selection among flat-headed bats (*Tylonycteris* spp.). **Journal of Zoology (London) 161**: 237-245. doi: 10.1111/j.1469-7998.1970.tb02038.x

O'DONNELL CFJ (2002) Influence of sex and reproductive status on nocturnal activity of long-tailed bats (*Chalinolobus tuberculatus*). **Journal of Mammalogy 83**: 794-803. doi: 10.1644/1545-1542

O'SHEA TJ, VAUGHAN TA (1977) Nocturnal and seasonal activities of the pallid bat, *Antrozous pallidus*. **Journal of Mammalogy 58**: 269-284. doi: 10.2307/1379326

RACEY PA (1982) Ecology of bat reproduction, p. 335-427. In: T.H. KUNZ (Ed.). **Ecology of bats.** New York, Plenum Publishing. doi: 10.1007/978-1-4613-3421-7_2

RACEY PA, SWIFT SM (1985) Feeding ecology of *Pipistrellus pipistrellus* (Chiroptera: Vespertilionidae) during pregnancy and lactation. I. Foraging behavior. **Journal of Animal Ecology 54**: 205-215. doi: 10.2307/4631

RICHARDS GC (1989) Nocturnal activity of insectivorous bats relative to temperature and prey availability in tropical Queensland. **Australian Wildlife Research 16**: 151-158. doi: 10.1071/WR9890151

RYDELL J (1989) Feeding activity of the northern bat *Eptesicus nilssoni* during pregnancy and lactation. **Oecologica 80:** 562-565. doi: 10.1007/BF00380082

RYDELL (J) 1993. Variation in foraging activity of an aerial insectivorous bat during reproduction. **Journal of Mammalogy 74:** 503-509. doi: 10.2307/1382411

RYDELL J, A ENTWISTLE, PA RACEY (1996) Timing of foraging flights of three species of bats in relation to insect activity and predation risk. **Oikos 76:** 243-252. doi: 10.2307/3546196

SPEAKMAN JR (1991) Why do insectivorous bats in Britain not fly in daylight more frequently? **Functional Ecology 5:** 518-524. doi: 10.2307/2389634

SPEAKMAN JR (2005) Body size, energy metabolism and lifespan. **Journal of Experimental Biology 208:** 1717-1730. doi: 10.1242/jeb.01556

SPEAKMAN JR, STONE RE, KERSLAKE JE (1995) Temporal patterns in the emergence behavior of pipistrelle bats, *Pipistrellus pipistrellus*, from maternity colonies are consistent with an anti-predator response. **Animal Behavior 50:** 1147-1156. doi: 10.1016/0003-3472(95)80030-1

SWIFT SM (1997) Roosting and foraging behavior of Natterer's bats (*Myotis nattereri*) close to the northern border of their distribution. **Journal of Zoology (London) 242:** 375-384. doi: 10.1111/j.1469-7998.1997.tb05809.x

SWIFT SM, RACEY PA (1983) Resource partitioning in two species of vespertilionid bats (Chiroptera) occupying the same roost. **Journal of Zoology (London) 200:** 249-2593. doi: 10.1111/j.1469-7998.1983.tb05787.x

TAYLOR LR (1963) Analysis of the effect of temperature on insects in flight. **Journal of Animal Ecology 32:** 99-117.

WAI-PING V, FENTON MB (1988) Ecology of spotted bat (*Euderma maculatum*): roosting and foraging behavior. **Journal of Mammalogy 70:** 617-622. doi: 10.2307/1381434

ZHANG L-B, LIANG B, JONES G, PARSONS S, WEI L, ZHANG S-Y (2007) Morphology, echolocation and foraging behavior in two sympatric sibling species of bat (*Tylonycteris pachypus* and *T. robustula*) (Chiroptera: Vespertilionidae). **Journal of Zoology (London) 271:** 344-351. doi: 10.1111/j.1469-7998.2006.00210.x

ZHANG L-B, LIANG B, ZHOU S-Y, LU L-R, ZHANG S-Y (2004) Group structure of lesser flat-headed bat *Tylonycteris pachypus* and greater flat-headed bat *T. robustula*. **Acta Zoology Sinica 50:** 326-333.

ZHANG L-B, JONES G, ROSSITER S, ADES G, LIANG B, ZHANG S-Y (2005a) Diet of flat-headed bats, *Tylonycteris pachypus* and *T. robustula*, in Guangxi, South China. **Journal of Mammalogy 86:** 61-66. doi: 10.1644/1545-1542

ZHANG L-B, JONES G, PARSONS S, LIANG B, ZHANG S-Y (2005b) Development of vocalizations in the flat-headed bats, *Tylonycteris pachypus* and *T. robustula* (Chiroptera: Vespertilionidae). **Acta Chiropterologica 7:** 91-99. doi: 10.3161/1733-5329

20

A new species of *Masteria* (Araneae: Dipluridae: Masteriinae) from southeastern Brazil

Denis Rafael Pedroso[1], Renner Luiz Cerqueira Baptista[2] & Rogério Bertani[3,4]

[1]*Laboratório de Aracnologia, Museu Nacional, Universidade Federal do Rio de Janeiro, Brazil. E-mail: drbpedroso@gmail.com*
[2]*Laboratório de Diversidade de Aracnídeos, Instituto de Biologia, Universidade Federal do Rio de Janeiro, Brazil.*
E-mail: baptistr@gmail.com
[3]*Laboratório Especial de Ecologia e Evolução, Instituto Butantan. Avenida Vital Brazil 1500, 05503-900 São Paulo, SP, Brazil.*
[4]*Corresponding author. E-mail: rogerio.bertani@butantan.gov.br, rogerio.bertani@uol.com.br*

ABSTRACT. A new species of *Masteria* L. Koch, 1873 from iron ore caves at Caeté and Santa Bárbara, state of Minas Gerais, Brazil, *Masteria emboaba* **sp. nov.**, is described. It was collected inside caves and in the litter of nearby dry forests. It is the first masteriine species described from southeastern Brazil and the second masteriine species for the country. The new species is the only known *Masteria* with only two eyes. Additionally, the male of *M. emboaba* **sp. nov.** has only two regular, thin spines at the apex of tibia I, lacking the tibial apophysis found in most other *Masteria* species. The only other described *Masteria* species that has spines in the place of tibial apophysis is *M. aimeae* (Alayón, 1995) from Cuba; however, the last species has a longer and sinuous embolus, contrasting the embolus of *M. emboaba* **sp. nov.**, which is much smaller, less sinuous and transversally placed. The only other described Brazilian species, *M. manauara* Bertani, Cruz & Oliveira, 2013, has a double tibial apophysis, with both ends tipped by a strong, short spine, and a very long embolus, parallel to the bulb.

KEYWORDS. Iron ore; lateritic cave; Minas Gerais; Mygalomorphae; Serra da Gandarela.

Masteria L. Koch, 1873 belongs to Dipluridae, Masteriinae (Raven 1985), which contains many of the smallest mygalomorph species known (Raven 1979, 1981, Raven & Platnick 1981, Platnick & Foster 1982). The 23 described species are widely distributed in the Australasian Region and in the New World (Platnick 2014). The Australasian region has eight species, including the type species, *M. hirsuta* L. Koch, 1873, from the Fiji Islands and Micronesia. In the New World, there are 15 described species: *M. aimeae* (Alayón, 1995) and *M. golovatchi* Alayón, 1995 – Cuba; *M. lewisi* (Chickering, 1964) and *M. pecki* Gertsch, 1982 – Jamaica; *M. petrunkevitchi* (Chickering, 1964) – Puerto Rico; *M. modesta* (Simon, 1891) – Saint Vincent; *M. barona* (Chickering, 1966) and *M. simla* (Chickering, 1966) – Trinidad; *M. downeyi* (Chickering, 1966) and *M. spinosa* (Petrunkevitch, 1925) – Costa Rica and Panama; *M. colombiensis* Raven, 1981 – Colombia; *M. cyclops* (Simon, 1889), *M. lucifuga* (Simon, 1889) and *M. tovarensis* (Simon, 1889) – Venezuela. Finally, a single species is known from Brazil (state of Amazonas), *M. manauara* Bertani, Cruz & Oliveira, 2013. It inhabits the ground litter (Bertani et al. 2013), particularly amongst the leaves of small palm trees.

Besides *M. manaura*, there are records of an unidentified *Masteria* species from the state of Piauí (L.S. Carvalho unpubl. data) and several areas of the state of Minas Gerais: Nova Lima (AngloGold Ashanti 2009), Conceição de Mato Dentro (Leão & Auler 2012), Caeté and Santa Bárbara (Coelho et al. 2010). The records from Minas Gerais are based on material collected in and around caves in iron ore deposits.

There are more than 3,000 caves in iron ore deposits in Brazil (Auler et al. 2014). Most of these caves are located at the two major iron ore provinces: Carajás ridge, in the Amazon, and Iron Quadrangle, in Minas Gerais, southeastern Brazil. The Iron Quadrangle caves are found on iron-rich deposits topped by "canga", an iron-rich breccia surface cemented by ferruginous matrix (Auler et al. 2014). The vegetation of the "canga" areas is open grassland with scattered trees and is dominated by herbs and bushes. The alfa and beta diversity of plant species is high, including dozens of rare and endemic species (Carmo & Jacobi 2013). The plant species are a mix of elements from the Atlantic Forest, Cerrado and Serra do Espinhaço (Carmo & Jacobi 2013).

Despite the small average size of iron ore caves, they have "a high potential as habitat of troglobitic invertebrates in Brazil" (Trajano & Bichuette 2010). Several troglobitic species have been found in Brazilian iron ore caves, both in the Carajás area (Pellegrini & Ferreira 2011, Pedroso & Baptista 2014) and the Iron Quadrangle (Souza & Ferreira 2005, Coelho et al. 2010). Caves from iron ore areas are usually near the surface and have an extensive array of microchannels that house a highly diverse associated fauna (Ferreira 2005). The fauna of the "canga" areas of Minas Gerais also include many endemic taxa (e.g., BERNARDI et al. 2013, ÁZARA & FERREIRA 2013).

Herein we describe the second masteriine species from Brazil and the first from Southeast Brazil. In addition, it is the first Masteriinae from iron ore caves collected at Caeté and Santa Barbára, Minas Gerais.

MATERIAL AND METHODS

Specimens were collected inside and near iron ore caves located at the hilltops of "Serra da Gandarela" (Gandarela range), at Caeté and Santa Bárbara, in the Iron Quadrangle area, state of Minas Gerais, southeastern Brazil. The caves are located on "canga" plates; however, the collecting area also included dry patches of Atlantic forest, especially on the small river valleys near the tips of the hill range. Most specimens were collected during a trip of the "Projeto Mina Apolo", a project to study the environmental impact from the installation of a large iron ore mine. The first author participated in the project as a member of the environmental consulting "Amplo Consultoria" team. Additional specimens from the caves in the collecting area had been mentioned by COELHO et al. (2010).

The general description format follows RAVEN (1981) with some modifications; e.g., eye diameters are given in their real measurements, not in ratios. All measurements are in millimeters and were obtained with a Leica LAS Interactive Measurements module. Leg and palp measurements were taken from the dorsal aspect of the left side (unless appendages were lost or obviously regenerated). A Leica LAS Montage and LAS 3D module, mounted on a Leica M205C dissecting microscope, were used to capture images of the structures of spiders. Illustrations of spermathecae were drawn over images obtained with a Leica DM 2500 compound microscope. The spermathecae soft tissues were digested with trypsin for several days and subsequently cleared with clove oil before they were photographed.

Abbreviations. (ITC) inferior tarsal claw, (PLS) posterior lateral spinnerets, (PME) posterior median eyes, (PMS) posterior median spinnerets, and (STC) superior tarsal claws.

Specimens are deposited in the arachnological collection of the Museu Nacional, Universidade Federal do Rio de Janeiro (MNRJ).

TAXONOMY

Masteria emboaba **sp. nov.**

Figs. 1-16

Diagnosis. Males and females of *Masteria emboaba* **sp. nov.** have only two eyes (Figs. 5 and 14). Most other species have six to eight eyes and two, *M. pecki* and *M. caeca* (Simon, 1892), have no eyes at all. This new species is much smaller than the other known South American species, except for *M. manauara*, a tiny species from the state of Amazonas and the only other species described from Brazil. The male of *M.*

emboaba **sp. nov.** has only two regular, thin spines at the apex of tibia I, lacking the tibial apophysis found in most other *Masteria* species. The only other described *Masteria* species with spines in the place of tibial apophysis is *M. aimeae*, from Cuba; however, the last species has a longer and sinuous embolus (ALAYÓN 1995: figs. 1c-d), contrasting with the embolus of *M. emboaba* **sp. nov.**, which is much smaller, less sinuous and transversally placed (Figs. 6-8). The only other described Brazilian species, *M. manauara*, has a double tibial apophysis, with both ends tipped by a strong, short spine, and a very long embolus, parallel to the bulb (BERTANI et al. 2013: figs. 6-9). Table I summarizes the available information on the geographical range and characters of all species of *Masteria*.

Description. Male holotype (MNRJ 4540) (Fig. 1). Entirely pale yellow. Carapace 1.22 long, 0.96 wide, clothed with long (ca. 0.15) prostate gray bristles on interstrial ridges (Figs. 1 and 2). Two eyes on tubercle occupying 0.28 of head width (Fig. 5). Eye group 0.12 wide. Sizes and interdistances: PME 0.04, PME-PME 0.04. Chelicerae 0.29 long, 0.21 wide, with nine spaced teeth on promarginal furrow and six spinules mesobasally. Labium 0.10 long, 0.20 wide. Maxillae 0.31 long, 0.26 wide. Sternum 0.66 long, 0.63 wide; sigilla not evident (Fig. 3).

Palp with elongated cymbium, bearing seven spines on apical edge; bulb pear-shaped, tegulum 0.22 long, 0.12 wide, tapering and giving origin to a short (0.13) and relatively thin (0.012) embolus, almost transversally placed, due to a strong basal curve to retrolateral side, keeping its diameter and with a gentle curvature on apical half, not tapering to apex (Figs. 6-8). Leg lengths and midwidths in Table II.

Leg formula 4123. Tibia I lacking spur, with two large, thin spines ventrally on distal edge. Metatarsus I lacking spur (Fig. 9). Spines elongate: leg I tibia v3, metatarsus v2; leg 2, patella d1, tibia p1, v4, metatarsus v4; leg 3, patella d3, tibia d3, p2, r1, v5, metatarsus d6, p1, v3; leg 4, patella d2, tibia d4, p3, r1, v4, metatarsus d7, p2, r1, v3; palp, tarsus 7. Five to seven teeth on STC; zero to three teeth on ITC. Abdomen 1.42 long, 0.98 wide. Spinnerets (Fig. 4): PMS 0.20 long, 0.81 wide, 0.24 apart; basal, middle, and apical segments of PLS 0.34 long, 0.14 wide; 0.29 long, 0.13 wide; 0.29 long, 0.11 wide, respectively.

Female paratype (MNRJ 4540). As in male, except as noted. Abdominal tegument translucent, allowing recognition of internal structures (Figs. 10 and 13). Carapace 1.20 long, 0.97 wide, clothed with long (ca. 0.11) prostate gray bristles on interstrial ridges (Fig. 11). Two eyes on tubercle occupying 0.36 of head width (Fig. 14). Eye group 0.07 wide. Sizes and interdistances: PME 0.05, PME-PME 0.05. Chelicerae 0.31 long, 0.26 wide, with eleven widely spaced teeth on promarginal furrow and six spinules mesobasally. Labium 0.10 long, 0.22 wide. Maxillae 0.32 long, 0.28 wide. Sternum 0.72 long, 0.65 wide; sigilla not evident (Fig. 12). Leg lengths and midwidths in Table II. Leg formula 4123. Spines elongate: leg 1, metatarsus v2; leg 2, metatarsus v3; leg 3, tibia d4, v2, metatarsus d5, v4; leg 4, tibia d4, v4, metatarsus d6, v3; palpus, tarsus v4. Four

Figures 1-9. *Masteria emboaba* **sp. nov.** Holotype male (MNRJ 4540) (1) habitus; (2) carapace and chelicerae; (3) sternum, maxillae, chelicerae, coxae and labium; (4) abdomen, ventral; (5) eye tubercle; (6-8) left male palp, (6) retrolateral, (7) prolateral, (8) ventral; (9) left leg I, ventral. Scale bars: 1-2, 4 = 1 mm, 5-9 = 0.1 mm, 3 = 0.5 mm.

Table I. Comparison of geographical distribution and main characters of species of *Masteria*. (M) Male, (F) female, (I) immature, "(–)" non-applicable, "(?)" data not available.

Species	Sex	Type-Locality	Eyes	Carapace length	Spermathecae	Apex of male tibia I	Metatarsal apophysis	Paraembolic apophysis	Embolus
M. aimeae (Alayón, 1995)	M	Cuba: Holguín	6	1.3	–	four spines	no	no	long, curved
M. barona (Chickering, 1966)	MF	Trinidad: Arima Valley, Simla	6	M 1.69 F 1.63	?	holotype legs I missing	holotype legs I missing	no	long, thickened, sinuous
M. caeca (Simon, 1892)	F	Philippines: Morong, d'Antipolo Cave	0	?	?	–	–	–	–
M. cavicola (Simon, 1892)	F	Philippines: Manila, Montalvan, San-Mateo Cave	6	?	?	–	–	–	–
M. colombiensis Raven, 1981	MF	Colombia: Magdalena	8	M 1.8 F 2.1	two spermathecae, basally divided, wide stalk, rounded tip	three distal processes on tibia articulation.	yes	no	short, slightly curved
M. cyclops (Simon, 1889)	I	Venezuela: Caracas, Catuche Forest	6	?	?	–	–	–	–
M. downeyi (Chickering, 1966)	MF	Costa Rica: Turrialba	6	M 1.67 F 2.05	?	three distal processes on tibia articulation	no	no	very short, straight
M. emboaba **sp. nov.**	MF	Brazil: Minas Gerais	2	M 1.2 F 1.2	Two spermathecae, double thin stalk, wide flattened tip	two unmodified spines	no	no	short, slightly curved
M. franzi Raven, 1991	M	New Caledonia: Hienghène, Tiouandé	6	1.25	–	three distal processes on tibia articulation	?	yes	long, filiforme, almost straight
M. golovatchi Alayón, 1995	M	Cuba: Guantánamo, Paso Cuba	6	1.5	–	six spines	no	no	very short, curved
M. hirsuta L. Koch, 1873	F	Fiji, Micronesia: Ovalau	6	F 2.75	four spermathecae, the outer ones large and rounded, the inner short and coniform; all lobes lack circular ribbing (Raven, 1991)	–	–	–	–
M. kaltenbachi Raven, 1991	F	New Caledonia: Nékliai	6	1.0	four spermathecae, the outer ones with a spiral ribbing	–	–	–	–
M. lewsi (Chickering, 1964)	MF	Jamaica: St. Catherine Parish	6	1.34	?	three distal processes on tibia articulation	no	no	short, thickened, tapering strongly on its distal portion
M. lucifuga (Simon, 1889)	I	Venezuela: Aragua, Colonia Tovar	6	?	?	–	–	–	–
M. macgregori (Rainbow, 1898)	F	New Guinea: Neneba	6	1.7	four spermathecae	–	–	–	–
M. manauara Bertani, Cruz & Oliveria, 2013	MF	Brasil: Amazonas	6	M 0.7 F 0.8	two spermathecae, convoluted stalk, large rounded tip	one upper and one lower spur, both with an enlarged spine at tip	yes	no	long, very thin
M. modesta (Simon, 1891)	F	St. Vincent	6	?	?	–	–	–	–
M. pallida (Kulczyn'ski, 1908)	F	New Guinea	6	?	?	–	–	–	–
M. pecki Gertsch, 1982	F	Jamaica: Falling Cave	0	2.75	two low suboval spermathecae	–	–	–	–
M. petrunkevitchi (Chickering, 1964)	MF	Puerto Rico: Mayaguez	8	1.69	?	–	no	no	short, thickened, tapering to its tip
M. simla (Chickering, 1966)	MF	Trinidad: Arima Valley, Simla	8	M 1.52 F 1.78	?	three distal processes on tibia articulation	yes	yes	short, curved
M. spinosa (Petrunkevitch, 1925)	MF	Panama: San Lorenzo River	8	M 1.86 F 2.05	?	three distal processes on tibia articulation	no	yes	short, thin, curved
M. toddae Raven, 1979	MF	Australia: Queensland, Home Rule	6	M 1.56 F 2.26	four spermathecae	three distal processes on tibia articulation	no	yes	short, thin, straigh
M. tovarensis (Simon, 1889)	F	Venezuela: Aragua, Colonia Tovar	6	?	?	–	–	–	–

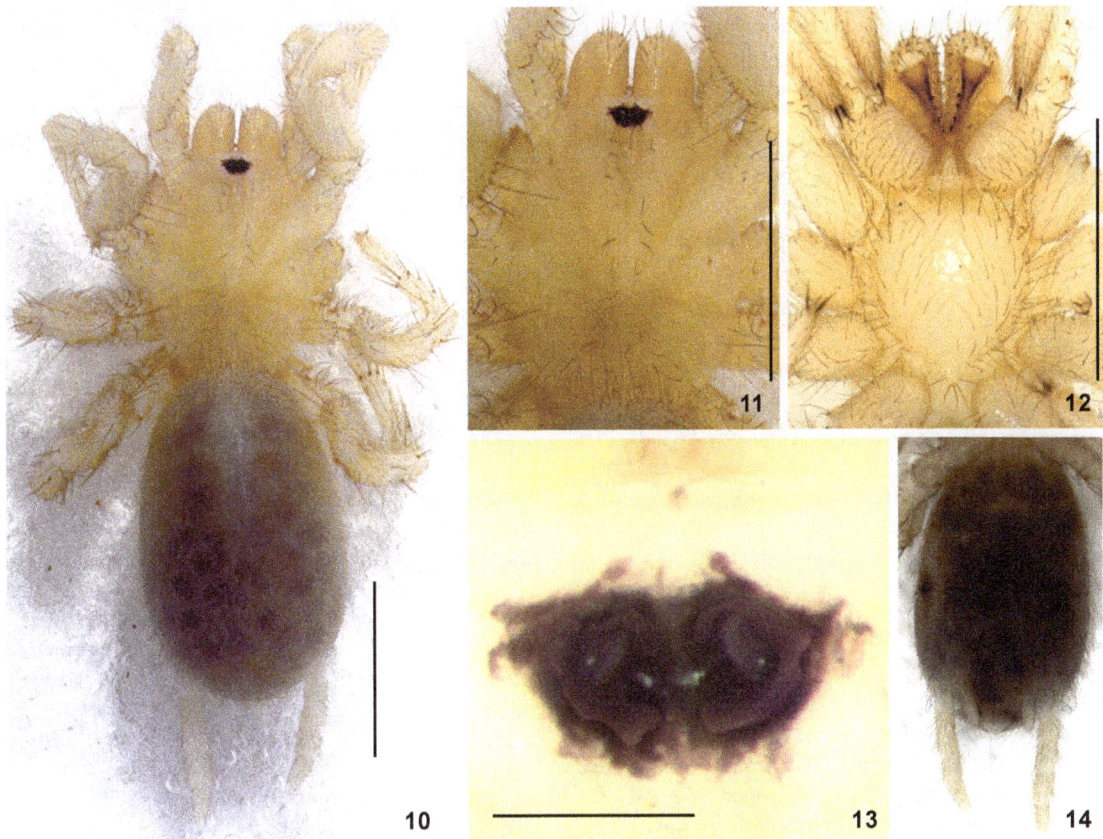

Figures 10-14. *Masteria emboaba* **sp. nov.** Paratype female (MNRJ 4540): (10) habitus; (11) carapace and chelicerae; (12) sternum, maxillae, chelicerae, coxae and labium; (13) eye tubercle; (14) abdomen and spinnerets, ventral. Scale bars: 10-12, 14 = 1 mm, 13 = 0.1 mm.

to eleven teeth on STC; two to four on ITC. Abdomen 1.86 long, 1.22 wide. Spinnerets (Fig. 13): PMS 0.23 long, 0.09 wide, 0.34 apart; basal, middle, and apical segments of PLS 0.33 long, 0.14 wide; 0.19 long, 0.14 wide; 0.22 long, 0.11 wide, respectively. Epigastric plate not posteriorly produced; two spermathecae, each one with one pair of long, thin stalks bearing a distal rounded and somewhat flattened receptacle. The outer stalk-receptacle set is smaller than the inner one (Fig. 15).

Type material. Male Holotype: BRAZIL, *Minas Gerais*: Caeté (inside a natural cavity, 20°01′40″S, 43°40′52″W, 1,484 m a.s.l., highlands of Serra da Gandarela), May 2011, Bichuettte, M.E. *leg.* (MNRJ 4540). Paratypes: BRAZIL, *Minas Gerais*: Caeté, same data as holotype (2 females, 1 immature, MNRJ 4540); near cave AP. 09 (20°01′33″S, 43°40′54″W), 1,439 m a.s.l., Projeto Mina Apolo, sifting forest litter, July 09 2011, Equipe Aracno *leg.* (2 females, 10 immatures, MNRJ 4388); Santa Bárbara: near cave AP. 31 (20°02′14″S, 43°40′38″W), 1,443 m a.s.l., Projeto Mina Apolo, sifting forest litter, July 08 2011, Equipe Aracno *leg.* (1 immature, MNRJ 4380).

Additional material. BRAZIL, *Minas Gerais*: Caeté: (near cave AP. 54, 20°01′40″S, 43°40′52″W, 1,484 m a.s.l.), Projeto Mina

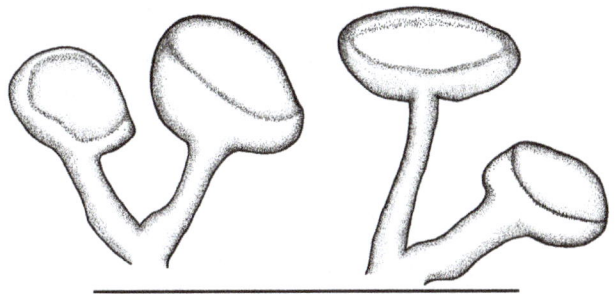

Figure 15. *Masteria emboaba* **sp. nov.** paratype (MNRJ 4540), spermathecae dorsal view. Scale bar: 100 μm.

Apolo, sifting forest litter, July 06 2011, Equipe Aracno *leg.* (1 immature, MNRJ 4378); same locality, under stones and rotten wood, forest, September 24 2011, Equipe Aracno *leg.* (1 female, 1 immature, MNRJ 4436; 1 immature, MNRJ 4437).

Distribution. Only known from small tracts of Atlantic Forest and caves on "canga" areas, on the hilltops of the Gandarela range, Caeté and Santa Bárbara, state of Minas Gerais, Brazil.

Table II. *Masteria emboaba* **sp. nov.** Male holotype and femlale paratype (MNRJ 4540). Length and midwidths of right legs and palpal segments.

Length/Midwidths	Male holotype						Female paratype					
	Femur	Patella	Tibia	Metatarsus	Tarsus	Total (length)	Femur	Patella	Tibia	Metatarsus	Tarsus	Total (length)
Pp	0.54/0.19	0.31/0.16	0.41/0.16	–	0.33/0.14	1.59	0.53/0.17	0.38/0.18	0.40/0.17	–	0.46/0.14	1.77
I	0.84/0.26	0.46/0.18	0.54/0.17	0.49/0.13	0.30/0.11	2.63	0.85/0.29	0.45/0.26	0.59/0.23	0.46/0.15	0.39/0.12	2.74
II	0.77/0.25	0.47/0.19	0.47/0.17	0.43/0.14	0.38/0.11	2.52	0.72/0.23	0.48/0.20	0.44/0.20	0.44/0.15	0.36/0.12	2.44
III	0.70/0.23	0.42/0.19	0.45/0.17	0.50/0.12	0.43/0.09	2.50	0.66/0.23	0.38/0.21	0.44/0.18	0.48/0.13	0.41/0.11	2.37
IV	0.85/0.23	0.50/0.21	0.64/0.18	0.70/0.11	0.48/0.08	3.17	0.88/0.21	0.55/0.23	0.65/0.17	0.64/0.14	0.52/0.10	3.24

Etymology. The specific name, "emboaba" is a noun in apposition and refers to the historical episode "Guerra dos Emboabas" (loosely translated as "War of the Emboabas"). This episode was a series of fights between gold miners from different regions of Brazil in the early 18th century throughout Minas Gerais, especially at the Caeté region (MELLO 1979).

Remarks. The spiders were whitish and almost translucent when alive (Fig. 16), but became yellowish and opaque in alcohol (Figs. 1 and 10). Additional specimens of *M. emboaba* **sp. nov.** were collected during the initial phase of the "Mina Apolo" project (COELHO et al. 2010), inside several of the "canga" caves, located between 20°01'33"S and 21°02'31"S to 43°40'25"W and 43°41'18"W. The *Masteria* specimens from a locality near Caeté, in Nova Lima, Minas Gerais, mentioned by ANGLOGOLD ASHANTI (2009), may also belong to *M. emboaba* **sp. nov.**

Natural history. *Masteria emboaba* **sp. nov.** was collected both inside "canga" caves and in the dry forested tracts near the "canga" area. However, they were not found in open grassland areas covering the "canga" around the caves (Fig. 17). In the dry forest (Fig. 18), they were found sieving through the litter, or were spotted under rotten wood and stones. Therefore, we may assume that this species is associated with forest litter, eventually invading nearby caves. Cave colonization would be an easy step for an animal that is already adapted to small cavities in litter. The small patches of dry forests covering the slopes of drainage valleys do not present a continuous and deep litter layer, but there are litter pockets amassed in suitable areas. The "canga" caves are placed just below the surface and are penetrated by many plant roots. They seem to offer a suitable environment for the species, with high humidity and plenty of spider food. In addition, there are many access points to the cave through the microchannels in the porous iron matrix (FERREIRA 2005). These access points may serve as a refuge for animals during dry periods, allowing the species to survive the long, dry winter of the area.

The *Masteria* species for which habitat information is available live in underground habitats, for instance in the depths of litter or in caves. Their pale color and some degree of eye reduction seem to be correlated with life in dark habitats. Most species of *Masteria* have only six eyes (e.g., SIMON 1889, 1892, RAVEN 1991), and those that have eight eyes have a pronounced reduction of the anterior median pair, e.g. *M. petrunkevitchi* and *M. simla* (CHICKERING 1964, 1966). The complete loss of eyes is found in two troglobite species: *M. caeca*, from Phillipines (SIMON 1892), and *M. pecki*, from Jamaica (GERTSCH 1982). The reduction to only two eyes in *Masteria emboaba* **sp. nov.** may point to a high degree of specialization to underground habitats.

Figures 16-18. (16) *Masteria emboaba* **sp. nov.**, living female, near AP-54; (17) view of typical open grassland vegetation found in "canga" areas; (18) view of dry forest patches covering drainage valleys near "canga" areas. Photos: D. Pedroso.

ACKNOWLEDGMENTS

We thank Norman Platnick for help with the literature. Support: FAPERJ PhD grant for D. Pedroso, FAPESP 2012/01093-0 and CNPq Research Fellow-Brazil for R. Bertani.

LITERATURE CITED

ALAYÓN GG (1995) La subfamilia Masteriinae (Araneae: Dipluridae) en Cuba. **Poeyana 453**: 1-8.

ANGLOGOLD ASHANTI (2009) **Diversidade da Mata Samuel de Paula.** Nova Lima, AngloGold Ashanti, 296p.

AULER AS, PILÓ LB, PARKER CW, SENKO JM, SASOWSKY ID, BARTON HA (2014) Hypogene cave patterns in iron ore caves: convergence of forms or processes? **Karst Waters Institute Special Publication 18**: 15-19.

ÁZARA LN, FERREIRA RL (2013) The first troglobitic *Cryptops* (*Trigonocryptops*) (Chilopoda: Scolopendromorpha) from South America and the description of a non-troglobitic species from Brazil. **Zootaxa 3709**(5): 432-44. doi: 10.11646/zootaxa.3826.1.10

BERNARDI LFO, KLOMPEN H, ZACARIAS MS, FERREIRA RL (2013) A new species of *Neocarus* Chamberlin & Mulaik, 1942 (Opilioacarida, Opilioacaridae) from Brazil, with remarks on its postlarval development. **Zookeys 358**: 69-89. doi: 10.3897/zookeys.358.6384

BERTANI R, CRUZ WR, OLIVEIRA MEES (2013) *Masteria manauara* **sp. nov.**, the first masteriine species from Brazil (Araneae: Dipluridae: Masteriinae). **Zoologia 30**(4): 437-440. doi: 10.1590/S1984-46702013000400010

CARMO FF, JACOBI CM (2013) A vegetação de canga no Quadrilátero Ferrífero, Minas Gerais: caracterização & contexto fitogeográfico. **Rodriguesia 64**(3): 527-541. doi: 10.1590/S2175-78602013000300005

CHICKERING AM (1964) Two new species of the genus *Accola* (Araneae, Dipluridae). **Psyche 71**: 174-180.

CHICKERING AM (1966) Three new species of *Accola* (Araneae, Dipluridae) from Costa Rica and Trinidad, W. I. **Psyche 73**: 157-164.

COELHO A, PILÓ LB, AULER AS, BESSI R (2010) **Espeleologia da área do Projeto Apolo, Quadrilátero Ferrífero, MG.** Belo Horizonte, Carste Consultores Associados, Relatório Técnico, 179p.

FERREIRA RL (2005) A vida subterrânea nos campos ferruginosos. **O Carste 3**(17): 106-115.

GERTSCH WJ (1982) The troglobitic mygalomorphs of the Americas (Arachnida, Araneae). **Association for Mexican Cave Studies Bulletin 8**: 79-94.

LEÃO MR, AULER AS (2012) **Pedido de Supressão da Cavidade ASS-01, Serra do Sapo – Conceição do Mato Dentro.** Belo Horizonte, Carste Consultores Associados, Relatório Técnico, 22p.

MELLO JS (1979) **Os Emboabas.** São Paulo, Governo do Estado de São Paulo, 295p.

PEDROSO DR, BAPTISTA RLC (2014) A new troglomorphic species of *Harmonicon* (Araneae, Mygalomorphae, Dipluridae) from Pará, Brazil, with notes on the genus. **Zookeys 389**: 77-88. doi: 10.3897/zookeys.389.6693

PELLEGRINI TG, FERREIRA RL (2011) *Coarazuphium tapiaguassu* (Coleoptera: Carabidae: Zuphiini), a new Brazilian troglobitic beetle, with ultrastructural analysis and ecological considerations. **Zootaxa 3116**: 47-58.

PLATNICK NI (2014) The World Spider Catalog, version 15. American Museum of Natural History, online at http://research.amnh.org/entomology/spiders/catalog/index.html. doi: 10.5531/db.iz.0001

PLATNICK NI, FOSTER RR (1982) On the Micromygalinae, A New Subfamily of Mygalomorph Spiders (Araneae, Microstigmatidae). **American Museum Novitates 2734**: 1-13.

RAVEN RJ (1979) Systematics of the mygalomorph spider genus *Masteria* (Masteriinae: Dipluridae: Arachnida). **Australian Journal of Zoology 27**: 623-636.

RAVEN RJ (1981) Three new mygalomorph spiders (Dipluridae, Masteriinae) from Colombia. **Bulletin of the American Museum of Natural History 170**: 57-63.

RAVEN RJ (1985) The spider infraorder Mygalomorphae (Araneae): Cladistics and systematics. **Bulletin of the American Museum of Natural History 182**: 1-180.

RAVEN RJ (1991) A revision of the mygalomorph spider family Dipluridae in New Caledonia (Araneae). In: CHAZEAU J, TILLIER S (Eds). Zoologia Neocaledonica. **Mémoires du Museum National d'Histoire Naturelle (A) 149**: 87-117.

RAVEN RJ, PLATNICK NI (1981) A revision of the American spiders of the family Microstigmatidae (Araneae, Mygalomorphae). **American Museum Novitates 2707**: 1-20.

SIMON E (1889) Arachnides. In: Voyage de M.E. Simon au Venezuela (décembre 1887-avril 1888), 4e Mémoire. **Annales de la Societe Entomologique de France 9**: 169-220.

SIMON E (1892) Arachnides. In: RAFFREY A, BOLIVAR I, SIMON E (Eds) Etudes cavernicoles de l'île Luzon. Voyage de M.E. Simon aux l'îles Phillipines (mars et avril 1890), 4e Mémoire. **Annales de la Société Entomologique de France 61**: 35-52.

SOUZA MFVR, FERREIRA RL (2005) *Eukoenenia* (Palpigradi: Eukoeneniidae) in Brazilian caves with the first troglobiotic palpigrade from South America. **Journal of Arachnology 38**: 415-424.

TRAJANO E, BICHUETTE ME (2010) Diversity of Brazilian subterranean invertebrates, with a list of troglomorphic taxa. **Subterranean Biology 7**: 1-16.

21

A new genus and new species of spittlebug (Hemiptera: Cercopidae: Ischnorhininae) from Southern Brazil

Andressa Paladini[1] & Rodney Ramiro Cavichioli[1,2]

[1]*Departamento de Zoologia, Universidade Federal do Paraná. Caixa Postal 19020, 81531-980 Curitiba, PR, Brazil.*
E-mail: andri_bio@yahoo.com.br
[2]*Corresponding author. E-mail: cavich@ufpr.br*

ABSTRACT. A new genus of spittlebug is described to include *Gervasiella oakenshieldi* **sp. nov.** (holotype male from Brazil, state of Paraná, municipality of Piraquara, Mananciais da Serra at 25°29'46"S, 48°58'54"W, 1000 m a.s.l., 15.XI.2008, P.C. Grossi *leg.*, deposited in DZUP). In addition, *Aeneolamia bucca* Paladini & Cavichioli, 2013 is transferred to *Gervasiella* **gen. nov.** based on the results of a cladistic analysis. *Gervasiella* **gen. nov.** can be distinguished from the other cercopid genera by the following: postclypeus inflated with upper portion black and basal one yellowish; color of tylus distinct from color of head and rostrum, barely reaching mesocoxae. *Gervasiella oakenshieldi* **sp. nov.** is diagnosed by having the head black with tylus white, postclypeus in profile inflated and convex with a prominent longitudinal carina; tegmina black with two elongate white maculae near costal margin, one on anterior third and the other on posterior third.

KEY WORDS. Auchenorrhyncha; Neotropical Region; phylogeny; taxonomy.

Insects belonging to Cercopidae are known as spittle-bugs due to the bubble nest produced by the nymphs. This family forms a large group of xylem feeding insects with approximately 1500 worldwide species included in 150 genera. Most species are distributed in the tropical and subtropical regions. Adults feed on leaves or stems of a wide variety of plants, nymphs can feed on roots and in some cases they complete their development above the ground (Carvalho & Webb 2005).

The Neotropical genera of Cercopidae have been usually defined by characters of the head and pronotum, and by the number of spines on the hind leg. The same set of characters also form the basis of the tribal classification proposed by Fennah (1968).

An ongoing study on Neotropical cercopids has revealed a new genus of Ischnorhininae. The new genus and the new species are described and illustrated. Also, we propose a new combination: *Aeneolamia bucca* Paladini & Cavichioli, 2013 is transferred to *Gervasiella* **gen. nov.** The species of *Gervasiella* **gen. nov.** are distinguished based on a comparative diagnosis.

MATERIAL AND METHODS

The specimens studied are deposited in the Coleção Entomológica Padre Jesus Santiago Moure, Departamento de Zoologia, Universidade Federal do Paraná, Curitiba, Paraná, Brazil (DZUP). Morphological terminology follows Fennah (1968) and Paladini & Cryan (2012). Techniques for preparation of genital structures follow Oman (1949). The dissected parts were stored in micro vials with glycerin. Photographs were obtained with a Leica DFC-550 digital camera attached to the stereomicroscope (Leica MZ16) and captured with the software IM50 (Image Manager; Leica Microsystems Imaging Solutions Ltd, Cambridge, UK), after montage using Auto-Montage Syncroscopy of Taxonline (Rede Paranaense de Coleções). Illustrations were made with the aid of a camera lucida and the final art were finalized using vectors with the software Corel-Draw version X5.

Terminal taxa. Besides the Neotropical cercopids present in the matrix analyzed by Paladini et al. (2015) two species of *Gervasiella* **gen. nov.** were also included, to test the validity of the new genus proposed here.

Characters were coded to include most of the morphological variation of the external morphology of the adult, and male and female genitalia. We included and reanalyzed 108 characters from the cladistics analysis of Paladini et al. (2015), and added two characters totaling 110 characters. Each character was considered a hypothesis of grouping. Primary homologies were proposed by similarity or topological correspondence (de Pinna 1991). The contingent coding was used when novel features appeared and evolved, and this feature shows variation (Sereno 2007). Multistate characters were treated as unordered (nonadditive) (Fitch 1971). Character state polarity was determined by outgroup rooting (Nixon & Carpenter 1993). Missing data were coded as '?' and nonapplicable characters were coded as '–'. The data matrix was built using Winclada v1.00.08 (Nixon 2002).

Analyses were performed using two character weighting schemes: equal weight and implied weight. Analyses were conducted using TNT version 1.1 (GOLOBOFF et al. 2008) using Traditional Search basing the heuristic search strategies on RAS + TBR (random addition sequences plus swap by tree bisection and reconnection), with 1,000 replications with 100 trees saved per replication. The choice for the best constant of concavity (K) values range for the data followed the methodology of PALADINI et al. (2015). The best K range for the data matrix presented here was 8-13. Branch support was calculated using the relative Bremer support (GOLOBOFF & FARRIS 2001). Nonparametric Bootstrap support values were computed running 1000 bootstrap pseudoreplicates (FELSENSTEIN 1985).

TAXONOMY

Gervasiella gen. nov.

Type species. *Gervasiella oakenshieldi* **sp. nov.** by original designation.

Diagnosis. *Gervasiella* **gen. nov.** can be distinguished from all other cercopids genera by the following combination of characters: 1) postclypeus inflated with upper portion black and basal one yellowish; 2) tylus with a distinct color from the head; 3) rostrum barely reaching mesocoxae; 4) subgenital plates shorter than pygofer, in ventral view quadrangular with apex truncate; 5) paramere long and slender, apex rounded, a unique subapical spine quadrangular; 6) paramere's spines located upon a lateral concavity similar to a hole; 7) aedeagus slender with quandrangular and wide base, one pair of dorsal processes long and slender turned upward.

Description. Head triangular with two deep impressions on the vertex near the median line; tylus quadrangular; ocelli near to each other than the compound eyes. Antennae with pedicel visible in dorsal view, flagellum normal in length, with ovoid basal body and an arista almost as long as pedicel; postclypeus inflated, convex in profile with a well-marked longitudinal carina; rostrum barely reaching mesocoxae. Pronotum hexagonal, surface smooth; anterior and lateral anterior margins straight; posterior and lateroposterior margins slightly sinuated; tegmina long and slender. Hindwings with Cu1 not thickened at base. Hind tibiae with two lateral spines and a row of apical spines; hind basitarsus with apical spines distributed in two irregular rows. Pygofer with one process between anal tube and subgenital plate in lateral view; subgenital plate quandrangular in ventral view. Aedeagus long, slender with one pair of dorsal processes; parameres slender, dorsal margin with two processes, one subapical spine quadrangular and sclerotized. First valvulae of ovipositor with two processes near the base; second valvulae with dorsal margin smooth.

Etymology. The genus is named in honor of Prof. Dr. Gervásio Silva Carvalho, a specialist of Neotropical cercopids, in recognition of his expertise and several contributions to the taxonomy of the group.

Remarks. Based on the tree resulting from the cladistics analysis (Figs. 12 and 13), *Gervasiella* **gen. nov.** is sister group of the clade including *Prosapia* Fennah, 1949, *Aeneolamia* Fennah, 1949 and *Isozulia* Fennah, 1953. In these genera the aedeagus presents a long and slender dorsal process inserted medially. *Gervasiella* **gen. nov.** is supported by two synapomorphies: paramere with a concavity under the spine (Figs. 14 and 15) and aedeagus base quadrangular (Figs. 16 and 17).

Gervasiella oakenshieldi **sp. nov.**
Figs. 1-11, 19

Measurements. Length, male 6.8 mm; females 6.7-7.8 mm.

Diagnosis. Head black with tylus white, postclypeus in profile, inflated and convex with a prominent longitudinal carina; tegmina black with two elongate white maculae near the costal margin, one on the anterior third and the other on the posterior third.

Description. Head triangular black; rostrum yellowish with the third segment black; compound eyes black, rounded, arranged transversely; vertex smooth, rectangular, with a prominent median carina; ocelli reddish near to each other than to compound eyes; tylus white, smooth, quadrangular, lacking a median carina; antennae black, pedicel scarcely setose, basal body of flagellum ovoid with an arista almost as long as the pedicel; postclypeus inflated, convex in profile with a wide longitudinal carina, lateral grooves slightly marked, apical portion black and two basal thirds yellowish. Thorax black; pronotum black, flattened, hexagonal, lacking median carina, anterior margin straight, lateral-anterior margins straight, lateral-posterior margin slightly sinuous, posterior margin with a light groove; scutellum black, with a slight central concavity and transversal grooves. Tegmina black with two white maculae: the first one elongated, located near the costal margin, extending from its base until the median third of the tegmina; the second one rounded located between the median and apical third; apical plexus of vein poorly developed; hindwings hyaline with brown venation; vein Cu1 not thickened at base; legs brownish; metathoracic tibia with two lateral spines (basal spine equal in size to spines in apical crown; apical spine larger than spines in apical crown); apical crown of spines on tibia consisting of two rows; basitarsus with one row of spines covered by sparse setae; subungueal process absent.

Male genitalia. Pygofer with one quadrangular process between the anal tube and subgenital plates; subgenital plates short, quadrangular with a rounded apex, dorsal margin produced in a rectangular process (Fig. 3); parameres long and slender with a quadrangular sclerotized spine turned backwards located over a concavity on the external side, dorsal margin with a finger like process turned to the inner side (Figs. 7 and 8); aedeagus cylindrical with a pair of dorsal processes long and slender turned upward, aedeagus base quadrangular and wide, apex quadrangular (Figs. 5 and 6).

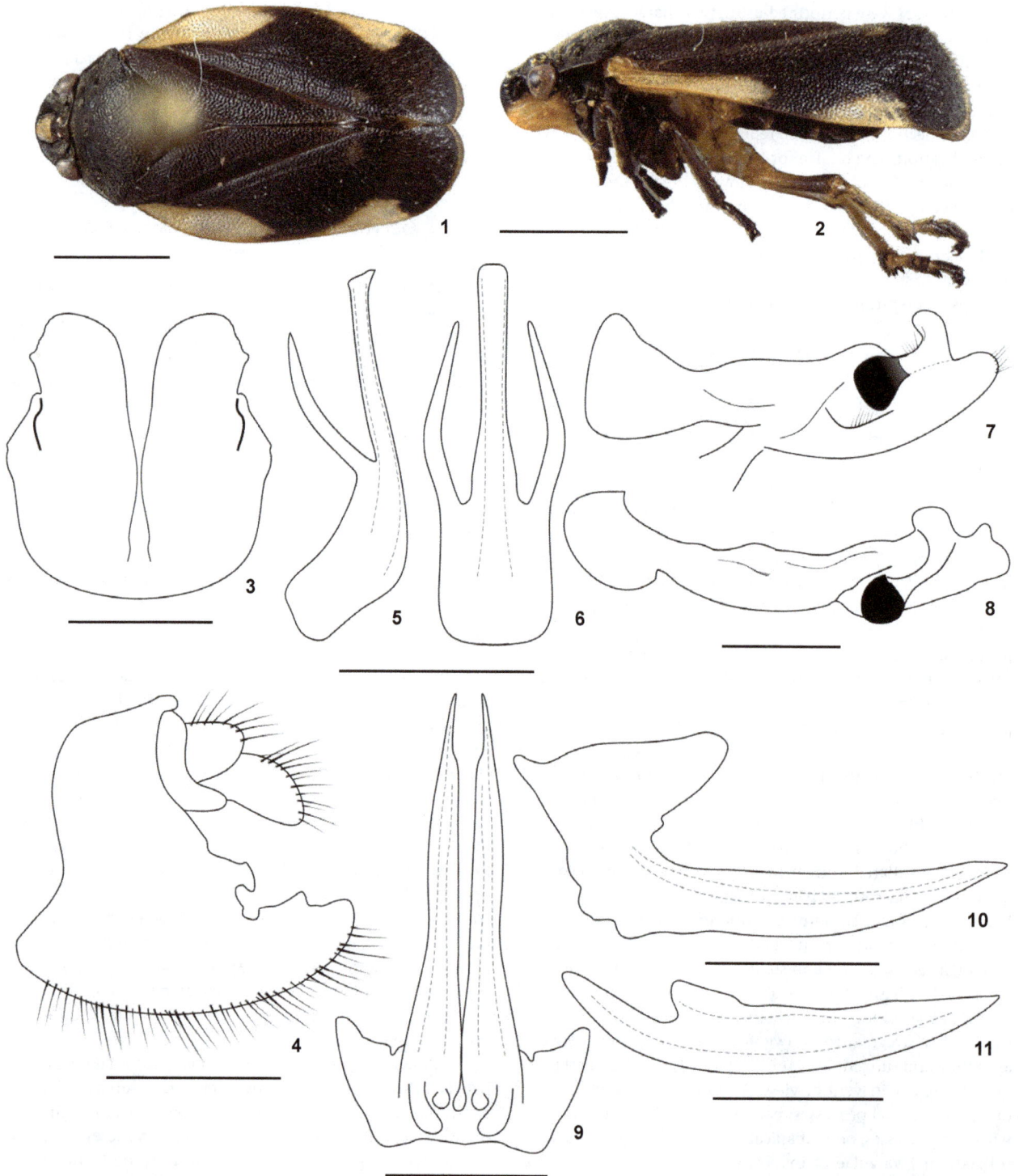

Figures 1-11. *Gervasiella oakenshieldi* **sp. nov.**: (1-2, 9-11) female paratype, (3-8) male holotype: (1) habitus, dorsal view; (2) habitus, lateral view; (3) subgenital plates and pygofer, ventral view; (4) pygofer and subgenital plates, lateral view; (5) aedeagus, lateral view; (6) aedeagus, dorsal view; (7) paramere, lateral view; (8) paramere, dorsal view; (9) first valvulae of ovipositor, ventral view; (10) first valvulae of ovipositor, lateral view; (11) second valvulae of ovipositor, lateral view. Scale bars: 1-2 = 2 mm, 3, 4, 9-11 = 0.5 mm, 5-8 = 0.25 mm.

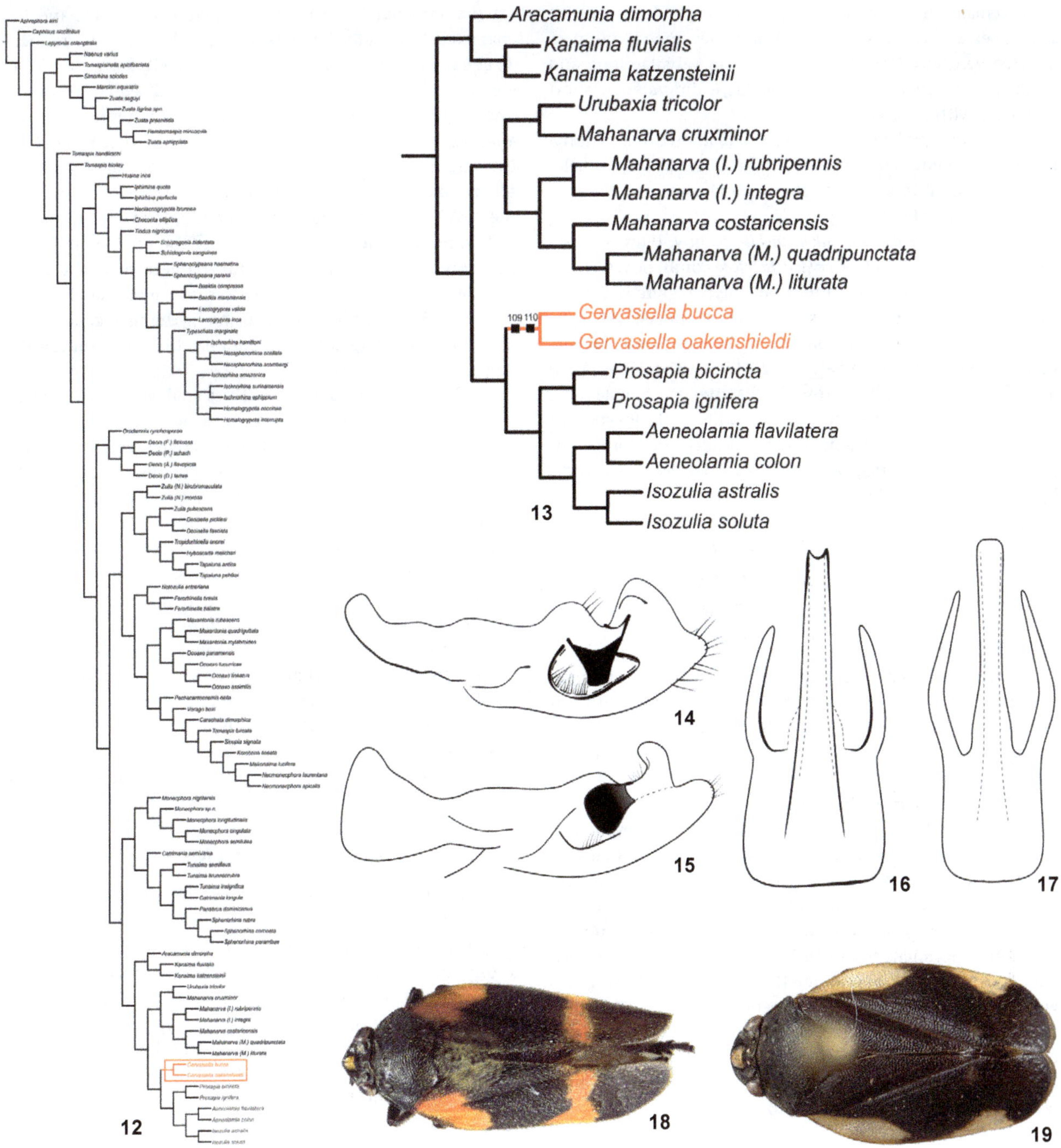

Figures 12-19. Phylogenetic relationships of Ischnorhininae and diagnostic characters of *Gervasiella* **gen. nov.**: (12) unique most parsi-monious tree resulting from the analysis of morphological data with the implied weighting scheme using the optimal constant of concavity interval of K8-12, highlighting the position of *Gervasiella* **gen. nov.**; (13) clade including *Gervasiella* **gen. nov.**; (14) *Gervasiella bucca* **comb. nov.** paramere in lateral view; (15) *Gervasiella oakenshieldi* **sp. nov.** parameres in lateral view; (16) *Gervasiella bucca* **comb nov.** aedeagus in dorsal view; (17) *Gervasiella oakenshieldi* **sp. nov.** aedeagus in dorsal view; (18) *Gervasiella bucca* dorsal habitus; (19) *Gervasiella oakenshieldi* **sp. nov.** dorsal habitus.

Female. First valvulae of ovipositor long and slender with acute apex and two basal process poorly developed, rounded, directed ventrally (Figs. 9 and10); second valvulae long and slender, dorsal margin smooth (Fig. 11), third valvulae short and wide, with long setae ventrally.

Etymology. Noun in genitive singular after to a fictional character surname of the novel The Hobbit, Thorin Oakenshield; in honor to J.R.R Tolkien an English writer known as the author of classic fantasy books.

Remarks. *Gervasiella oakenshieldi* **sp. nov.** (Figs. 1-2 and 19) superficially resembles *Gervasiella bucca* **comb. nov.** (Fig. 18) in having the same color pattern but the paramere is slender, with the concavity under the spine less pronounced (Fig. 15).

Examined material. Holotype male from BRAZIL, *Paraná*: Piraquara (Mananciais da Serra 25°29'46"S, 48°58'54"W, 1000 m a.s.l., 15.XI.2008), P.C. Grossi *leg*. Paratypes: 1 female same data as holotype; 1 female, same locality as holotype but 25.III.2012, light trap; 1 female same locality as holotype but flight interception [trap], XI.2007, P.C. Grossi & D. Parizotto *leg*. All deposited in DZUP.

Gervasiella bucca
(Paladini & Cavichioli, 2013) **comb. nov.**
Figs. 14, 16, 18

Aeneolamia bucca Paladini & Cavichioli, 2013: 353.

Diagnosis. General coloration black; tegmina with basal red macula and one apical red stripe; Pygofer short with finger-like process between anal tube and subgenital plates; aedeagus with one pair of dorsal, slender processes directed upward.

Remarks. *Gervasiella bucca* was originally described in *Aeneolamia* due to a superficial resemblance in the morphology of the male genitalia, although other features indicated that those species were not congeneric. Subsequently, *Gervasiella* **gen. nov.** was erected to accommodate *G. bucca* and the newly described *G. oakenshieldi* **sp. nov.**, based on the examination of additional specimens from Southern Brazil (from the municipality of Piraquara, Paraná State). Diagnostic traits of this genus were previously mentioned in the generic diagnosis. These features were later recovered as synapomorphies validating the monophyly of *Gervasiella* **gen. nov.**, as inferred in our morphology-based phylogenetic analysis of Ischnorhininae that included both *G. bucca* and *G. oakenshieldi* **sp. nov.**

CLADISTIC ANALYSIS

List of new characters. The complete list of characters can be found in PALADINI et al. (2015) and the full data matrix is available in Appendix S1[1]. The two new characters included in the analysis are: Male genitalia: (109) Paramere, lateral view, concavity under the main spine: (0) absent; (1) present; and (110) Aedeagus, shape of base in dorsal view: (0) rectangular;

(1) quadrangular. The data matrix included 102 taxa and 110 characters (Appendix S1[1]). The equal weights analysis produced 30 equally parsimonious trees (length = 1061 steps, CI = 14, RI = 61). The strict consensus cladogram had 59 collapsed nodes. The relationships among genera were poorly resolved. In the analysis with implied weighting scheme the best K range for the data matrix presented here was 8-12; this range was chosen based on PALADINI et al. (2015). The five trees obtained with the best K range were had the same topology,(Fig. 12) which will be used as the hypothesis to infer the phylogenetic relationship and monophyly of *Gervasiella* **gen. nov.** The topology obtained in the present analysis is similar to that of PALADINI et al. (2015) except for the inclusion of the new genus. Only unambiguous characters were optimized in the resultant cladogram.

The main goal of this cladistics analysis was to evaluate and to support the description of a new genus. The clade *Gervasiella* **gen. nov.** (Fig. 13) has a relative Bremer support of 71 and a Bootstrap support of 99. The genus is supported by two synapomorphies: paramere with a concavity located under the main spine (109_1) and aedeagus with a quadrangular base (110_1). and 10 a homoplasious character-state transformations: vertex shape narrow (3_0); antennae with basal body of flagellum ovoid (6_1); posterior margin of pronotum slightly grooved (31_0); tegmina venation almost indistinct (37_2); tegmina with apical plexus of veins reduced (40_1); basitarsus of the posterior leg with two rows of spines (47_1); subgenital plates short compared to pygofer (53_0); apex of subgenital plates truncated (50_2); spine of paramere oriented vertically (69_1); ovipositor with two basal processes (103_1).

Gervasiella **gen. nov.** is included in the clade Tomaspidini and is sister group to *Prosapia*, *Aeneolamia*, and *Isozulia*.

ACKNOWLEDGEMENTS

We thank Paschoal C. Grossi (UFRPE) and Daniele Parizzoto for collecting and generously providing specimens; Olivia Envangelista (MZUSP) for her valuable suggestions; the anonymous reviewers and associate editor for their constructive comments and significant improvement on an earlier version of this manuscript. This work was supported by a CNPq postdoctoral grant (process 150163/2013-4) to the senior author. This research is also partially funded by the advisor's grant (RRC) from PROTAX/CNPq (processes 561298/2010-6 and 303127/2010-4). This paper is the contribution number 1917 of the Departamento de Zoologia, Universidade Federal do Paraná.

LITERATURE CITED

CARVALHO GS, WEBB MD (2005) **Cercopid Spittlebugs of the New World (Hemiptera, Auchenorrhyncha, Cercopidae).** Sofia, Pensoft, 271p.

DE PINNA MCC (1991) Concepts and tests of homology in the

cladistics paradigm. **Cladistics** 7(4): 367-394.

FENNAH RG (1968) Revisionary notes on the new world genera of cercopid froghoppers (Homoptera, Cercopoidea). **Bulletin of Entomological Research 58**: 165-190.

FELSENSTEIN J (1985) Confidence limits on phylogenies: an approach using the bootstrap. **Evolution 39**: 783-791.

FITCH WN (1971) Toward defining the course of evolution, minimum change for a specified tree topology. **Systematic Zoology 20**: 406-416.

GOLOBOFF PA, JS FARRIS (2001) Methods for quick consensus estimation. **Cladistic 17**(1): S26-S34. doi: 10.1111/j.1096-0031.2001.tb00102.x

GOLOBOFF PA, FARRIS JS, NIXON KC (2008) TNT, a free program for phylogenetic analysis. **Cladistics 24**(5): 774-786. doi: 10.1111/j.1096-0031.2008.00217.x

NIXON KC (2002) **Winclada**. New York, Published by the Author, v. 1.00.08.

NIXON KC, CARPENTER JM (1993) On outgroups. **Cladistics 9**(4): 413-426. doi: 10.1111/j.1096-0031.1993.tb00234.x

OMAN PW (1949) The Nearctic leafhoppers (Homoptera: Cicadellidae). A generic classification and check list. **Memoirs of the Entomological Society of Washington 3**: 1-253.

PALADINI A, CAVICHIOLI RR (2013) A new species of *Aeneolamia* (Hemiptera: Cercopidae: Tomaspidinae) from the Neotropical Region. **Zoologia 30**(3): 353-355. doi: 10.1590/S1984-46702013000300016

PALADINI A, CRYAN JR (2012) Nine new species of Neotropical spittlebugs. **Zootaxa 3519**: 53-68.

PALADINI A, TAKIYA DM, CAVICHIOLI RR, CARVALHO GS (2015) Phylogeny and biogeography of Neotropical spittlebugs (Hemiptera: Cercopidae: Ischnorhininae): revised tribal classification based on morphological data. **Systematic Entomology 40**(1): 82-108. doi: 10.1111/syen.12091

SERENO PC (2007) Logical basis for morphological characters in phylogenetics. **Cladistics 23**(6): 565-587.

22

A new species of *Kingsleya* (Crustacea: Decapoda: Pseudothelphusidae) from the Xingu River and range extension for *Kingsleya junki,* freshwater crabs from the southern Amazon basin

Manuel Pedraza[1], José Eduardo Martinelli-Filho[2] & Célio Magalhães[3,4]

[1]*Programa de Pós-Graduação, Museu de Zoologia, Universidade de São Paulo. Avenida. Nazaré 481, Ipiranga, 04263-000 São Paulo, SP, Brazil. E-mail: manupedrazam@gmail.com*
[2]*Faculdade de Oceanografia, Instituto de Geociências da Universidade Federal do Pará. Campus Universitário do Guamá, 66075-110 Belém, PA. Brazil.*
[3]*Instituto Nacional de Pesquisas da Amazônia. Caixa Postal 2223, 69080-971 Manaus, AM, Brazil.*
[4]*Corresponding author. E-mail: celiomag@inpa.gov.br*

ABSTRACT. *Kingsleya castrensis* **sp. nov.**, a pseudothelphusid crab is described and illustrated from the Xingu River, state of Pará, southern Amazon region, Brazil. The new species is characterized by the male first gonopod bearing a large, well-developed apical plate, with a broadly rounded, thick distal lobe. New records of *Kingsleya junki* Magalhães, 2003 extend the distribution of this species eastward to the Tocantins River basin, in the state of Pará, Brazil.

KEY WORDS. Amazon; Brachyura; Kingsleyini; Neotropical region; taxonomy.

Kingsleya Ortmann, 1897 currently comprises seven species that are all distributed in the highlands of the Guyanan and Central Brazilian Shields. This area encompasses a large portion of northern South America from southern Venezuela, Guyana, Suriname, and French Guiana to the northern Brazilian states of Amazonas, Pará and Roraima (MAGALHÃES 2003a, MAGALHÃES & TÜRKAY 2008). In Brazil, species of this genus occurs in tributaries of the Amazon River draining the Guyana Shield where it is represented by *K. latifrons* (Randall, 1840) (in Rio Branco, Rio Negro, and Rio Trombetas), and by *K. siolii* Bott, 1967 (in Rio Trombetas and Rio Paru do Oeste); and the Central Brazilian Shield: *Kingsleya gustavoi* Magalhães, 2005 (Rio Tocantins), and *K. junki* Magalhães, 2003 (Xingu River). *Kingsleya ytupora* Magalhães, 1986 (found in the Rio Uatumã, Rio Trombetas, Rio Curuá-Una, Rio Xingu) is the only species known to occur on both sides of the Amazon valley (MAGALHÃES 1986, 2003a, b, MAGALHÃES & TÜRKAY 2008).

Although pseudothelphusids living in high altitude localities typically have restricted distributions, this may not be the case for species living in the Amazon basin which have wide distributions (although much of the southern Amazon River tributaries are still poorly surveyed for decapods). Crab samples sporadically collected during ichthyological and entomological expeditions to the middle and lower course of the Xingu River were studied by MAGALHÃES (2003b) and indicated that five species occur in this stretch of the river basin: two pseudothelphusids (*K. junki* and *K. ytupora*) and three trichodactylids – *Sylviocarcinus devillei* H. Milne Edwards, 1853,

S. pictus (H. Milne-Edwards, 1953), and *Trichodactylus ehrhardti* Bott, 1969. Recent collections from southern tributaries of the Amazon river revealed the presence of *Kingsleya*, including an undescribed species of this genus from the surroundings of the city of Altamira, on the left bank of the middle course of the Xingu River. The new species is herein described and illustrated, and a range extension for *K. junki* is reported.

MATERIAL AND METHODS

Specimens are deposited at the Instituto Nacional de Pesquisas da Amazônia, Manaus, Brazil (INPA), Museu Nacional, Universidade Federal do Rio de Janeiro, Rio de Janeiro (MNRJ), Museu Paraense Emilio Goeldi, Belém, Brazil (MPEG), Museu de Zoologia, Universidade de São Paulo, São Paulo, Brazil (MZUSP), and Senckenberg Research Institute and Natural History Museum (SMF). The following abbreviations are used: carapace width (cw), measured across the carapace at its widest point; carapace length (cl), measured along the midline, from the frontal to the posterior margin; carapace height (ch), the maximum height of the cephalothorax, measured as the distance between the dorsal and ventral edges of the shell; frontal width (fw), the width of the front measured along its upper border; male first (G1) and second (G2) gonopods; third maxilliped (Mxp3); cheliped (P1); pereiopods 2 to 5 (P2-P5); and sternal sulcus (s). Geographic coordinates inserted between brackets were taken from Google Earth. Illustrations were made using a Leica M8 stereomicroscope with a camera lucida; the computerized photographs were taken using a stereomicroscope Zeiss Discovery V12 (Automontage® system). Measurements of carapace width and carapace length, in millimeters, were made with a calipers and are given in parentheses after the number of specimens examined. Terminology for describing the morphology of the G1 was adapted from SMALLEY (1964) and MAGALHÃES & TÜRKAY (2008).

TAXONOMY

Kingsleya castrensis **sp. nov.**
Figs. 1-4, 7-14

Diagnosis. G1 with large, roughly rounded, thick apical plate, widest medially; proximal lobe of apical plate subtriangular, well developed, situated on mesio-caudal side; distal margin straight, stretching diagonally over the distal lobe, fusing to mesiodistal portion of apical plate; distal lobe of apical plate broad, with lateral margin angulate in mesial view, caudal margin straight, distal margin slightly concave, mesial margin rounded, thick.

Description. Carapace outline ellipsoid, widest medially (cb/cl 1.68); dorsal surface smooth, slightly convex, regions partially defined (Fig. 7). Two distinct gastric pits, close to each other, on metagastric region. Cervical grooves deep, narrow, nearly straight, faint proximally, distal end failing to reach anterolateral margin. Postfrontal lobules small, quite distinct; median groove indistinct. Surface of carapace between front and postfrontal lobules smooth and slightly inclined anteriorly and medially. Upper border of front smooth, angulate, slightly convex in dorsal view, median notch absent; lower border carinate, slightly sinuous in both frontal and dorsal view, more projected anteriorly than upper one, except medially. Upper orbital margin smooth, lower orbital margin slightly crenualte; exorbital angle low, obtuse (Fig. 13). Anterolateral margin of carapace nearly smooth, with very shallow depression just behind exorbital angle, followed by a set of faint, minute teeth increasing in size from the anterior to posterior portion; posterolateral margin smooth, barely defined. Epistome narrow; epistomial tooth triangular, deflexed, with carinate, smooth borders. Suborbital and subhepatic regions of carapace sidewall smooth; pterygostomial regions with narrow pilose patches along outer borders of bucal cavity (Figs. 8 and 13).

Endopod of Mxp3 with outer margin of ischium slightly convex, inner margin straight; outer margin of merus rounded, inner surface of palp covered with large setae; exopod of Mxp3 short, narrow, 0.18 times length of outer margin of ischium (Fig. 14). Aperture of efferent branchial channel wide, upper margin subquadrate, lacking setae (Fig. 13).

First pereiopods heterochelous in both males and females, similarly armed, right P1 usually largest (holotype left P1 major). Major cheliped merus subtriangular in cross section; superior margin rounded with irregular row of tubercles, fainter distally; medial margin lined by longitudinal row of rounded,

low teeth, slightly increasing in size distally; inferior lateral margin marked by row of faint tubercles, smooth distally; distal margin arched, smooth laterally, with straight row of faint tubercles mesially. Carpus with inner margin granular proximally, with prominent median spine, smooth distally; outer margin rounded, smooth. Palm narrow (length/breadth 1.61 in holotype), smooth on both sides. Fingers moderately gaping, tips not crossing; both fingers with large triangular teeth sometimes interspaced with small ones, smaller distally. Dactylus distinctly arched, longer then palm (dactylus/palm 1.36 in holotype, measured dorsally), upper, outer surface of dactylus smooth, distomedian portion darker than proximal. Propodal finger with smooth surfaces. P2-5 slender, ratios dactylus/propodus, dactylus/merus (left side measurements in holotype), respectively, as follows: P2 = 1.80 and 0.90, P3 = 1.73 and 0.88, P4 = 1.66 and 0.88, P5 = 1.64 and 0.92. P2-5 with dactyli shorter than propodi, bearing five longitudinal rows of sharp, corneous spines, increasing in size distally, 2 faint grooves on the proximal external surface.

Thoracic sternum slightly longer than broad. Thoracic sternites of Mxp3 and P1 completely fused, except for small notches at lateral edges of sternum; s4/s5, s5/s6, s6/s7 distinct, interrupted medially, just failing to reach midline of thoracic sternum; s7/s8 complete, reaching midline. Midline of thoracic sternum marked by deep groove between sternites VII, VIII, deeper at interception with sternal suture 7/8. Episternites 4-6 triangular posteriorly, episternite 7 posteriorly truncate. Sterno-abdominal cavity strongly concave, with few, scattered pubescence. Penis noticeably long, emerging from nearby coxosternal condyle articulation, located in shallow depression on sternite 8, proximally thick, abruptly tapering distally.

All abdominal segments free. Lateral margins of male telson slightly concave, slightly cranulate, tip rounded (Fig. 8).

G1 (Figs. 1-4 and 9-12) sinuous, broadened distally, with strong median curvature on caudal surface in mesial view, bearing well-developed mesial process. Marginal suture sinuous, displaced to mesial side in distally, bearing several setae proximally. Lateral suture deep, extending 2/3 of gonopod length from proximal portion. Marginal process short, broad, subrectangular in mesial view, not projecting distally beyond field of apical spines area, distal notch in latero-caudal surface. Mesial process well developed, roughly subretangular, approximately 1.8 times longer than apical plate in mesial view, proximal portion rounded, distal portion produced into sharp conical spine pointing in mesial direction; mesial process juxtaposed to the apical plate, both structures clearly separated by a deep incision. Apical plate well developed, large, thick, expanded along caudo-cephalic axis, with 2 juxtaposed lobes; proximal lobe subtriangular, well developed, situated subdistally on mesio-caudal side, narrower than distal lobe, distal margin straight, stretching diagonally over the distal lobe, gradually merging to mesiodistal portion of the apical plate; distal lobe enlarged, caudal margin rather angulate in mesial

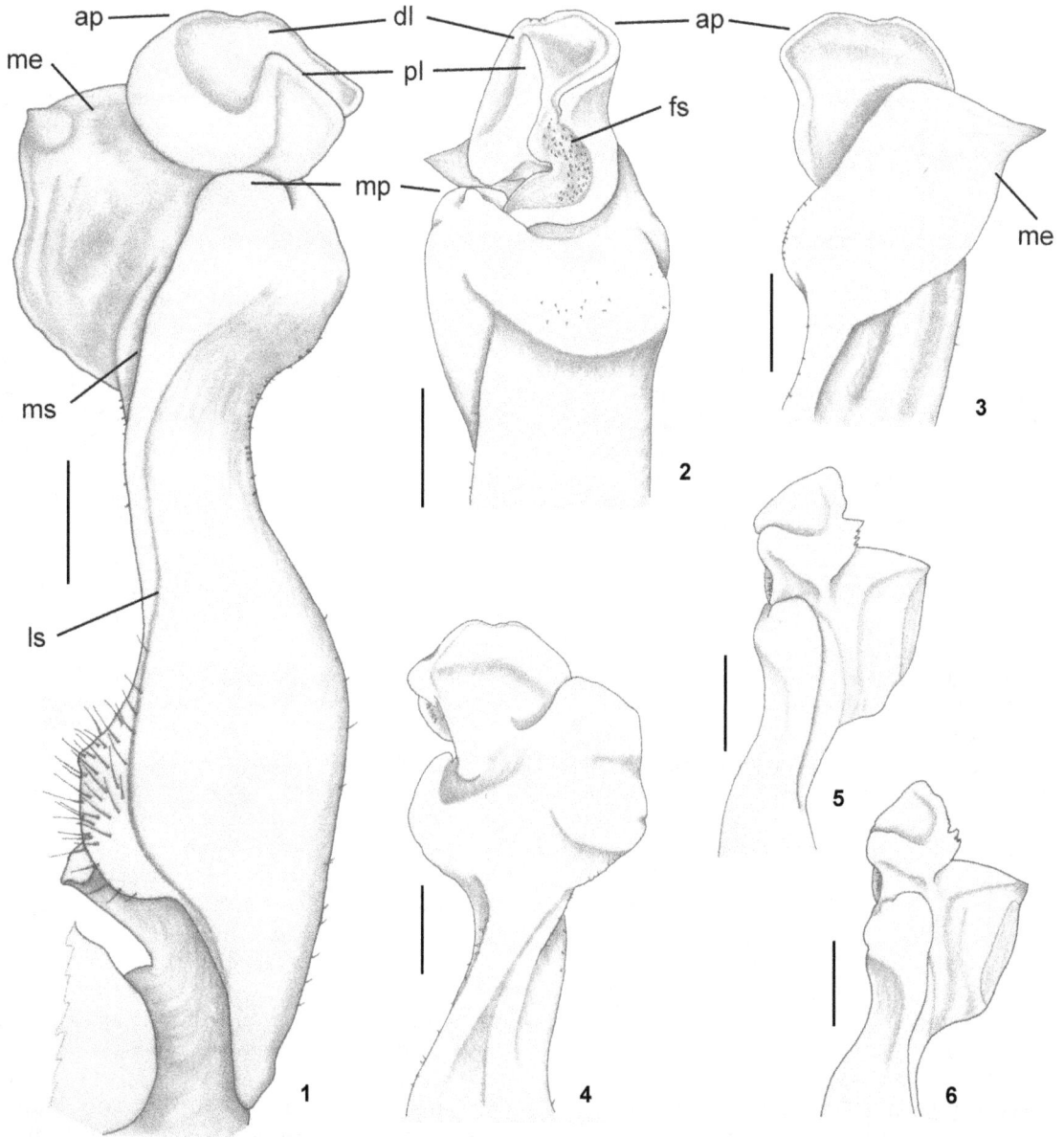

Figures 1-6. (1-4) *Kingsleya castrensis* **sp. nov.**, male, holotype (cw 46.4mm, cl 27.7mm), left first gonopod, INPA 2010: (1) whole limb, mesial view; (2) distal part, caudal view; (3) distal part, cephalic view; (4) distal part, lateral view. (5-6) *Kingsleya junki*, male, right first gonopod: (5), distal part, caudal-mesial view, INPA 1708; (6) distal part, caudal-mesial view, INPA 2012. (ap) Apical plate, (dl) distal lobe of apical plate, (fs) field of apical spines, (ls) lateral suture, (me) mesial process, (mp) marginal process, (ms) marginal suture. Scale bars: 1 mm.

view, distal margin slightly concave, mesial margin rounded. Apical spine field well developed, curved, narrow patch of minute spines, longitudinally directed along caudal side of apical plate, delimited by mesial, lateral borders of apical plate, distally opened by distinct notch at apex of apical plates proximal lobe. Sperm channel opening proximally at base of apical spine field.

G2 straight, almost as long as G1 (ca. 0.8 times length of G1), flagellum slender, strongly tapering after distal quarter, tip flattened, with short spinules on sternal surface.

Type material. Brazil, *Pará*: Altamira (51° B.I.S. – Batalhão de Infantaria de Selva camp area, 3°1147"S, 52°0958"W), male (cw 46.4, cl 27.7, ch 18.3, fw 13.7), holotype, 16.VIII.2011, José E. Martinelli Filho and Cléber S. de Sousa *leg.*, INPA 2010; same

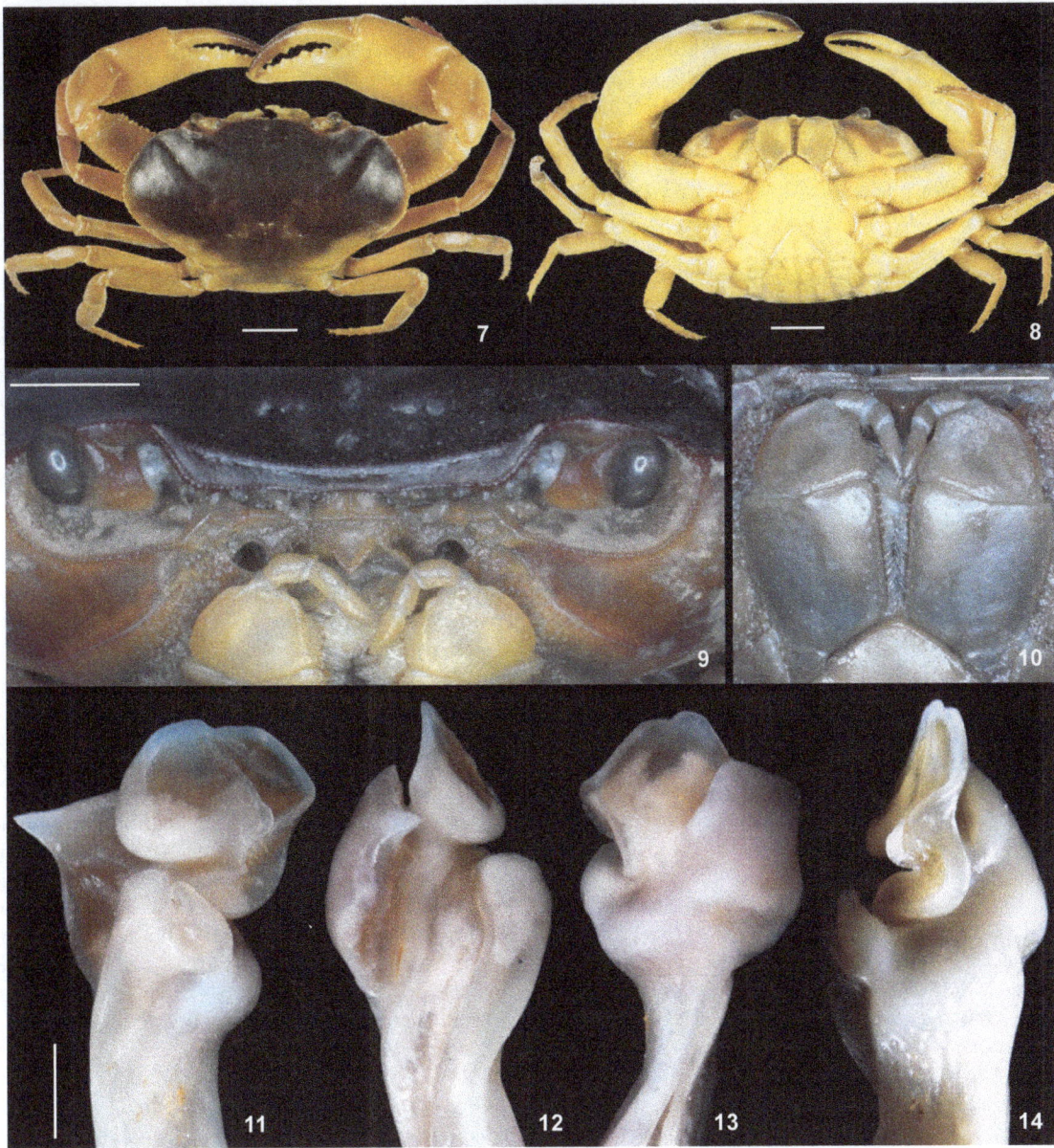

Figures 7-14. *Kingsleya castrensis* **sp. nov.**, male, paratype, MZUSP 26394: (7) habitus, dorsal view; (8) habitus, ventral view; (9) frontal view; (10) pair of third maxillipeds, frontal view. Male, paratype, left first gonopod, MZUSP 23393: (11) caudal-mesial view; (12) mesial-cephalic view; (13) lateral view; (14) idem, caudal view. Scale bars: 7, 8 = 10 mm, 9, 10 = 5 mm; 11-14 = 1 mm.

data as holotype, 1 male (cw 40.8, cl 25.1), 3 females (cw 24.7, cl 16.6; cw 46.3, cl 28.4; cw 21.5, cl 14.5), paratypes, 16.VIII.2011, José E. Martinelli Filho and Cléber S. de Sousa *leg.*, INPA 2011; same data as holotype, 4 males (cw 37.2, cl 22.7; cw 37.2, cl 23.3; cw 41.9, cl 25.2; cw 44.1, cl 26.5), paratypes, 26.VIII.2011, Cléber S. de Sousa *leg.*, MPEG 1013; same data as holotype, 2 males (cw 39.7, cl 24.8; cw 42.1, cl 25.7), 1 female (cw 39.0, cl 25.0), paratypes, 26.VIII.2011, Cléber

S. de Sousa *leg.*, SMF 47661; same data as holotype, 1 male (cw 32.2, 20.4), paratype, 28.VIII.2011, José E. Martinelli Filho and Cléber S. de Sousa *leg.*, MNRJ 25117; Altamira (Recanto Cardoso, 3°0920"S, 52°1533"W), 1 male (cw 42.9, cl 27.1), paratype, 9.IV.2012, Ronan Santos *leg.*, INPA 2058; Altamira (Princesa do Xingu road, 3°0935"S 52°1439"W), 1 male, (cw 32.9, cl 21.2), paratype, 6.I.2012, Cléber S. de Sousa *leg.*, INPA 2056; Altamira (Princesa do Xingu road, 3°0942"S, 52°1433"W), 1 male (cw

34.3, 22.2), paratype, 06.I.2012, Cléber S. de Sousa *leg.*, INPA 2057; Altamira (Abrigo Pedra do Navio, 3°1706.4"S, 52°13 42.2"W), 1 male (cw 32.57, cl 21.08), 11.IV.2012, R. Pinto-da-Rocha *leg.*, MZUSP 26943; Brasil Novo (Travessão 16, 3°1820"S, 52°3540"W), 1 male (cw 31.2, cl 19.8), 1 female (40.2, 25.2), paratypes, 13.I.2012, Cléber S. de Sousa *leg.*, MZUSP 32737; Altamira (Pedra da Cachoeira cave, 3°1914.8"S, 52°1953.1"W), 1 male (cw: 49.8, cl: 29.7), paratypes, 20.VI.2012, R. Pinto-da-Rocha *leg.*, MZUSP 26394.

Additional material examined: BRAZIL, *Pará*: Altamira (Princesa do Xingu road, 3°0935"S, 52°1439"W), 1 female (cw 39.8, cl 25.6), 20.I.2012, Cléber S. de Sousa *leg.*, INPA 2061; Altamira (Princesa do Xingu road, 3°0935"S, 52°1439"W), 1 male (cw 28.8, cl 18.4), 21.VIII.2011, Cléber S. de Sousa *leg.*, MPEG 1014; Altamira (Princesa do Xingu road, 3°1004"S, 52°2156"W), 1 male (cw 38.7, cl 24.2), 19.VI.2011, Cléber S. de Sousa *leg.*, MZUSP 32738; Altamira (51° B.I.S. – Batalhão de Infantaria de Selva camp area, 3°11478"S, 52°0956"W), 1 male (cw 41.5, cl 24.7), 10.XII.2010, José E. Martinelli Filho and Cléber S. de Sousa *leg.*, INPA 2060; Altamira (Arapujá island, 3°1330"S, 52°1213"W), 1 female (cw 37.0, cl 24.0), 10.V.2010, Anderson Prates *leg.*, INPA 2059; Brasil Novo (Travessão 8, 3°2157"S, 52°3218"W), 1 male (cw 38.1, cl 24.2), 2 females (cw 28.7, cl 18.9; cw 35.8, cl 28.2), 15.VIII.2012, Cléber S. de Sousa *leg.*, INPA 2062; Altamira (Planaltina cave, 3°2239"S, 52°3431"W), 1 female (cw 27.1, cl 17.4), 17.VIII.2011, Cléber S. de Sousa *leg.*, MPEG 1141.

Type locality and distribution. Brazil, state of Pará, city of Altamira. Most specimens were collected within the camp area of the "51° Batalhão de Infantaria de Selva" (B.I.S.), a unit of the Brazilian Armys Battalion of Jungle Infantry, headquartered in the city of Altamira. Additional specimens were also collected in in Altamira and Brasil Novo cities.

Ecological notes. Most of the crabs were collected on the margins of a 1-3m wide, third order tributary stream of the Xingu river. The stream is located inside a secondary forest fragment dominated by palm trees: *Euterpe oleracea* Mart. and *Attalea phalerata* Mart. ex Spreng in the flooded area, and *Astrocaryum gynacanthum* Mart. and *A. aculeatum* G. Mey in the well-drained soil area (*terra firme*). A floodplain along the borders of the stream varies from a few meters wide to almost a hundred meters according to topography and season (SALM et al. 2015). Adult crabs dig holes in the mud or hide between the aerial roots of the palm trees. Occasionally adult male and female crabs were found together in the same hole. Juvenile crabs were found on palm leaves and trunks or beneath leaf litter. A few specimens were collected outside the flooded margins of the stream, in the *terra firme* area.

Female crabs carried young crabs under their abdominal brood pouch. Morning field investigations revealed that 68.5% of the 108 observed crabs were males. The species is probably aggressive and territorial, since one of the chelipeds was lacking in 20% of the males and 13% of the females. The loss of

chelipeds may also be attributed to autotomy as a response to predators, since skeletal remains and pereiopods of *Kingsleya* were frequently found in the studied area.

Etymology. The specific epithet refers to *castra*, the Latin word for military camp, in reference to the Brazilian Army battalion camp where this species was found.

Remarks. The new species is attributed to *Kingsleya* since its G1 shows the diagnostic characters of the genus, namely the marginal process distally enlarged, not overreaching the apical field of spines, the bi-lobed apical plate, the mesial process clearly separated from the apical plate and standing out from the cephalic surface of the stem; the apical plate with two partially superimposed lobes; and the field of apical spines distally divided by a terminal notch (MAGALHÃES & TÜRKAY 2008).

Kingsleya castrensis **sp. nov.** can be easily distinguished from *K. junki* Magalhães, 2003 and *K. ytupora* Magalhães, 1986, the other two species of the genus that occur in the Xingu River (MAGALHÃES 2003b) by characters of the G1s apical plate. In both *K. junki* and *K. ytupora* (see MAGALHÃES 2003b: 384, figs. 1B-D, and 385, fig. 2B, respectively) the apical plate is narrow and produced distally in relation to the mesial process, whereas the apical plate of *K. castrensis* **sp. nov.** is distinctly enlarged and short in relation to the mesial process (Figs. 1, 3, 5 and 6). Moreover, in *K. castrensis* **sp. nov.** the mesial margin of the apical plates distal lobe is smooth (Fig. 1), whereas in *K. junki* the distal plate is clearly indented (Figs. 5 and 6). The distal margin of the apical plate, in mesial view, is rounded and rather narrow in *K. ytupora* (see MAGALHÃES 2003: 385, fig. 2B), whereas in *K. castrensis* **sp. nov.** this margin is much broader rounded and enlarged (Figs. 1, 3 and 4). Another character that readily separates these two species is the presence (in *K. ytupora*) or absence (in *K. castrensis* **sp. nov.**) of a set of six to seven large, sharp teeth on the anterolateral margin of the carapace; in the latter species, this margin is fringed with a set of faint, minute teeth that lends to this margin an almost smooth appearance.

Kingsleya castrensis **sp. nov.** is unique among the species of the genus because of the distinctly enlarged, broadly rounded apical plate of its G1. All other species of *Kingsleya* have a G1 with an apical plate that is, in spite of their specific differences, much narrower, tapering and roughly subtriangular in shape (MAGALHÃES 1986, 1990, 2003b, 2005, MAGALHÃES & TÜRKAY 2008).

Kingsleya junki Magalhães, 2003
Figs. 5-6

Kingsleya junki Magalhães, 2003b: 378, fig. 1.

Material examined. 1 male (cw 26.7, cl 16.7), INPA 1708, Brazil, Pará, 2 km south of Jacundá [4°27S 49°07W], right bank of Tocantins River, 7.V.1984, W. Overal *leg.*; 1 male (cw 30.5, cl 19.5), 1 female (cw 53.2, cl 33.7), INPA 2012, Brazil, Pará, Altamira, Leonardo da Vinci stream, 3°0908"S, 52°0432"W, 26.III.2012, C.S. Souza *leg.*; 1 male (cw: 46.1, cb: 28.7), MZUSP 32818, Brazil, Pará, Altamira, Abrigo do Chuveiro cave, IV.2009, leg. unknow.

Distribution. The species was known only from its type locality, Vitória do Xingu, downstream from Altamira, on the left bank of Xingu River (MAGALHÃES 2003b). The present record from Jacundá extends its distribution to the eastern Amazon region, in the middle course of the Tocantins River basin.

Ecological notes. The specimens of *K. junki* were found in the vegetated margins of the Leonardo da Vinci stream, a small, clear-water tributary on the left bank of the Xingu River. All of the crabs were collected inside holes or beneath rocks. Two couples were found, and males and females apparently shared the same hole.

Remarks. The G1 of the specimens reported herein (Figs. 5 and 6) is similar to that of the holotype of *K. junki* (see MAGALHÃES 2003b: 384, fig. 1B), although some variability can be noticed in the morphology of the apical plate, particularly in the mesial margin of the distal lobe. In the holotype, this margin is indented both proximally and distally, whereas it is more distinctly indented only in the proximal portion of this margin in the present specimens. Since such a situation can be verified in the specimens from both the Xingu River and Tocantins River basins, this might be due to intraspecific variability and, therefore, the specimen from Tocantins River was considered to be conspecific with those from the Xingu River basin.

ACKNOWLEDGMENTS

MP thanks FAPESP (Fundação de Amparo à Pesquisa do Estado de São Paulo) for providing financial support through a doctoral fellowship (2012/01334-7). JEMF is grateful to Rodolfo Salm and to biologists Anderson Prates and Cleber Sousa for their assistance during field observations. JEMF was supported by Universidade Federal do Pará (research grants PROPESP #04/2014 and PROPESP/FADESP #09/2014). CM thanks the Conselho Nacional de Desenvolvimento Científico e Tecnológico for an ongoing Research Grant (Proc. 303837/2012-6). We also thank Barbara Robertson, Michael Türkay, and two anonymous reviewers for corrections and suggestions that greatly improved the manuscript.

LITERATURE CITED

MAGALHÃES C (1986) Revisão taxonômica dos caranguejos de água doce brasileiros da família Pseudothelphusidae (Crustacea, Decapoda). **Amazoniana 9**(4): 609-636.

MAGALHÃES C (1990) A new species of the genus *Kingsleya* from Amazonia, with a modified key for the Brazilian Pseudothelphusidae (Crustacea: Decapoda: Brachyura). **Zoologische Mededelingen 63**(21): 275-281.

MAGALHÃES C (2003a) Brachyura: Pseudothelphusidae e Trichodactylidae, p. 143-297. In: MELO GAS (Ed.). **Manual de Identificação dos Crustacea Decapoda de Água Doce do Brasil.** São Paulo, Edições Loyola.

MAGALHÃES C (2003b) The occurrence of freshwater crabs (Crustacea: Decapoda: Pseudothelphusidae, Trichodactylidae) in the Rio Xingu, Amazon Region, Brazil, with description of a new species of Pseudothelphusidae. **Amazoniana 17**(3/4): 377-386.

MAGALHÃES C (2005) A new species of freshwater crab (Crustacea: Decapoda: Pseudothelphusidae) from the southeastern Amazon Basin. **Nauplius 12**(2): 99-107.

MAGALHÃES C, TÜRKAY M (2008) A new species of *Kingsleya* from the Yanomami Indians area in the Upper Rio Orinoco, Venezuela. (Crustacea: Decapoda: Brachyura: Pseudothelphusidae). **Senckenbergiana biologica 88**(2): 1-7.

SALM R, PRATES A, SIMÕES NR, FEDER L (2015) Palm community transitions along a topographic gradient from floodplain to terra firme in the eastern Amazon. **Acta Amazonica 45**(1): 65-74. doi: 10.1590/1809-4392201401533

SMALLEY A (1964) A terminology for the gonopods of the American river crabs. **Systematic Zoology 13**: 28-31.

23

Description of the first species of *Metharpinia* (Crustacea: Amphipoda: Phoxocephalidae) from Brazil

Luiz F. Andrade[1], Rodrigo Johnsson[2] & André R. Senna[2]

[1]*Programa de Pós-graduação em Biologia Animal, Universidade Federal Rural do Rio de Janeiro. Rodovia BR 465, km 7, 23890-000 Seropédica, RJ, Brazil. E-mail: lzflp.andrade@hotmail.com*
[2]*Laboratório de Invertebrados Marinhos: Crustacea, Cnidaria & Fauna Associada, Instituto de Biologia, Universidade Federal da Bahia. Rua Barão de Jeremoabo 147, Ondina, 40170-290 Salvador, BA, Brazil. E-mail: r.johnsson@gmail.com; senna.carcinologia@gmail.com*

ABSTRACT. A new amphipod species of *Metharpinia* Schellenberg, 1931 is described from Campos Basin, southeastern Brazilian coast. The material was collected with van Veen grab from unconsolidated substratum, off the mouth of the Paraíba do Sul River. The new species can be distinguished from its congeners by presenting a strongly constricted rostrum and a slender palp of maxilla 1. There are four species in *Metharpinia* from the South Atlantic: *M. dentiurosoma* Alonso de Pina, 2003, *M. grandirama* Alonso de Pina, 2003 and *M. iado* Alonso de Pina, 2003, and *Metharpinia taylorae* **sp. nov.** This is the first record of a species of the genus from Brazilian waters.

KEY WORDS. Amphipod; Campos Basin; Habitats Project; *Metharpinia taylorae* **sp. nov.**; taxonomy.

Phoxocephalidae Sars, 1895, one of the most diverse amphipod taxa in terms of taxonomic characters, is characterized by the following: antennae 1 and 2 haustorioid in shape and with multiarticulate accessory flagellum; gnathopods 1 and 2 subchelated or chelated; pereopod 7 distinct from pereopod 5-6, shortened, article 2 expanded posteriorly; uropod 3 biramous; telson deeply cleft (BARNARD & DRUMMOND 1978, 1982). The Phoxocephalidae are benthic-burrowing amphipods (HURLEY 1954). They are widely distributed from shallow to deep waters. According to BARNARD & DRUMMOND (1978), Australia is the evolutionary center of the Phoxocephalidae, and there are two areas of dispersion of shallow water phoxocephalids, the Magellanic region plus the Falkland islands, and the Antarctica (including South Geogia Islands). However, this hypothesis needs to be tested by modern phylogenetic methods.

Phoxocephalidae currently includes more than 460 species around the world, grouped in 11 subfamilies and 74 genera (HORTON & DE BROYER 2014).Most species are found exclusively in the deep sea of the southern hemisphere (BARNARD & DRUMMOND 1978). According to SENNA & SOUZA-FILHO (2011) there are 13 Phoxocephalidae species recorded from Brazilian waters: *Bathybirubius margaretae* Senna, 2010, *Coxophoxus alonso* Senna, 2010, *Harpiniopsis galera* Barnard J.L., 1960, *Hererophoxus videns* Barnard KH, 1930, *Leptophoxoides marina* Senna, 2010, *Microphoxus breviramus* Bustamante, 2002, *M. cornutus* (Schellenberg, 1931), *M. moraesi* Bustamante, 2002, *M. uroserratus* Bustamante, 2002, *Phoxocephalus homilis* Barnard JL, 1960, *Pseudharpinia berardo* Senna, 2010, *P. ovata* Senna, 2010, and *P. tupinamba* Senna & Souza-Filho, 2011.

Metharpinia Schellenberg, 1931 and its sister-group, *Microphoxus* Barnard, 1960, are among the most primitive genera in the birubiin-parharpiniin group of the Americas. *Metharpinia* has nine species, all distributed along the west and east coasts of North and Central America, and Argentina (BARNARD & DRUMMOND 1978, BARNARD & KARAMAN 1991, ALONSO DE PINA 2001, 2003a, b). According to ALONSO DE PINA (2003a), species of *Metharpinia* are characterized by the following characters: antenna 1, article 2 with ventral setae placed proximally; maxilliped with dactylar nail partially fused and immersed; gnathopods 1 and 2, palms acute and propodus poorly setose anteriorly; and pereopods 3 and 4, propodus with facial setal formula composed of stout setae and dactyli with inner acclivity sharp, produced as tooth.

We describe a new species of *Metharpinia*. This is the first record of the genus from Brazilian waters, increasing the Phoxocephalidae diversity in Brazil to 14 species in nine genera.

MATERIAL AND METHODS

The material examined was collected during the Habitats Project (Environmental Heterogeneity of Campos Basin), coordinated by the Brazilian Oil Company (CENPES/PETROBRAS). Collecting trips were conducted at the Campos Basin, off the mouth of the Paraíba do Sul River, between the states of Rio de Janeiro and Espírito Santo, southeastern Brazil, in the Summer and Winter of 2009. Collections were made aboard the R/V Gyre, from unconsolidated substratum, using a van Veen grab.

The specimens were dissected under a stereoscopic microscope Motic K-401L and mounted in glycerine gel slides. The illustrations were produced under an optic microscope with a camera lucida Motic BA-310. The type material is deposited at the Crustacea Collection of the Museu Nacional, Universidade Federal do Rio de Janeiro (MNRJ). It is preserved in 70% ethanol or glycerine gel slides. The setal classification adopted in this paper follows WATLING (1989). Nomenclature of the gnathopod palm is based on POORE & LOWRY (1997).

TAXONOMY

Metharpinia Schellenberg, 1931

Diagnosis. See BARNARD & KARAMAN (1991).

Composition of the genus: *M. coronadoi* Barnard, 1980; *M. dentiurosoma* Alonso de Pina, 2003; *M. floridana* (Shoemaker, 1933); *M. grandirama* Alonso de Pina, 2003; *M. iado* Alonso de Pina, 2003; *M. jonesi* (Barnard, 1963); *M. longirostris* Schellenberg, 1931; *M. oripacifica* Barnard, 1980; *M. protuberantis* Alonso de Pina, 2001; *M. taylorae* **sp. nov.**

Metharpinia taylorae **sp. nov.**
Figs. 1-25

Diagnosis. Rostrum strongly constricted. Right mandible, incisor with two spines, one large, apically bifid, and one small and subrounded. Left mandible, incisor with three teeth, one large and two small, one of them apically bifid, lacinia mobilis well developed, apically smooth and subrounded. Maxilla 1, palp very slender and setose. Maxilliped, inner plate with three apical plumose setae. Gnathopods 1-2 poorly setose. Pereopod 7, basis with posteroventral lobe rounded. Epimeral plate 3, ventral margin with five submarginal pectinate setae, posterior margin slightly serrate, with 13 long slender setae, posteroventral corner broadly rounded. Urosomite 3 without dorsal hook. Uropod 3, inner ramus bearing few plumose setae, outer ramus with one plumose seta. Telson, deeply cleft, about 90% of its length, apical margin sinuous, each lobe with one lateral small plumose seta, subapical margin bearing five slender setae on each lobe, inner teeth of apex naked.

Description. Based on the holotype (MNRJ 477) and allotype (MNRJ 479). Head (Figs. 1 and 2), eyes present, rostrum strongly constricted, narrow, spatulate, elongate, about 1.3X longer than antenna 1 peduncular article 1. Antenna 1 (Fig. 3), peduncle article 1 about 1.4X longer than wide, without setae; article 2 anterodorsal corner with one small slender seta, ventral margin with seven slender setae, 1.3X longer than wide; article 3 shortened with two ventral slender setae, 1.2X wider than long; flagellum 16-articulate, poorly setose; accessory flagellum 14-articulate, elongate, poorly setose. Antenna 2 (Fig. 4), peduncle, article 4 about 1.5X longer than wide ventral margin with a row of 10 slender setae, 11 stout facial setae arranged in three rows, anterodorsal corner with one stout seta and one

simple seta; article 5, ventral margin setose, facial row of setae with nine medium to small stout setae, apical margin with two medium slender setae; flagellum 20-articulate, poorly setose. Right mandible (Fig. 5), incisor with two teeth, one large, apically bifid, and one small and blunt; accessory setal row with seven stout multi-cuspidate setae; molar not triturative with four slender setae; palpar hump small, lacinia mobilis absent, palp 3-articulate, article 3 apically setose. Left mandible (Fig. 6), incisor with three teeth, one large and two small, one of them apically bifid; accessory setal row with eight stout multi-cuspidate setae; molar not triturative, with nine apical slender setae; lacinia mobilis well developed, apically smooth and subrounded. Maxilla 1 (Fig. 7), inner plate 1.1X wider than long with five apical plumose setae; outer plate 1.3X longer than wide with 10 multi-cuspidate robust apical setae; palp 2-articulate, very slender, article 2, outer margin with five slender setae, three proximal and two distal, inner margin with four slender setae, and apical margin with three slender setae. Maxilla 2 (Fig. 8), inner plate about 1.6X longer than wide and slightly shorter than outer plate, apical margin with nine slender setae; outer plate about 2.6X longer than wide, apical margin with eight slender setae. Maxilliped (Fig. 9), inner plate subrectangular, with three apical plumose setae; outer plate lanceolate with seven medial setae, two apical setae, and one small lateral setae; palp, 4-articulate, article 2, suboval, about 2.2X longer than wide medially setose; article 3 suboval, medially setose, with one lateral slender setae in notch, about 1.9X longer than wide; article 4 simple, curved and slender.

Gnathopod 1 (Fig. 10) poorly setose, coxa weakly expanded anteriorly, posteroventral corner with six slender setae; basis about 3.5X longer than wide, subrectangular, posterior margin with 13 short slender setae, ventral margin without setae; ischium, small, posteroventral corner with three slender setae; merus, small, subtriangular, posterior margin with three slender setae; carpus, about 1.4X longer than wide posterior margin medially setose; propodus, about 1.4X longer than wide, anterior margin without setae, anterodistal corner with four slender setae, posterior margin straight with 10 slender setae, palm almost transverse, palmar corner defined by a small and slightly upwards curved spine with one lateral stout seta with accessory seta; dactylus, curved, simple, subequal in length to palm. Gnathopod 2 (Fig. 11) poorly setose, coxa weakly expanded anteriorly, posteroventral corner with 12 slender setae, posterior margin with two pairs of small slender setae, anterior margin with one pair of small slender setae; basis about 4.1X longer than wide, subrectangular, anterior margin with two slender setae, anteroventral corner with six slender setae, posterior margin with two slender setae; ischium, small, without setae; merus, small, subtriangular, posterior margin with six slender setae; carpus, elongate, about 3X longer than wide posterior margin setose; propodus, about 1.9X longer than wide, anterodistal corner setose, posterodistal margin setose, palm almost transverse, palmar corner defined by a small spine

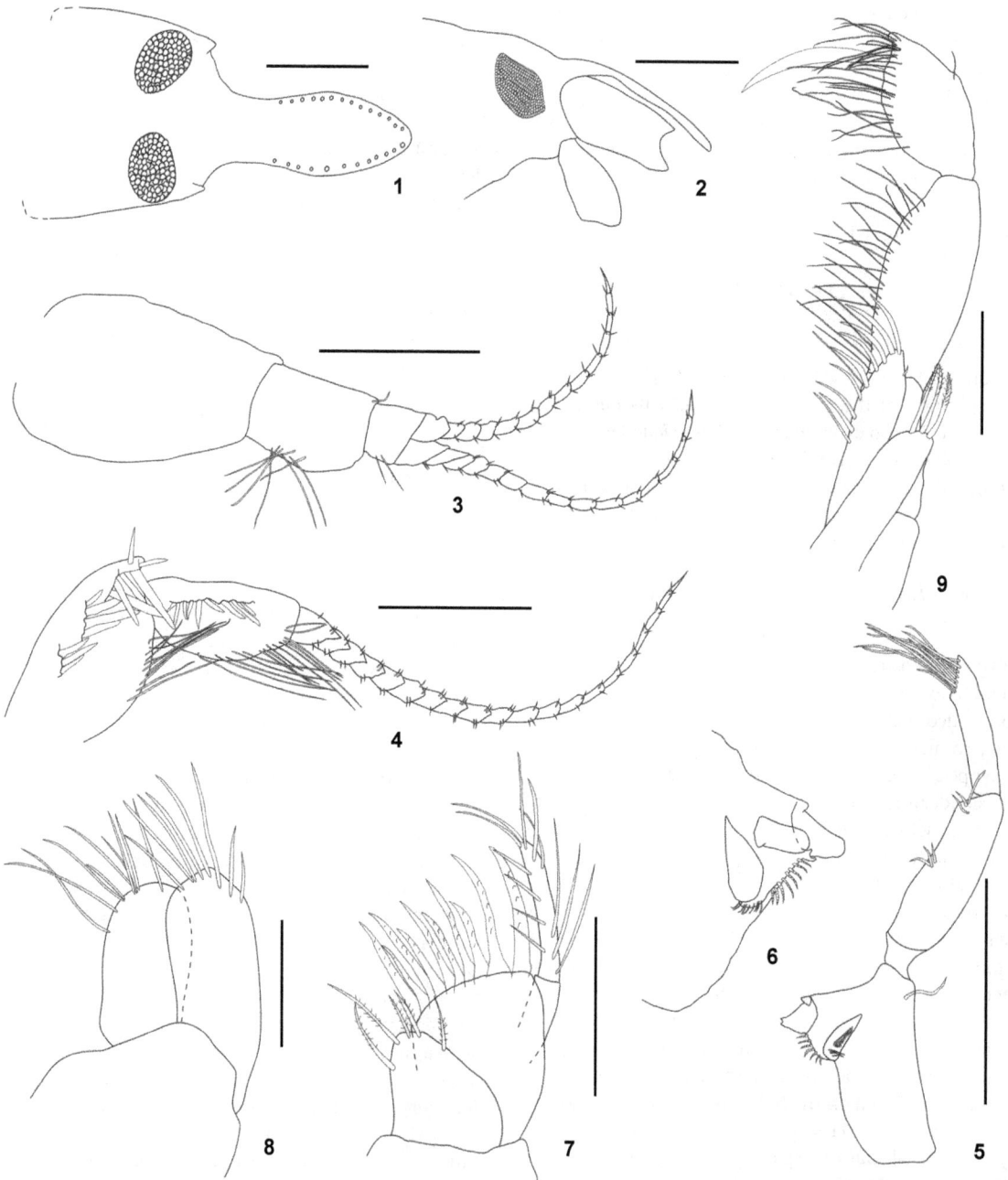

Figures 1-9. *Metharpinia taylorae* **sp. nov.**, holotype, female: (1) head, dorsal view; (2) head, lateral view; (3) antenna 1; (4) antenna 2; (5) right mandible; (6) left mandible; (7) maxilla 1; (8) maxilla 2; (9) maxilliped. Scale bars: 0.2 mm for maxilla 1-2; 0.5 mm for the remainder.

slightly curved upwards; dactylus curved, simple, slightly longer than palm. Pereopod 3 (Fig. 12), coxa subrectangular, about 1.7X longer than wide, ventral margin subrounded, without setae; basis, about 3X longer than wide, subrectangular, posterior margin with five long slender setae, anterior margin with seven small setae; ischium small, posteroventral corner with

one slender setae; merus elongate, about twice longer than wide, subrectangular, posterior margin with three sets of long setae (2-1-2), posterodistal corner with a row of seven long setae, anterodistal corner with two small setae; carpus broad, about 1.2X longer than wide ventral margin setose; propodus elongate, about 3.8X longer than wide, posterior margin with eight sets of setae (6-2-3-2-5-2-2-1); dactylus simple, about 0.4X as long as propodus. Pereopod 4 (Fig. 13), coxa suboval, ventral margin rounded, posteroventral corner with two small slender setae; basis, 3.2X longer than wide, posterior margin with five long slender setae, posteroventral corner with three slender setae; ischium, small, posterior margin with five slender setae; merus, elongate, about 2.1X longer than wide, subrectangular, anterodistal corner with two slender setae, posterior margin with eight pairs of long slender setae, posterodistal corner with four slender setae; carpus, broad, about 1.4X longer than wide, posterior margin setose; propodus elongate, about 3.9X longer than wide, posterior margin with four sets of slender setae (4-2-4-2); dactylus robust, simple, about half length of propodus. Pereopod 5 (Fig. 14), coxa wider than long, with two lobes, posterior lobe with sinuous ventral margin, deeply produced, bearing four slender setae; basis, about twice longer than wide, subrectangular, posteriorly slightly expanded, anterior margin setose, anterodistal corner with five slender setae, posterior margin naked and slightly concave; ischium, small, naked; merus, about 1.7X longer than wide posteriorly expanded, anterior margin with one slender seta, posterior margin setose; carpus, about 1.4X longer than wide, anterior margin setose, posterior margin with three sets of slender setae (5-1-11), posteroventral corner with two long plumose setae; propodus, about 4.1X longer than wide, anterior margin with two sets of slender setae (4-5), anteroventral corner with five slender setae, posteroventral corner with six setae; dactylus slender, simple, about half length of propodus. Pereopod 6 (Fig. 15), coxa with a subacute posterior lobe, posterior margin setose; basis, about 1.2X longer than wide expanded posteriorly, anterior margin setose, posterior margin naked; ischium, small, naked, about twice wider than long; merus, wide, about 1.6X longer than wide anterior margin with four sets of setae (2-4-2-4), posterior margin with four sets of setae (3-4-5-5); carpus, about 2.2X longer than wide, anterior margin with three sets of setae (2-2-4), posterior margin with four sets of setae (1-2-2-6); propodus, elongate, about 8.2X longer than wide, anterior margin with two sets of setae (3-2), anterodistal corner with two slender setae and one stout setae with accessory seta, posterior margin with six sets of setae (3-2-2-2-3-1), posteroventral corner with five slender setae; dactylus, about 0.3X as long as propodus. Pereopod 7 (Fig. 16), coxa, posterior margin minutely setose; basis strongly expanded posteriorly, posterior margin serrate, posteroventral lobe rounded, smooth, and naked; ischium small, anterodistal corner with one slender seta; merus, anterior margin with two sets of setae (2-2), posterior margin with four sets of setae (1-2-2-2); carpus, about 1.5X longer than wide, anterior margin with four slender setae, anterodistal corner with a set of four slender setae, posterior margin with two sets of setae (3-2), posteroventral corner with a set of four slender setae; propodus, about 4X longer than wide, anterior margin with one slender seta, anterodistal corner with one slender seta, posterior mar-

gin with two sets of setae (2-4), posteroventral corner with three slender setae; dactylus slightly robust, about 0.6X as long as propodus.

Epimeral plate 1 (Fig. 17), anterior margin with one slender seta, ventral margin with two slender and 13 plumose setae, posterior margin slightly serrate, with 11 long slender setae, posteroventral corner rounded. Epimeral plate 2 (Fig. 18), anterior margin naked, ventral margin with seven medium and four long plumose submarginal setae, posterior margin with slightly serrate, with 10 long slender setae, posteroventral corner subrounded. Epimeral plate 3 (Fig. 19), anterior margin with seven small slender setae, ventral margin with five submarginal pectinate setae, posterior margin with slightly serrate, with 13 long slender setae, posteroventral corner broadly rounded. Uropod 1 (Fig. 20), peduncle elongated, about 2.1X longer than wide, dorsal margin with four stout setae; outer ramus slightly longer than inner ramus, about 7.9X longer than wide, dorsal margin with 11 stout setae, plus one subapical long stout seta; inner ramus subequal in length to peduncle, about 6X longer than wide, dorsal margin with three stout setae, plus one subapical long stout seta. Uropod 2 (Fig. 21), peduncle about 1.7X longer than wide, dorsal margin with three stout setae, dorsoapical corner with one stout setae, apicolateral corner with one stout setae; outer ramus, slightly longer than inner ramus, about 6.7X longer than wide, dorsal margin with nine stout setae, plus one subapical long stout seta; inner ramus, about 1.2X longer than peduncle, about 5.7X longer than wide, dorsal margin naked, with one subapical long stout seta. Urosomite 3 (Fig. 22) without dorsal hook. Uropod 3 (Fig. 23), peduncle short, about 1.3X longer than wide, apicolateral corner with five stout setae; outer ramus 2-articulate, about 1.6X longer than inner ramus, about 2.7X longer than peduncle, article 1 elongated, about 4.7X longer than wide, about 3.4X longer than article 2, dorsal margin bearing two stout setae and one small distal seta, ventral margin with one long distal plumose seta; article 2, about 4.7X longer than wide, bearing two apical slender setae; inner ramus, about 1.7X longer than peduncle, about 4.7X longer than wide, bearing seven apical and one subapical stout plumose setae. Male uropod 3 (Fig. 24), peduncle short, about 2.2X longer than wide, apicolateral corner with three stout setae with accessory setae, lateral margin with one stout setae with accessory setae and one short setae, facial margin with one slender setae; outer ramus 2-articulate, about 1.1X longer than inner ramus, about 2,2X longer than peduncle, article 1 elongated, about 4.6X longer than wide, about 4.4X longer than article 2, dorsal margin bearing two sets of stout setae with accessory setae (2-2) and three stout setae with accessory setae distally, ventral margin with five long plumose setae; article 2, about 3.4X longer than wide, bearing two apical slender setae; inner ramus, about 1.9X longer than peduncle, about 5.6X longer than wide, dorsal margin bearing two long plumose setae, ventral margin bearing five long plumose setae and one short simple setae, apical margin bearing two long plumose setae. Telson (Fig.

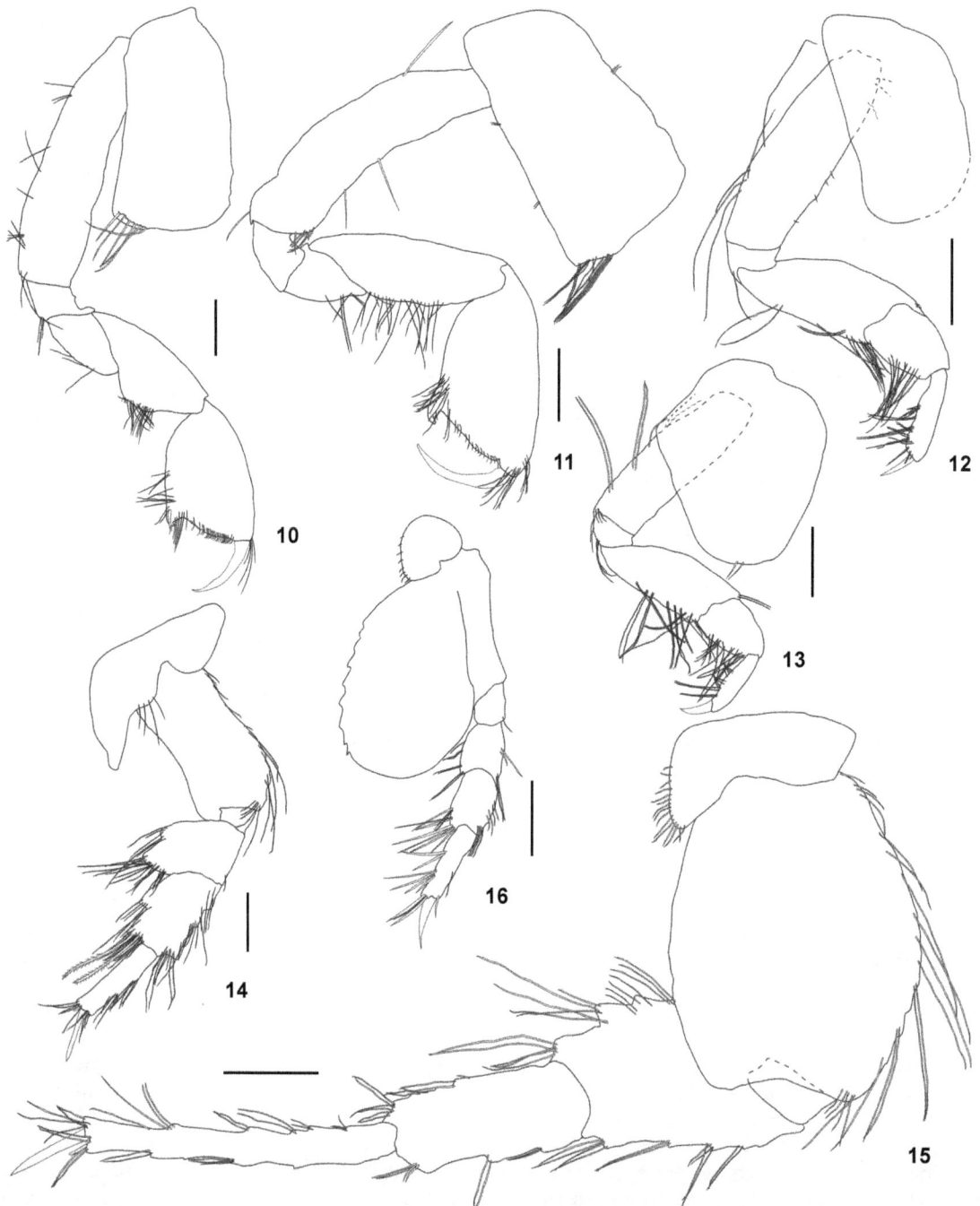

Figures 10-16. *Metharpinia taylorae* **sp. nov.**, holotype, female: (10) gnathopod 1; (11) gnathopod 2; (12) pereopod 3; (13) pereopod 4; (14) pereopod 5; (15) pereopod 6; (16) pereopod 7. Scale bars: 1.0 mm for gnathopods 1-2; 0.5 mm for the remainder.

25), about 1.3X longer than wide, deeply cleft, about 90% of its length, apical margin truncate with blunt cusp, each lobe with one lateral small plumose seta, subapical margin bearing five slender setae on each lobe, apex naked.

Material examined. Holotype female, 8.5 mm, BRAZIL, *Rio de Janeiro*: Campos Basin (21°33'52.574"S, 40°42'53.900"W, 22 m depth), 10 March 2009, R/V Gyre *leg.*, MNRJ 477. Allotype male, BRAZIL, *Espírito Santo*: Campos Basin (21°11'0,850"S,

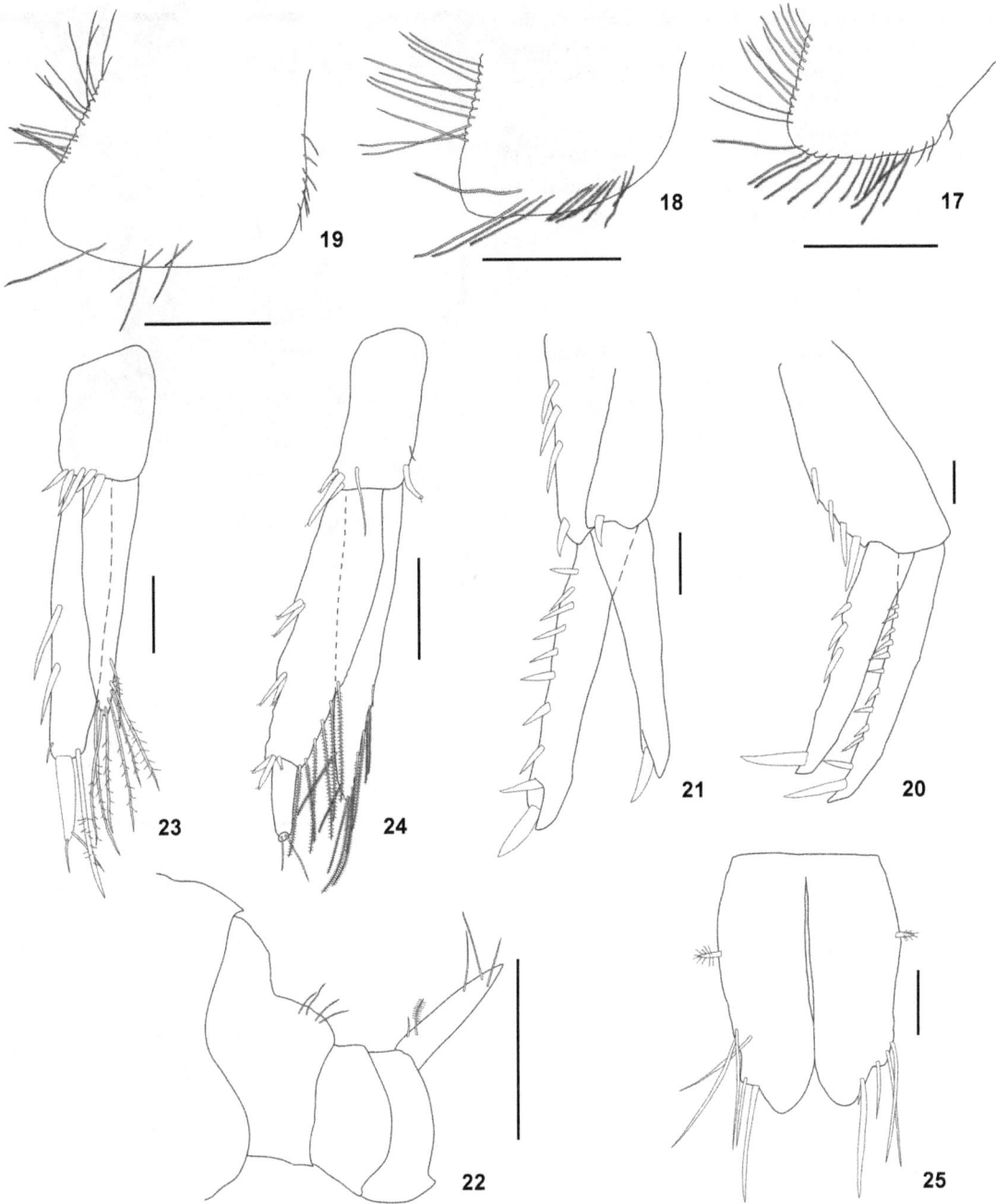

Figures 17-25. *Metharpinia taylorae* **sp. nov.**, holotype, female: (17) epimeral plate 1; (18) epimeral plate 2; (19) epimeral plate 3; (20) uropod 1; (21) uropod 2; (22) urosomite 3; (23) uropod 3; (25) telson; paratype, male: (24) uropod 3. Scale bars: 0.2 for male and female uropod 3; 0.5 mm for the remainder.

40°28′27.125″W, 26 m depth); 5 March 2009, R/V Gyre *leg.*, MNRJ 479. Paratypes: 1 male, Brazil, *Espírito Santo*: Campos Basin (21°17′51.743″S, 40°30′59.011″W, 29 m depth), 07 March 2009, R/V Gyre *leg.*, MNRJ 478; 1 male and 12 juveniles, *Rio de Janeiro*: Campos Basin (21°33′53.096″S, 40°42′55.466″W, 21 m depth, van Veen), 10 March 2009, R/V Gyre *leg.*, MNRJ 480; 2 females and 13 juveniles, (21°39′11.066″S, 40°48′49.898″W, 21 m depth, van Veen), 11 March 2009, R/V Gyre *leg.*, MNRJ 481; 1 ovigerous female, 1 male and 8 juveniles, (21°39′9.790″S, 40°48′50.234″W, 22 m depth, van Veen), 19 July 2009, R/V Gyre *leg.*, MNRJ 482.

Geographic distribution. Brazil, north coast of Rio de Janeiro State and south coast Espírito Santo State, Campos Basin, off the mouth of the Paraíba do Sul River (Fig. 26). Type locality: 21°33'52.574"S, 40°42'53.900"W.

Bathymetric range. Collected from 21 to 29 m depth.

Etymology. The species epithet, *taylorae*, is dedicated to Dr. Joanne Taylor, from the Museum Victoria, Australia, to honor her important contributions to the knowledge on the amphipod family Phoxocephalidae.

DISCUSSION

Metharpinia taylorae **sp. nov.** shares the diagnostic characters of the genus, such as the constricted, narrow, spatulated and elongated rostrum, and uropod 3 with one of rami longer than peduncle, bearing article 2 on outer ramus, with two apical setae (BARNARD & KARAMAN 1991). Although *M. taylorae* **sp. nov.** shares some characters with species of *Microphoxus* (see comparison between the two genera in ALONSO DE PINA 2003a), we placed the new species in *Metharpinia* because, for the most part, it fits the diagnosis of this genus.

The new species is easily distinguished from *M. dentiurosoma* and from *M. grandirama* by in lacking the dorsal hook on urosomite 3, a unique feature of *M. dentiurosoma* and *M. grandirama* (ALONSO DE PINA 2003a).

Metharpinia taylorae **sp. nov.** differs from *M. protuberantis* by the following combination of characters (*M. protuberantis* characters in parenthesis): rostrum strongly constricted and highly developed (weakly constricted, poorly developed); gnathopod 2, basis, slightly elongate (strongly elongate); pereopod 5, coxa posterior lobe deeply produced (slightly produced); epimeral plate 3, posteroventral corner broadly rounded (strongly produced into a large tooth) (ALONSO DE PINA 2001).

Metharpinia taylorae **sp. nov.** differs from *M. coronadoi* by the following characters (*M. coronadoi* characters in parenthesis): coxa 1 weakly expanded anteriorly (anterior margin straight); right lacinia mobilis absent (present); left lacinia mobilis well developed, apically smooth and subrounded (with 2-3 teeth plus 1-2 accessory teeth); epimeral plate 3, posteroventral margin rounded (rounded-quadrate) (BARNARD 1980).

Metharpinia taylorae **sp. nov.** differs from *M. jonesi* by the following characters (*M. jonesi* characters in parenthesis): pereopods 3 and 4 very similar in shape (pereopod 4 stouter and longer than pereopod 3); epimeral plate 3, ventral margin with five submarginal pectinate setae (with large tooth, ventral margin with 4 setae) (BARNARD 1963).

Metharpinia taylorae **sp. nov.** differs from *M. floridana* by the following characters (*M. floridana* characters in parenthesis): epimeral plate 2 rounded (rounded-subquadrate); telson, both male and female, each lobe with 1 lateral small plumose seta (each lobe with 1 lateral and 1 subapical plumose setae) (SHOEMAKER 1933).

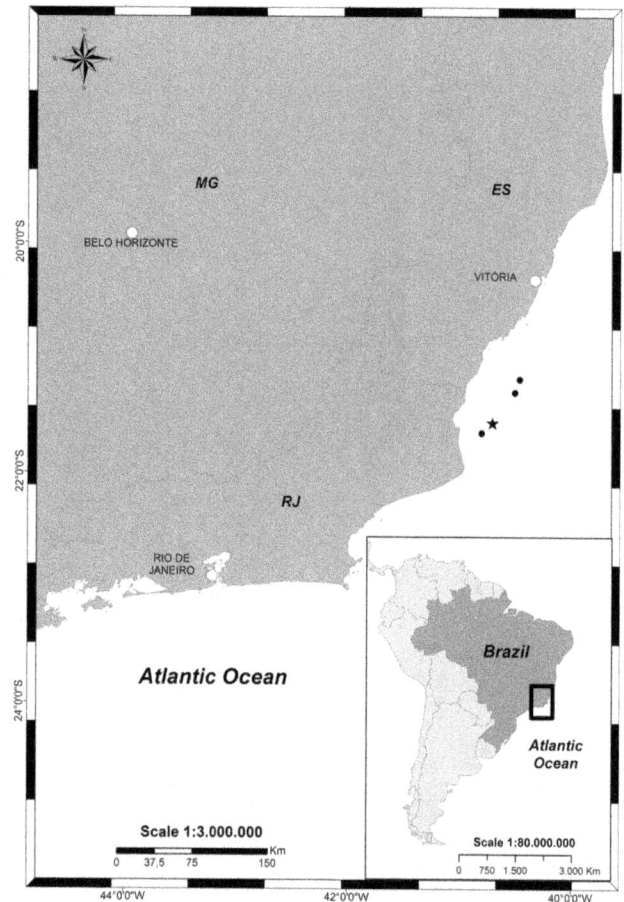

Figure 26. Distribution of *Metharpinia taylorae* **sp. nov.** Star: type locality, 21°33'52.574"S, 40°42'53.900"W; Circle: ocurrence of paratypes. RJ: Rio de Janeiro State; ES: Espírito Santo State; MG: Minas Gerais State (Distribution map by Danielle P. Cintra).

Metharpinia taylorae **sp. nov.** differs from *M. oripacifica* by (*M. oripacifica* characters in parenthesis): right lacinia mobilis absent (present); left lacinia mobilis well developed, apically smooth and subrounded (with five teeth, middle teeth scarcely shortened); telson, each lobe bearing 5 slender setae (dorsolateral brush of 7 setae) (BARNARD 1980).

Metharpinia iado, recorded from Argentina, is probably the most different from *M. taylorae* **sp. nov.**, due to the following characters (*M. iado* characters in parenthesis): left mandible, lacinia mobilis apically smooth and blunt (multi-cuspidate); maxilla 2, inner plate without plumose setae (with apical and medial plumose setae), outer plate, outer margin naked (setose); maxilliped, inner plate, with three apical plumose setae (eight apical plumose setae, plus one stout seta); gnathopods 1-2 weakly setose (strongly setose), palm of gnathopods 1-2 sinuous (almost straight); pereopod 5, coxa posterior lobe produced and subacute (not produced, round), pereopods 5-6, merus and carpus without facial setae (with facial sets of stout setae); telson, deeply cleft, about 90% (three-quarters cleft) (ALONSO DE PINA 2003b).

The description of the type-species of *Metharpinia*, *M. longirostris*, is insufficient and the species is poorly illustrated. However, we can distinguish the new species from it by the following characters (characters of *M. longirostris* within parenthesis): pereopods 5-6 without facial setae (with facial setae); posteroventral lobe of basis of pereopod 7 round (truncated); coxa 4, posteroventral corner with two small slender setae (ventral and posterior margins setose); each telson lobe with one lateral small plumose seta, apex naked (without lateral plumose setae, apex setose) (SCHELLENBERG 1931, BARNARD 1980).

Two characteristics are exclusive of the new species among the representatives of *Metharpinia*: palp of maxilla 1 slender, article 2 bearing slender setae on both sides and at apical margin; and the lacinia mobilis absent in right mandible and smooth and subrounded in the left mandible.

ACKNOWLEDGMENTS

We thank Coordenação de Aperfeiçoamento de Pessoal de Nível Superior (CAPES) and Fundação Carlos Chagas Filho de Amparo à Pesquisa do Estado do Rio de Janeiro (FAPERJ) for the financial support and fellowships. The material examined was provided by Centro de Pesquisas e Desenvolvimento Leopoldo Américo Miguez de Mello (CENPES-PETROBRAS). The distribution map was made by Danielle P. Cintra from Instituto de Geociências, Universidade Federal do Rio de Janeiro (IGEO-UFRJ).

LITERATURE CITED

ALONSO DE PINA, GM (2001) Two new phoxocephalids (Crustacea: Amphipoda: Phoxocephalidae) from the south-west Atlantic. **Journal of Natural History 35**: 515-537.

ALONSO DE PINA GM (2003a) Two new species of *Metharpinia* Schellenberg (Amphipoda: Phoxocephalidae) from the southwest Atlantic. **Journal of Natural History 37**: 2521-2545.

ALONSO DE PINA GM (2003b) A new species of Phoxocephalidae and some other records of sand-borrowing Amphipoda (Crustacea) from Argentina. **Journal of Natural History 37**: 1029-1057.

BARNARD JL (1963) Relationship of benthic Amphipoda to invertebrate communities of inshore sublittoral sands of southern California. **Pacific Naturalist 3**(15): 437-467.

BARNARD JL (1980) Revision of *Metharpinia* and *Microphoxus* (marine phoxocephalid Amphipoda from the Americas). **Proceedings of the Biological Society of Washington 93**(1): 104-135.

BARNARD JL, DRUMMOND MM (1978) Gammaridean Amphipoda of Australia, part III: The Phoxocephalidae. **Smithsonian Contributions to Zoology 245**: 1-551.

BARNARD JL, DRUMMOND MM (1982) Gammaridean Amphipoda of Australia, Part V: Superfamily Haustorioidea. **Smithsonian Contributions to Zoology 360**: 1-148.

BARNARD JL, KARAMAN GS (1991) The Families and Genera of Marine Gammaridean Amphipoda (Except Marine Gammaroidea). **Records of the Australian Museum 13**: 1-866.

HORTON T, DE BROYER C (2014) Phoxocephalidae Sars, 1891. In: HORTON T, LOWRY J, BROYER C DE (Ed.). **World Amphipoda Database.** Available online at: http://www.marinespecies.org/aphia.php?p=taxdetails&id=101403 [Accessed: 22 May 2014]

HURLEY DE (1954) Studies on the New Zealand Amphipodan Fauna 3. The Family Phoxocephalidae. **Transactions of the Royal Society of New Zealand 81**: 579-599.

POORE AGB, LOWRY JK (1997) New ampithoid amphipods from Port Jackson, New South Wales, Australia (Crustacea: Amphipoda: Ampithoidae). **Invertebrate Taxonomy 11**: 897-941.

SENNA AR, SOUZA-FILHO JF (2011) A new species of *Pseudharpinia* (Amphipoda: Haustorioidea: Phoxocephalidae) from Southeastern Brazilian continental shelf. **Nauplius 19**(1): 7-16.

SCHELLENBERG A (1931) Gammariden und Caprelliden des Magellangebietes, Südgeorgiens und der Westantarktis. **Further Zoological Results of the Swedish Antarctic Expedition 1901-1903 2**(6): 1-290.

SHOEMAKER CR (1933) Amphipoda from Florida and the West Indies. **American Museum Novitates 598**: 1-24.

WATLING L (1989) A classification of crustacean setae based on the homology concept, p. 15-26. In: FELGENHAUER BE, THISTLE AB, WATLING L (Ed.) **Functional Morphology of Feeding and Grooming in Crustacea.** New York, CRC Press, Crustacean Issues, vol. 6.

Permissions

List of Contributors

Mingming Zhang
College of Wildlife Resources, Northeast Forestry University, No.26 Hexing Road, Xiangfang District, Harbin 150040, P.R. China

Zhensheng Liu and Liwei Teng
College of Wildlife Resources, Northeast Forestry University, No.26 Hexing Road, Xiangfang District, Harbin 150040, P.R. China
Key Laboratory of Conservation Biology, State Forestry Administration, No.26 Hexing Road, Xiangfang District, Harbin 150040, P.R. China

Jorge L. Ramirez and Rina Ramírez
Departamento de Malacología y Carcinología, Museo de Historia Natural, Universidad Nacional Mayor de San Marcos, Apartado 14-0434, Lima-14, Perú

Marília A. S. Barros and Daniel M. A. Pessoa
Departamento de Fisiologia, Centro de Biociências, Universidade Federal do Rio Grande do Norte. Campus Universitário Lagoa Nova, 59078-970 Natal, RN, Brazil

Ana Maria Rui
Departamento de Ecologia, Zoologia e Genética, Instituto de Biologia, Universidade Federal de Pelotas. Campus Universitário Capão do Leão, Caixa Postal 354, 96001-970 Pelotas, RS, Brazil

Marcelo M. Dalosto, Alexandre V. Palaoro, Davi de Oliveira and Sandro Santos
Núcleo de Estudos em Biodiversidade Aquática, Programa de Pós-Graduação em Biodiversidade Animal, Centro de Ciências Naturais e Exatas, Universidade Federal de Santa Maria. Avenida Roraima 1000, 97105-900 Santa Maria, RS, Brazil

Évelin Samuelsson
Programa de Pós-Graduação em Ecologia, Departamento de Ciências Biológicas, Universidade Regional Integrada do Alto Uruguai e das Missões. Avenida Sete de Setembro 1621, 99700-000 Erechim, RS, Brazil

Ingrid Mattos and José Ricardo M. Mermudes
Laboratório de Entomologia, Departamento de Zoologia, Universidade Federal do Rio de Janeiro. Caixa Postal 68044, 21941-971 Rio de Janeiro, RJ, Brazil

Mauricio O. Moura
Departamento de Zoologia, Universidade Federal do Paraná. Caixa Postal 19020, 81531-980 Curitiba, PR, Brazil

Gilmar Perbiche-Neves and Carlos E. F. da Rocha
Departamento de Zoologia, Instituto de Biociências, Universidade de São Paulo. Rua do Matão, travessa 14, 321, 05508-900 São Paulo, SP, Brazil

Marcos G. Nogueira
Departamento de Zoologia, Instituto de Biociências, Universidade Estadual Paulista. Distrito de Rubião Júnior, 18618-970 Botucatu, SP, Brazil

Débora de S. Silva-Camacho, Joaquim N. de S. Santos, Rafaela de S. Gomes and Francisco G. Araújo
Laboratório de Ecologia de Peixes, Universidade Federal Rural do Rio de Janeiro. Antiga Rodovia Rio-SP km 47, 23851-970 Seropédica, RJ, Brazil

João P. Vieira and Michelle N. Lopes
Laboratório de Ictiologia, Universidade Federal do Rio Grande. Avenida Itália km 8, 96201-900 Rio Grande, RS, Brazil

Amilcar Brum Barbosa and Sonia Barbosa dos Santos
Laboratório de Malacologia Límnica e Terrestre, Departamento de Zoologia, Instituto de Biologia Roberto Alcantara Gomes, Universidade do Estado do Rio de Janeiro. Rua São Francisco Xavier 524, PHLC sala 525-2, 20550-900 Rio de Janeiro, RJ, Brazil

Alessandra da Fonseca Viana and Marcelo Vianna
Laboratório de Biologia e Tecnologia Pesqueira, Instituto de Biologia, Universidade Federal do Rio de Janeiro. Avenida Carlos Chagas Filho 373, Bloco A, 21941-902 Rio de Janeiro, RJ, Brazil

Austin L. Hughes
Department of Biological Sciences, University of South Carolina, Columbia SC 29205 USA

Regiane Saturnino, Bruno V.B. Rodrigues and Alexandre B. Bonaldo
Laboratório de Aracnologia, Coordenação de Zoologia, Museu Paraense Emílio Goeldi. Avenida Perimetral 1901, Terra Firme, 66077-830 Belém, Pará, Brazil

Diane Nava, Rozane M. Restello and Luiz U. Hepp
Programa de Pós-graduação em Ecologia, Universidade Regional Integrada do Alto Uruguai e das Missões. Avenida Sete de Setembro 1621, 99709-910 Erechim, RS, Brazil

Silvio T. da Costa, Luíza Loebens and Rafael Lazzari
Departamento de Zootecnia e Ciências Biológicas, Centro de Educação Norte do Rio Grande do Sul, Universidade Federal de Santa Maria. 98300-000 Palmeira das Missões, RS, Brazil

Luciane T. Gressler and Fernando J. Sutili
Programa de Pós-graduação em Farmacologia, Universidade Federal de Santa Maria. 97105-900 Santa Maria, RS, Brazil

Bernardo Baldisserotto
Departamento de Fisiologia e Farmacologia, Universidade Federal de Santa Maria. 97105-900 Santa Maria, RS, Brazil

Ana Lucia Henriques-Oliveira and Leandro Lourenço Dumas
Universidade Federal do Rio de Janeiro, Laboratório de Entomologia, Departamento de Zoologia, Instituto de Biologia, Caixa Postal 68044, Cidade Universitária, 21941-971, Rio de Janeiro, RJ, Brazil

Diego Astúa
Departamento de Zoologia, Universidade Federal de Pernambuco. Avenida Professor Moraes Rêgo, s/n, Cidade Universitária, 50670-901 Recife, PE, Brazil

Raul Fonseca
Departamento de Zoologia, Universidade Federal de Pernambuco. Avenida Professor Moraes Rêgo, s/n, Cidade Universitária, 50670-901 Recife, PE, Brazil
Departamento de Zoologia, Universidade do Estado do Rio de Janeiro. 20550-013 Rio de Janeiro, RJ, Brazil

Aline S. Maciel, Thereza de A. Garbelotto, Ingrid C. Winter, Talita Roell and Luiz A. Campos
Departamento de Zoologia, Universidade Federal do Rio Grande do Sul. Avenida Bento Gonçalves 9500, Agronomia, 91501-970 Porto Alegre, RS, Brazil

Ana Paula S. Dornellas and Luiz R. L. Simone
Museu de Zoologia, Universidade de São Paulo. Caixa Postal 42494, 04218-970 São Paulo, SP, Brazil

Li-Biao Zhang and Qi Liu
Guangdong Public Laboratory of Wild Animal Conservation and Utilization & Guangdong Key Laboratory of Integrated Pest Management in Agriculture, Guangdong Entomological Institute, Guangzhou 510260, China

Fu-Min Wang
Guangdong Provincial Wildlife Rescue Center, Guangzhou 510520, China

Li Wei
College of Ecology, Lishui University, Lishui 323000, China

Denis Rafael Pedroso
Laboratório de Aracnologia, Museu Nacional, Universidade Federal do Rio de Janeiro, Brazil

Renner Luiz Cerqueira Baptista
Laboratório de Diversidade de Aracnídeos, Instituto de Biologia, Universidade Federal do Rio de Janeiro, Brazil

Rogério Bertani
Laboratório Especial de Ecologia e Evolução, Instituto Butantan. Avenida Vital Brazil 1500, 05503-900 São Paulo, SP, Brazil

Andressa Paladini and Rodney Ramiro Cavichioli
Departamento de Zoologia, Universidade Federal do Paraná. Caixa Postal 19020, 81531-980 Curitiba, PR, Brazil

Manuel Pedraza
Programa de Pós-Graduação, Museu de Zoologia, Universidade de São Paulo. Avenida. Nazaré 481, Ipiranga, 04263-000 São Paulo, SP, Brazil

José Eduardo Martinelli-Filho
Faculdade de Oceanografia, Instituto de Geociências da Universidade Federal do Pará. Campus Universitário do Guamá, 66075-110 Belém, PA. Brazil

Célio Magalhães
Instituto Nacional de Pesquisas da Amazônia. Caixa Postal 2223, 69080-971 Manaus, AM, Brazil

Luiz F. Andrade
Programa de Pós-graduação em Biologia Animal, Universidade Federal Rural do Rio de Janeiro. Rodovia BR 465, km 7, 23890-000 Seropédica, RJ, Brazil

Rodrigo Johnsson and André R. Senna
Laboratório de Invertebrados Marinhos: Crustacea, Cnidaria & Fauna Associada, Instituto de Biologia, Universidade Federal da Bahia. Rua Barão de Jeremoabo 147, Ondina, 40170-290 Salvador, BA, Brazil

Index

www.ingramcontent.com/pod-product-compliance
Lightning Source LLC
Chambersburg PA
CBHW082015190326
41458CB00010B/3194